China's e-Science Blue Book 2018

Chinese Academy of Sciences ·
Cyberspace Administration of China ·
Ministry of Education of the PRC ·
Ministry of Science and Technology of the PRC ·
Chinese Academy of Social Sciences ·
National Natural Science Foundation of China ·
Chinese Academy of Agricultural Sciences
Editors

China's e-Science Blue Book 2018

中国工信出版集团　　電子工業出版社·
PUBLISHING HOUSE OF ELECTRONICS INDUSTRY
http://www.phei.com.cn　 Springer

Editors

Chinese Academy of Sciences
Beijing, China

Cyberspace Administration of China
Beijing, China

Ministry of Education of the PRC
Beijing, China

Ministry of Science and Technology
of the PRC
Beijing, China

Chinese Academy of Social Sciences
Beijing, China

National Natural Science Foundation
of China
Beijing, China

Chinese Academy of Agricultural Sciences
Beijing, China

ISBN 978-981-13-9392-1   ISBN 978-981-13-9390-7   (eBook)
https://doi.org/10.1007/978-981-13-9390-7

Jointly published with Publishing House of Electronics Industry
The print edition is not for sale in China. Customers from China please order the print book from:
Publishing House of Electronics Industry.

This Springer imprint is published by the registered company Springer Nature Singapore Pte Ltd.
The registered company address is: 152 Beach Road, #21-01/04 Gateway East, Singapore 189721, Singapore

Contents

Introduction

Summary on the Development of e-Science in China

Yang Wang, Beibei Gu and Xuehai Hong

Abstract Over the past two years, the domestic experts and scholars had completed a lot of research on the development of e-Science in China, and received a series of outstanding achievements. This chapter mainly summarized the fruitful achievements which supported by e-Science and information technology, so as to provide a reference for the researchers in this field for the further e-Science research.

Keywords e-Science · Development of e-Science · e-Science applications

1 Introduction

During the "13th Five-Year Plan" period, the "Outline of the National Informatization Development Strategy" [1] issued by the General Office of the CPC Central Committee and the General Office of the State Council emphasized that efforts should be made to improve the level of informatization application while centering on the "Five-in-One Overall Layout" and the "Four-Pronged Comprehensive Strategy" and focusing on informatization to drive modernization.

To deeply implement the national science and technology policies and strategic planning, the Chinese Academy of Sciences and relevant departments issued *China's e-Science Blue Book (2017)*, which effectively reflected China's major achievements and progress in informatization and scientifically guided the future of e-Science. The Blue Book is published every two years, and it is the fourth public release this year. As a typical representative report that is highly authoritative, the Blue Book comprehensively displays the development trend and achievements of China's e-Science,

Y. Wang (✉) · B. Gu
Computer Network Information Center, Chinese Academy of Sciences, Beijing, China

B. Gu
University of Chinese Academy of Sciences, Beijing, China

X. Hong
Institute of Computing Technology, Chinese Academy of Science, Beijing, China

© Publishing House of Electronics Industry 2020
China's e-Science Blue Book 2018,
https://doi.org/10.1007/978-981-13-9390-7_1

3

and it provides benefits in many aspects as a record of the major scientific and technological progress in the application of e-Science in the past two years.

2   Overview

China's e-Science Book 2018 contains 21 chapters. It gives a summary of the major progress of China's e-Science in the past two years from the perspective of "three orientations" while focusing on China's informatization trend and major technological achievements.

2.1   Application in the Frontier Research of Science and Technology

"It is better to ride the wave than to be wise." We must take advantage of the situation to approach the frontier research of science and technology by accelerating our steps of innovation in order to build China into a world SciTech power, and forward-looking research should be carried out to strengthen the research and development on future mainstream technologies. General Secretary Xi Jinping pointed out that, "China's scientific and technological community should be confident in innovation and dare to dream big. We shall work hard in achieving originality and ingenuity, dare to challenge the most cutting-edge scientific problems, propose more original theories, and make more original discoveries, so as to achieve leapfrog development in key fields of science and technology, keep up with the new direction of scientific and technological development, and take a strategic initiative in a new round of global SciTech competition."

With the implementation of the innovation-driven development strategy, China has achieved fruitful results in building itself into an innovative country, and a number of major scientific and technological achievements have come out successively, such as "China's Eye of Heaven," gravitational wave detection, and space science satellite system. In the first section of "Application in the Frontier research of Science and Technology," there are a total of five chapters that mainly elaborate on the major scientific and technological achievements and progress in achieving e-Science.

Four chapters are selected here.

1. The development of the Internet has attracted worldwide attention, and a leap has been made in system architecture and key technologies.

In the chapter titled "Trend and Prospects of the Internet" written by Wu Jianping, Academician of the Chinese Academy of Engineering and Director of the Department of Computer Science at Tsinghua University, we can see that the Internet has achieved numerous remarkable achievements after half a century of rapid development and evolution. Under the development trend of multi-network interconnection and integration, the network architecture and key technologies are in a critical period of change. This chapter analyzes the current situation of the Internet and its main features, with a summary on the development of the system architecture and key technologies related to Internet at home and abroad from the perspective of research and practice, and suggestions on integration, high performance, security, reliability, intelligence, and so on are proposed for the future development of China's Internet.

2. Major informatization infrastructure of astronomy such as "Eye of Heaven" was built and put into use.

In the chapter titled "Informatization Needs and Construction of Major Astronomical Science and Technology Infrastructure Such as 'Eve of Heaven'" written by Cui Chenzhou, Director of the Information and Computing Center of the National Astronomical Observatory of the Chinese Academy of Sciences, it can be seen that the astronomical research has entered an era of data-intensive scientific discoveries along with the advancing technology. Whether it is a major astronomical infrastructure for technology or general astronomical research, it is inseparable from the support of information infrastructure. This chapter introduces five representative large-scale astronomical projects to show the demand and construction of the information infrastructure for major astronomical programs. These projects include the Five-hundred-meter Aperture Spherical Radio Telescope (FAST, see Fig. 1), the Large Sky Area Multi-Object Fiber Spectroscopic Telescope (LAMOST), the Mingantu Ultra-wide Spectral Radioheliograph (MUSER), the Space Variable Objects Monitor (SVOM)/Ground-based Wide Angle Cameras (GWACs), and the Square Kilometer Array (SKA). Besides, the chapter introduces how the Chinese Virtual Astronomical Observatory is established and applied as the public supporting platform for China's astronomy informatization, and at last, it puts forward suggestions for developing the infrastructure of astronomical e-Science in China.

3. The information system has promoted China's space science research into the forefront of the world.

The "Strategic Pilot Project for Space Science" series of satellite programs have ushered China's space science into a booming new era and provided a solid support of infrastructure and system platform services for the successful launch of dark matter particle detection satellite ("Wukong"), "Practice 10" returning scientific experimental satellite (SJ-10), quantum science experimental satellite ("Mozi," see Fig. 2) and hard X-ray modulation telescope satellite ("Huiyan") in the Pilot Project for Space Science (Phase I). The success of these four satellite programs has promoted China's space science research into the forefront of the world. The chapter titled "Application of e-Science for Space Science Satellite Program" is written by Zou Ziming, Deputy Director of the National Space Science Center of the Chinese Academy of Sciences and Chief Conductor and Chief Designer of the ground support system for the space science pilot satellite project. Firstly, the chapter analyzes the needs, opportunities, and challenges of informatization for space science in the new era, and then, it takes the construction of "12th Five-Year Plan" e-Science Project ("Technology Cloud for

Fig. 1 Five-hundred-meter Aperture Spherical Radio Telescope known as "Tianyan"

Space Science" Program) as an example to introduce the Project's progress, system architecture, and construction achievements with details showing the application and effect of the solar-terrestrial space system research network in supporting the satellite mission coordination, satellite on-orbit operation and scientific research. It sets the future direction of informatization for space science and focuses on improving the core elements of e-Science, so as to establish a deep-integrated and ubiquitous cloud-platform for e-Science.

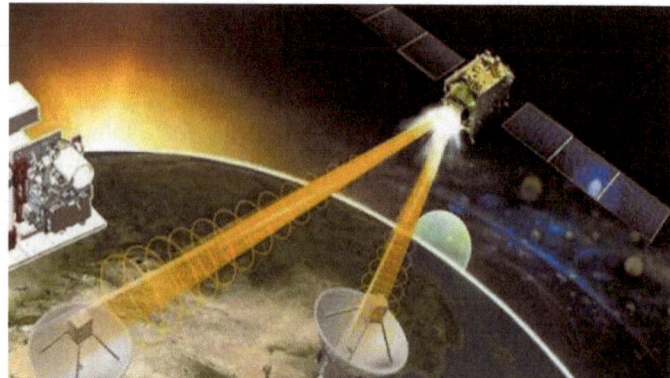

Fig. 2 An illustration of the quantum communication experiment of "Mozi"

4. The self-developed fully automatic "Survey" telescope provides powerful data support for the observation of gravitational waves.

The Chinese Antarctic Astronomical Cooperative Team led by the Purple Mountain Observatory and the Antarctic Astronomical Center of the Chinese Academy of Sciences has independently developed the "Survey" Telescope AST3 by utilizing the extremely good observation conditions of the Antarctic polar night. The telescope has realized long-distance automatic survey observation with continuous high reliability under extremely cold conditions. This is the first fully automatic survey telescope for Antarctica. It has adopted our self-developed software of informatization customization for data processing and cooperatively developed the system of AST3 operation and data processing auxiliary software. In this way, it provides the support of informatization software for the development of a series of time-domain astronomical frontier observation studies concerning gravitational wave source, fast radio burst, gamma-ray burst, supernova and exoplanets and has realized the observation of the AT2017gfo/SSS17a, which is the optical counterpart of the gravitational wave source GW170817 for the first-seen neutron star pair collision (see Fig. 3). The major observations were recorded in details in the chapter titled "Application of Antarctic Survey Telescopes AST3 in the Detection of Gravitational Wave Electromagnetic Counterparts" written by Wang Lifan, Director of the Antarctic Astronomical Research Center at Purple Mountain Observatory of Chinese Academy of Sciences.

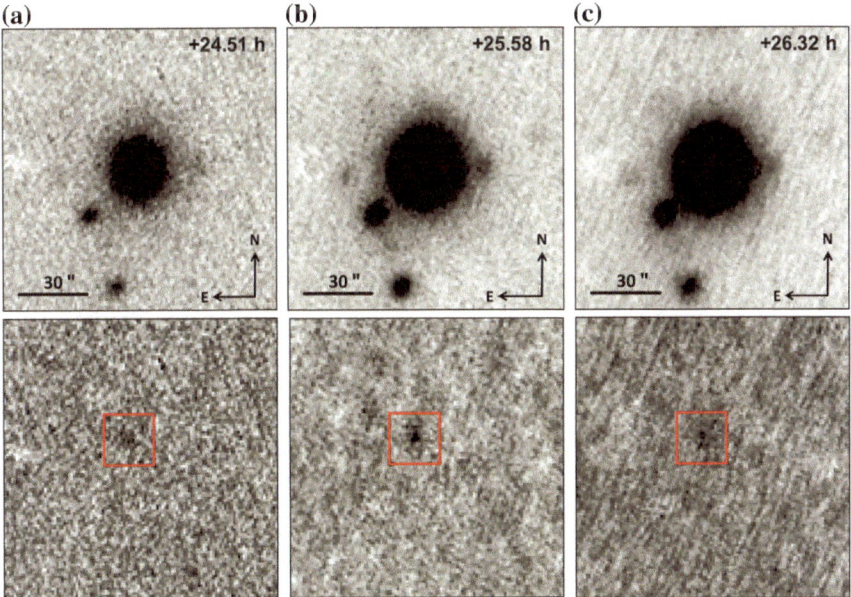

Fig. 3 Observation of the AT2017gfo/SSS17a which is the optical counterpart of the gravitational wave source GW170817

2.2   Progress of e-Science in Major Projects

At present, China has developed in a high speed, a lot of major projects facing a great demand of information technology and e-Science. It is necessary for the experts and scholars to solving the information technology problems in practical engineering and projects and should focus on the development of key core technologies and seize the long-term and overall commanding heights of technological strategy.

In the second section of "Progress of e-Science in Major Projects," there are eight chapters that mainly elaborate on the application in major projects and the analysis of the development trend of informatization in the technological area.

Among them, the chapters include "Current Status and Prospects of the Acoustic System of the 'Jiaolong' Manned Submersible," "Construction and Application of the Beidou Satellite Time System," "Efficient Access to Big Data—A Domestic Three-line Array Stereo Aerial Camera System," etc.

With a focus on the general needs, there is a summary on China's e-Science trends and typical cases, including "Construction and Application of Information System for National Science and Technology Management," "Current Status and Prospects of China's Science and Technology Cloud," "Precise Management and Open-sharing Enabled by Information Technology for the Natural Science Foundation of China," "Construction and Prospect of the National Philosophy and Social Science Documentation Center," "Scientific Data Shared for Accelerating Technology Innovation," "Current Status and Prospects of China's Technology Cloud for Security," etc.

Six chapters are selected here.

1. The "Jiaolong" manned submersible makes China an advanced country in deep diving.

In the chapter titled "Current Status and Prospects of the Acoustic System of the Manned Submersible" written by Zhu Min, Director of the Ocean Acoustics Technology Center of the Institute of Acoustics at the Chinese Academy of Sciences and Deputy Chief Designer of "Jiaolong" manned submersible (see Fig. 4), it is pointed out that the deep sea is a strategic space concerning the development of national security as well as an important battlefield for superpower games because it contains the resources for the sustainable development of human society. Mastering the key technologies in this field is the only way for China to enter the deep sea for research and development, ensuring the security of deep sea. The acoustic system of the "Jiaolong" submersible has achieved breakthrough in underwater acoustic communication (see Fig. 5) and detection, which is a highlight in the development of manned submersible technology in the world.

Fig. 4 Jiaolong manned submersible

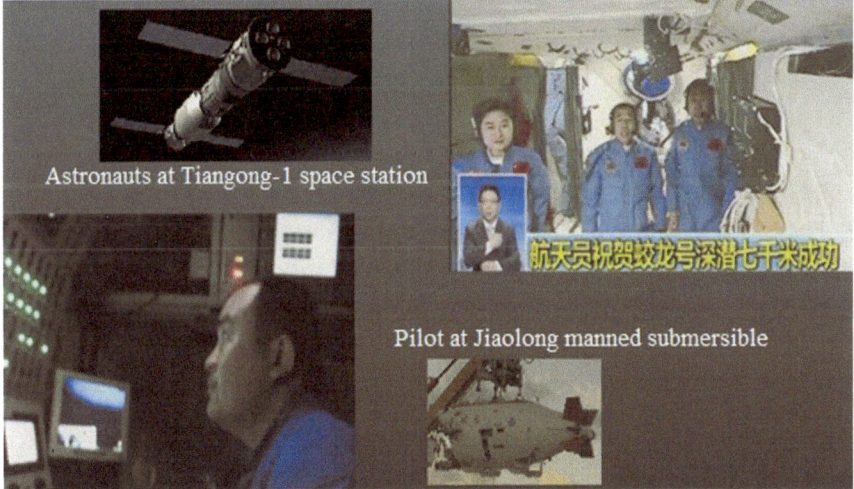

Fig. 5 Jiaolong dialogue with TIANGONG-1 space station with its acoustic communication system

2. Beidou satellite navigation system meets the development needs of the "Belt and Road Initiative" and it aims to provide services to the world.

In the chapter titled "Construction and Application of the Beidou Satellite Time System" written by Dong Shaowu, Director and Chief Researcher of the time-frequency reference laboratory of the National Time Service Center at the Chinese Academy of Sciences, it is pointed out that China's Beidou global satellite

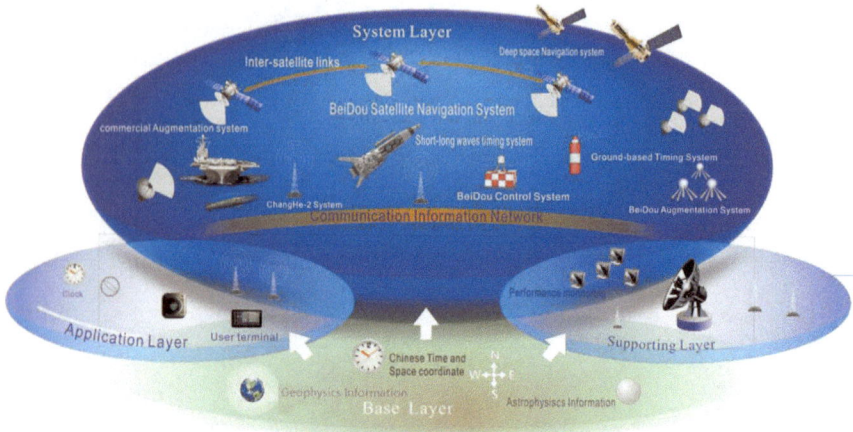

Fig. 6 Time and informatization system of integrated terrestrial and space

navigation system under construction (see Fig. 6) has already had the functions of navigation, positioning, and precise timing in the Asia-Pacific region. It is expected that a global system including 35 satellites will be built in 2020 to realize global coverage. This chapter briefly introduces the time system of China's Beidou navigation satellite system, the international time comparison and cooperation based on the Beidou system, the relationship with international standard time (UTC), and the latest progress of time service in China.

3. There are significant achievements in the construction and application of national science and technology information system.

In order to implement the innovation-driven development strategy, the State Council has launched a series of measures to deepen the reform of the science and technology system. It is required to construct and improve the National Science and Technology Information System (NSTIS), and realize the declaration, review, execution, acceptance, and management of the central financial technology projects (special programs, funds, etc.) through a unified information system. In the chapter titled "Construction and Application of Information System for National Science and Technology Management" written by Hu Shaohua, Executive Deputy Director of the Information Center of the Ministry of Science and Technology of the People's Republic of China, we can deeply understand the system architecture that meets the requirements of NSTIS and the overall architecture of the four-tier NSTIS system. The author deeply explored the core functions and key technologies of the system and made a comprehensive analysis of the overall effect of the system construction. Finally, it summarized the relevant work related to NSTIS and made an expectation to the direction of future research.

4. Informatization construction helps the Natural Science Foundation of China in its precise management and open-sharing.

In the chapter titled "Precise Management and Open-sharing Enabled by Information Technology for the Natural Science Foundation of China (NSFC)" written by Li Dong, Deputy Director of the NSFC Information Center, it focuses on the current status of the e-Science development as well as the historical process and development strategy of NSFC informatization. It also discusses the achievements of the NSFC's refined management platform "Science Fund Network Information System" as well as its sharing services and open-access platforms.

5. Social development and network security are gradually improved.

In recent years, our society has undergone the process of being reshaped in the rapid development of information technology, especially the wide application of technologies related to Internet, big data, and artificial intelligence. The research methods of philosophy and social sciences with human society as the research object have also gradually changed, and new requirements have been put forward for the informatization of philosophy and social sciences. The chapter titled "Construction and Prospect of the National Philosophy and Social Science Documentation Center" written by Wang Lan, Secretary of the Party Committee and Curator of the Library at the Chinese Academy of Social Sciences, it reviews the practice of the Chinese Academy of Social Sciences for informatization in recent years and analyzes the challenges in the process of philosophy and social science informatization. At last, it puts forward strategies to accelerate the construction of the National Philosophy and Social Science Documentation Center.

Besides, in the chapter titled "Current Status and Prospects of China's Technology Cloud for Security" written by Long Chun, Director of the Network Space Security Technology and Application Development Department of the Computer Network Information Center at the Chinese Academy of Sciences, it is pointed out that China's technology cloud is the key infrastructure to carrying forward the rapid development of China's e-Science, and the technology cloud is an important part to guarantee the system of "13th Five-Year Plan" special security project of the Chinese Academy of Sciences. Its goal is to build a security system integrated with cloud computing environment to provide multidimensional and high-precision network information security for research and production activities. This chapter gives an overview of the current status and support of China's technology cloud security technology and discusses the characteristics of security technology in the cloud computing environment and the next direction of development.

2.3   *Achievements of Informatization in Interdisciplinary*

As the saying goes, we should "get to the bottom of things to gain knowledge with further practice in person." Similarly, technology must be practiced if it aims at further development. The level of science and technology has become one of the most important variables affecting the world economic cycle, and it is also the most essential factor for the increase in economic aggregates. Scientific and technological revolution will invariably expand the economic aggregate and bring a golden period for economic development. To maintain the sustained and healthy development of China's economy, we should accelerate scientific and technological innovation and inject new impetus for economic growth to provide strong a guarantee for the coordinated development of economy and society. Therefore, it is necessary to promote the integration of technology and economic society in the development and open up the channel leading from strong technology to strong industry and strong economy.

The third section of "Achievements of informatization in interdisciplinary" contains a total of eight chapters, which mainly elaborate on important aspects of national economy such as agriculture, industry, medical care, and education promoted by technological innovation and guaranteed by network security, including "Key Technologies and Application Platforms of Spatial Information Services for Agricultural Economy—China's Electronic Map for Agricultural Economy," "Design of the China Academy of Agricultural Sciences for the Infrastructure of Technology Cloud," "Development and Prospects of Marking Service Technology for Advanced Manufacturing Industry," "Augmented Reality Assistive Technology for Complex Equipment Installation and Maintenance," "Clinical Oncology Research Supported by Big Data," "Application of Information Technology in Medical Ultrasound Engineering," "Application of Information Technology in Research and Management of Major Scientific Research Projects for Instrument Development," "Current Status and Prospects of China's Education and Research Computer Network".

Four chapters are selected here.

1. Innovations in the construction of agricultural informatization vigorously drive the development of the agricultural economy.

In the chapter titled "Key Technologies and Application Platforms of Spatial Information Services for Agricultural Economy—China's Electronic Map for Agricultural Economy" written by Liu Shengping, Deputy Director of the Intelligent Agricultural Technology Research Office of the Institute of Agricultural Information at the Chinese Academy of Agricultural Sciences, it is pointed out that after more than 30 years of information accumulation with diligent research, the largest database of rural economic basic data in China has been established to realize long-term interdisciplinary and inter-departmental technology development and application, with a focus on agricultural economic data management, analysis, and decision-making services. Research was carried out from perspectives of data management, information analysis, and decision-making services. It has provided effective information services for agricultural decision-making departments, scientific research units, and agricultural production departments and promoted the coordinated development of agriculture, society, economy, and environment. It plays a positive role in the development of China's agriculture and informatization in rural areas.

2. The development of intelligent industrial informatization has achieved remarkable results.

It is an urgent need for all countries to develop the manufacturing industry in order to cope with economic globalization and improve international competitiveness. It is also an inevitable choice to achieve industrialization through informatization and promote the adjustment and optimization of traditional manufacturing structures. The realization of advanced manufacturing must be based on the interconnection of information, especially the inter-connectivity and association mapping between cross-domain and multi-link applications. However, due to the existence of various heterogeneous marks in advanced manufacturing, there are information islands formed between application links, and the long-term coexistence of various heterogeneous marks is the normal state of future development. In the chapter titled "Development and Prospects of Identification Service Technology Facing the Advanced Manufacturing" written by Tian Ye, Executive Deputy Director of the Information Technology of Internet of Things and Application Laboratory of the Computer Network Information Center at the Chinese Academy of Sciences, it describes in details about how to build a heterogeneously compatible, multi-level, distributed, peer-to-peer identity service architecture (see Fig. 7) in various

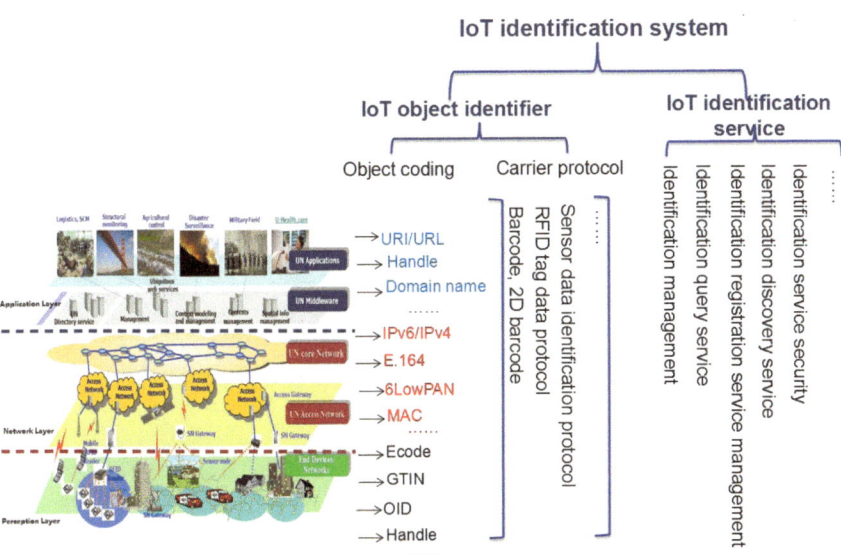

Fig. 7 Identification service system facing to the advanced manufacturing

manufacturing fields, such as the raw material supply chain management, person-alized customization, networked collaboration, remote operation and maintenance, providing a reference for the development of innovative models for advanced manufacturing.

3. The research and application of precision medicine have gradually benefited the public.

In the chapter titled "Clinical Oncology Research Supported by Big Data" written by Xu Ruihua, Director of the Cancer Center of Sun Yat-sen University and First Director of the Cancer Targeting Therapy Committee of China Anti-Cancer Association, we can see that population health is the most important social issue on livelihood, and it is related to the development of national economy and social progress, so that it is an issue that governments of all countries are trying to solve. There are millions of new cancer cases in China every year, and it is increasing year by year. Tumors have become a common disease that gradually surpasses the cardiovascular and cerebrovascular diseases to become the "No. 1 killer" as a cause of death in China. With the rapid development of big data and cloud computing technologies, the application of big data in basic medicine, clinical medicine, and public health is now in full swing (see Fig. 8). This chapter mainly shows the significance of big data on the development of oncology, the key technologies for the development of oncology big data, and the opportunities and challenges during this process.

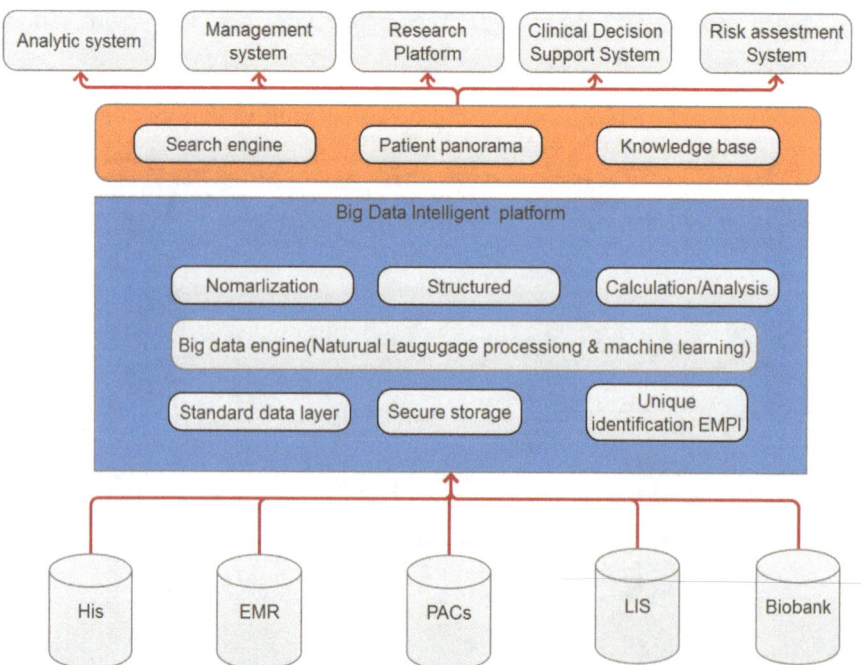

Fig. 8 Framework of the big data analysis and application system for oncology

In addition, in the chapter titled "Application of Information Technology in Medical Ultrasound Engineering" written by Chen Siping, Professor of Shenzhen University and director of the National Local Associated Engineering Laboratory of Medical Ultrasound Key Technology, it is mentioned that medical ultrasound imaging has become one of the most widely used diagnostic tools in modern clinical medicine. Each discovery of ultrasound echo implicit information can lead to a breakthrough of medical ultrasound imaging technology, whose development is a history of continuously discovering implicit information. Medical ultrasound has evolved from A-mode ultrasound of one-dimensional information at first to B-mode ultrasound of two-dimensional image information and the color ultrasound combined with blood flow motion information. It is developing toward the direction of early diagnosis and precision medicine. Through the complementation of ultrasound, electromagnetic and elastic and biochemical information, it can obtain important information simultaneously about the anatomical structure, mechanical properties, and electrical properties of human tissues. These diversified information discoveries are of great value in the early diagnosis of diseases and have achieved significant medical effect (see Fig. 9).

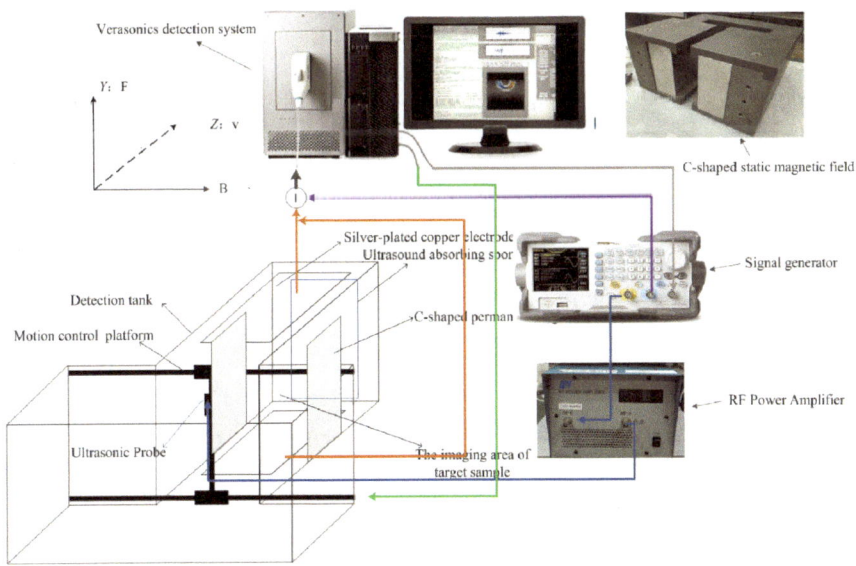

Fig. 9 Dual-mode imaging experimental platform

3   Summary and Outlook

Science and technology are worldwide and era-oriented, so we must have a global vision and grasp the pulse of the times for SciTech development. Historical experience shows that the scientific and technological revolution can always change the pattern of world development in a profound way. In today's world, a new round of scientific and technological revolution is waiting in its wings. The original breakthroughs in some major scientific issues such as material structure, evolution of the universe, origin of life, and the nature of consciousness are getting us to new frontiers and new directions. Some major subversive technological innovations are leading to new industry and new business. Information technology, biotechnology, manufacturing technology, new materials technology, and new energy technology have penetrated into almost all fields, driving a group of major technological changes featured as green, intelligent, and ubiquitous. A new generation of information technology such as big data, cloud computing, mobile Internet is more integrated with robots and intelligent manufacturing technologies. The chain of technological innovation is more flexible, with quickening technology updates and achievement transformation. Industrial upgrading is stepping up to transform social production and consumption from industrialization to automation and intelligentization. Social productivity will once again be greatly improved, and labor productivity will have another leap.

E-Science [2] is the engine and distinctive feature of the new round of scientific and technological revolution. In the face of the ascendant informatization and the increasingly fierce competition in science and technology, e-Science could not be neglected; otherwise, our high-end informatization applications will continue to fall behind or it means that we are impossible to grasp the historic opportunity of innovation leap in the new round of scientific and technological revolution. In order to solve the problems in China's e-Science process, we should seize the current historic opportunities, vigorously develop the technology of e-Science, and direct the scientific research and innovation activities to be more characterized as individualization, openness, networking, and clustering. We should constantly optimize and innovate the ecology, transform China's scientific research, innovation organizations and activity models and greatly improve China's capabilities of scientific and technological innovation, so as to stimulate unprecedented innovation and vitality with support for the innovation-driven development strategies in China.

The continuous improvement of China's e-Science is inseparable from the innovation-driven development of science and technology. *China's e-Science Blue Book (2017)* further explores the construction of a more complete e-Science system based on a summary of the major scientific research achievements in the past two years. With innovative research thinking and progress as its reference, it is highly cutting-edge, forward-looking, and authoritative. We believed that it can provide some guidance and reference for the majority of scientific research staff.

References

1. The "Outline of the National Informatization Development Strategy" issued by Xinhua News Agency, the General Office of the CPC Central Committee and the General Office of the State Council. [EB/OL]. http://www.gov.cn/xinwen/2016-07-27/content_5095297.htm, 2016-07-27.
2. Chen Mingqi, Chu Dawei, Hong Xuehai, and Cao Ning. Development Trend and Thinking about e-Science [J]. Bulletin of the Chinese Academy of Sciences, 2016, 31(6): 608–613

Yang Wang Ph.D., Senior Engineer, Head of the Information Development Strategy and Evaluation Center, Computer Network Information Center of the Chinese Academy of Sciences. He mainly focuses on the development strategy of information technology and e-Science. He has published more than 10 academic papers in SCI/EI and Chinese journals. He participated in the "13th Five-Year Plan for Informatization Development of the Chinese Academy of Sciences" and "Data and Computing Platform Planning". He also participated in the research work on "The promotion of scientific research informatization", "National Informatization Development Strategy Research", "Information Development Strategy in Interdisciplinary Research".

Application in the Frontier Research of Science and Technology

Internet Development Status and Future Trends

Jianping Wu, Mingwei Xu and Yuan Yang

Abstract This chapter analyzes the overall trend of Internet development, and discusses the research and development effort on new network architectures. Further, the chapter summarizes the development trend of Internet in China and proposes some suggestions on the future Internet development.

Keywords Internet · Network architecture · Trend analysis · Development suggestion

1 Introduction

After the fast development and evolution for more than half a century, the Internet core technologies have made considerable strides. In retrospect, fiber-optic communication was developed in the 1970s, Internet appeared in 1980s, mobile communication arose in 1990s, and telecommunication networks and Internet integrated in 2000s. After these dramatic revolutions, an adjustment period has come for the Internet and communication technologies, in which evolution and spiraling are the main characteristics. Emerging technologies such as ubiquitous perception, artificial intelligence, and big data are bringing new challenges to the Internet architecture.

Currently, Internet has become a critical factor that promotes economic growth and improves national competitiveness. It has also been the foundational infrastructure of information sharing and cooperation for human society. It is never exaggerated to highlight the historical significance of Internet. The Internet of things (IoT) will create a harmonic world of human, machines, and things, extending the application field of information technologies greatly. The mobile Internet will create new areas for economic growth. With emerging smartphone applications, Internet traffic is considered to increase explosively in the next decade. Internet promotes the integration and development of a number of emerging technologies, including clouding computing, IoT, and big data. Internet penetrates

J. Wu (✉) · M. Xu · Y. Yang
Department of Computer Science and Technology, Tsinghua University, Beijing, China

© Publishing House of Electronics Industry 2020
China's e-Science Blue Book 2018,
https://doi.org/10.1007/978-981-13-9390-7_2

and combines various areas, such as information, industry, and culture, bringing a better life that is more comfortable, convenient, and economical. Internet takes a great responsibility of industrialization and informatization, following the trend of mass entrepreneurship and innovation.

Internet has grown to the cyberspace, which is considered as the fifth territory besides land, sea, air, and space. There is a trend of integration of various networks, such as mobile Internet, energy Internet, industrial Internet, broadcasting networks, ocean networks, and space networks, to name but a few. Many networks are developing separately in the current stage, showing a "chimney-like" growing. Most networks are dedicated, independent, and heterogeneous. This introduces many challenges to the integration. However, great efforts are taken by both academia and industry.

In this chapter, we will review the overall trend of Internet development and discuss the research and development effort on new network architectures. Then, we will summarize the development trend of Internet in China and present our vision on the future Internet.

2 Analysis on Internet Development Trend

We summarize the features of Internet and communication technology development, since 2010, based on which we analyze the future trends as follows.

(1) IPv4/IPv6 will still be the core of the global Internet in the next decade. However, challenges including scalability, security, real time, and manageability are becoming more and more critical.

(2) The fast development of Internet promotes the integration with other networks, such as mobile networks and space networks. The network architectures are evolving toward a unified one. While Internet Plus and Industry 4.0 will become new drivers to stimulate the economic growth, emerging applications pose new challenges on developing core devices that are advanced, independent, and controllable.

(3) Internet traffic is growing rapidly, and data are transferred redundantly. Characteristics of network topology and traffic distribution are changing, which lead to adaption problems between service providers and content providers. Developing a "content-centric" broadband network becomes a new trend. ITnization, cloudification, and emerging technologies such as SDN and NFV will accelerate the evolution of network architecture, device form, and operation model.

(4) The inflexibility of traditional networks hastens the application of cloud computing and datacenters. In the meantime, billions of access devices introduce higher QoS requirements. Internet is facing the problem of sustainable development. More and more attentions are paid on a network architecture that is recognizable, scalable, and definable.

(5) Capacity and distance of data transmission are increasing continuously with the development of optical networking and core devices. The global IP traffic in 2018 is four times of that in 2013, and the peak traffic reaches 1 Pbps. Eighty percent of the traffic is introduced by video content. Optical transmission with a data rate of 100 Gbps is already in service, and a prototype of 400 Gbps transmission system has been realized. Next-generation 100 Tbps WDM optical transceivers are developed. In the future, increasing of multimedia traffic requires a transmit capacity of Pbps and a switch capacity of Ebps. The network capacity needs to increase by at least 1000 times in the next decade. There are a number of emerging technologies, such as gigabit optical access, software-defined optical networks, all-optical networks, photonic integrated circuit, and optoelectronic integrated circuit.

(6) Programmability, virtualization, and open resources are receiving more interests from the industry. Network functions in the Internet can be implemented and deployed in a virtualized manner, in industry standard high-volume servers, storages, and switches. Such a cloudified architecture can improve the utilization of network resources, reduce maintenance costs, and accelerate service deployment.

(7) Photonic integrated circuit (PIC) technology allows a single chip to hold various optically active and passive functions. The technology provides a promising direction to address the issues of data rate and energy consumption faced by current optical communications. In the future, Indium phosphide (InP)-based PIC, Si-based PIC, integrated optical waveguides such as lithium niobate-based PIC and hybrid PIC will become important technologies.

(8) Cellular mobile communications after 4G have become a global hot topic. Commercial deployment of 5G will be completed in 2020, satisfying the traffic increment in the next decade. The key technology beyond 5G is under preparing. The mobile communication in the future aims to bring new experience to users. Emerging services such as interactive games, hologram, and VR are taken into consideration when developing mobile communication systems. The application of mobile communication technologies will extend to a wider field, including densely connected IoT, IoV, industry Internet.

(9) A boom in broadband wireless access technologies such as Wi-Fi has arisen. The IEEE 802.11 family develops wireless LAN technologies with higher data rate and less latency and develops a set of technologies in new application areas, such as 802.11ah for IoT, 802.11p for IoV, 802.11ad for low latency and high bandwidth. Wi-Fi is also considering new functions including networking and voice roaming.

3 Review of New Network Architecture Research

The development of the Internet has stepped into a critical period of theoretical and technological revolution. The architecture is the core element of Internet development. Generally, there are two methods to develop a new Internet architecture, i.e., revolution and evolution. The revolutionary method aims to design a brand-new architecture from the very beginning. This method needs a high cost to deploy in practice. On the other hand, the evolutionary method aims to patch on the current architecture and make incremental updates. This method is limited on solving the fundamental problems. A good solution should combine the advantages of both methods. It should enhance the ability to support future applications, while the development should be based on actual conditions. It should maintain the stability of Internet core and the design principles, while enabling efficient innovations by loosening some basic limits that restrain the Internet extension.

3.1 Projects in the USA

The USA launched Future Internet Design (FIND) program in 2005 and started Global Environment for Network Innovations (GENI) program in 2006, which is subsequent to FIND and provides experiment environment for the future network research and development. By 2012, the management architecture and a new end-to-end working prototype of GENI had been developed and tested. GENI had entered the fifth stage. Based on these explorative programs, the US NSF launched Future Internet Architecture (FIA) program in 2010, which began to design and verify new architectures of the future Internet. The plan includes a number of projects, including MobilityFirst, NDN, NEBULA, XIA, ChoiceNet. These projects try to "materialize" the possible forms of future networks, with respect to content routing, mobility, evolutionability, etc. Some progresses had been made. After reviewing and analyzing such progresses, the NSF determined the further direction in 2014 and started the second stage of FIA. The focuses of this stage include information unit networks, scalable network architecture, mobility, and cloud computing, which aim at exploring the fundamental theory and methods further for the future Internet. Typical projects and their main features are summarized as follows.

(1) **MobilityFirst** aims to overcome the challenges brought by wireless access of large number of mobile devices and achieve smooth and seamless mobility support. MobilityFirst considers network nodes moving as an ordinary state instead of a special connection form. MobilityFirst achieves communication stability of mobile nodes by generalized delay-tolerant networking (GDTN) and considers the balance between mobility and scalability. Besides, MobilityFirst enables efficient communication between mobile nodes by scheduling network resources properly.

(2) **Named Data Networking (NDN)** uses content name (i.e., URL) to replace IP address as content identification. NDN develops content name-based routing protocol and introduces content caching mechanism at router nodes, which improves quality of service effectively by reducing redundant traffic and source server load. Meanwhile, NDN proposes to support mobility by separating content name and content address. NDN also proposes to improve information security by signatures and encryptions on contents.

(3) **NEBULA** is a secure and elastic network architecture, which leverages cloud computing (i.e., data centers) to complete data storage and computing. NEBULA constructs a high-speed, secure, and reliable backbone network, so as to connect data centers for cloud computing and distributed communication. The key technologies of NEBULA include reliable data control under cloud computing model, cloud computing-centric network architecture, etc.

(4) **eXpressive Internet Architecture (XIA)** aims to design an evolutionable network architecture with security mechanism embedded. The architecture is not limited to any specific network scenario. XIA is constructed based on the concept of internal security principle. XIA allows the application to specify requirements on data transmission and uses a secure and efficient transmission mechanism to satisfy the requirements.

(5) **ChoiceNet** utilizes economic theory for network architecture development. The purpose is to keep the activity of Internet on core innovations. ChoiceNet makes a revolutionary change on network development: combining Internet technology innovation and economic theory. ChoiceNet applies game theory and economic incentives on Internet technologies and proposes a competitive market. The design, development, and other aspects of the new-generation Internet architecture are driven by user choices and competitions. In such a way, the innovations and revolutions in every layer of the protocol stack are motivated.

3.2 Projects in Europe

Under the plan of FP7/Horizon 2020, the European Union (EU) started research and development on a network architecture, which aims at breaking through the limits of current Internet and supporting integrated future networks. In 2008, the EU launched the Future Internet Research and Experiment (FIRE) program to provide an experiment environment for research on future networks. Standardization of technologies is carried out by ITU-T. China launched the EU China Future Internet Common Activities and Opportunities (ECIAO) program and cooperates with the EU FIRE project. Typical projects in the FIRE program are summarized as follows.

(1) The **Publish-Subscribe Internet Routing Paradigm (PSIRP**, 2008–2010) project is one of the most influential projects of the EU FP7. PSIRP introduces the Pub/Sub communication paradigm to redefine Internet architecture and

operating mechanism. PSIRP makes multicast the basic paradigm and takes security and mobility into full consideration from the very beginning.

(2) The **Publish-Subscribe Internet Technology (PURSUIT**, 2010–2013) is a subsequent project of PSIRP. PURSUIT made a number of improvements to PSIRP. In particular, PURSUIT studied the role of cache in the Pub/Sub architecture, as well as the effect on congestion control and error control. PURSUIT studied the effect of the Pub/Sub architecture on underlying operating, including resource allocation and reservation. PURSUIT further studied how the new network architecture cooperates with underlying network resources, such as optical fiber reallocation, wireless resource binding. PURSUIT also studied the effect on social economic environment.

3.3 Other Architectures

(1) **Software-Defined Network (SDN)** follows a design philosophy of separating the control plane and the data plane. SDN uses a centralized controller to achieve more flexible routing policies. The controller exchanges control messages with routes/switches through standard communication protocols. SDN technology allows routers/switches to concentrate on forwarding, while enabling flexible network control by the controller. The SDN technology gives service providers a better control over their infrastructures and reduces the overall operating cost.

(2) **Dual-Architecture Network (DAN)** is proposed in 2005 by researchers from China, to solve the congestion and the content control problems of Internet. In DAN, Internet is considered as the primary architecture which applies a pull paradigm, and the broadcast storage system is seen as the secondary architecture which applies a push paradigm. DAN integrates the two architectures by using edge to assist core and using storage to assist routing. The two architectures complement each other and develop synchronically. DAN allows a transformation of the network architecture and enables efficient and coordinated delivery of information.

4 Research and Development of Internet in China

Internet in China has developed rapidly, under the overall situation of interconnection and integration of multiple networks. Especially during the "Twelfth Five-Year Plan" period, a number of projects have been launched around the overall demands of the construction and development of national broadband networks. These projects include 12 major and key projects of the National High Technology R&D Program of China (863 program), covering new network architecture, ultra-large capacity and ultra-high-speed optical communication, high-speed routing, next-generation broadcasting, integration of broadcasting/telecommunication networks and evolution, IPv6 monitor and management, end-to-end green network, etc. A series of important breakthroughs have been made with the execution of the broadband network technology key projects.

4.1 New Network Technologies

With respect to network development strategy, China is carrying out "Broadband China," "China's Next Generation Internet Demonstration Project," and "Infrastructure of Future Network Major Technology" strategies. There are also researches on development strategy, including "Innovation 2050: The Science and Technology Revolution and the Future of China," "Research on Medium and Long-term Development Strategy of Engineering Technology for 2030 China," and "Research on the Development Strategy of Future Network Technology, Platform and Mechanism."

With respect to research and development of new network architectures and key technologies, a number of 973 programs, 863 programs, and NSFC programs have been launched for the future Internet architecture. The key projects include "Research on new generation Internet architecture and protocol," "Research on service-oriented future Internet architecture and mechanism," and "Research on reconfigurable information communication network system." Various new network architectures were proposed during the projects. Tsinghua University proposed an evolutionable network architecture, and a real address-based secure and trusted network architecture. Beijing Jiaotong University proposed an integrated pervasive service network. Information Engineering University proposed a reconfigurable network. These studies consider different perspectives and layers of the Internet architecture and technologies.

With respect to experiment platforms, China has released an innovative experiment facility based on a new-generation network architecture, which covers 16 cities. Breakthroughs have been made in software-defined routing and switching devices, intelligent information resource scheduling system under cloud architecture, and big data-based network measurement and sensing. A series of achievements with independent intellectual property rights have been made. In 2013, China approved the China Environment for Network Innovations (CENI) as one of the 16 major scientific infrastructure projects in the "Twelfth Five-Year Plan."

With respect to Internet core devices, China has international leading enterprises such as Huawei and ZTE. High-end network devices have been developed. The 400G and T-class routing devices are global leading. The leadership is expanding from whole machines to network chips and protocol software. China has become one of the important providers of the IETF international standards. The impact on emerging international standardization organizations such as open networking foundation (ONF) is constantly increasing.

4.2 Interconnection and Integration of Networks

China carried out research on integration and evolution of broadcasting networks, telecommunication networks, and Internet. The core technologies of 10G optical line terminal (OLT) and 10G optical network unit (ONU) have been developed. The integrated broadcasting platform can provide video-on-demand services for about 500,000 users. The access network application demonstration area of the next-generation broadcasting (NGB)-LTE has covered 200,000 users, and the fiber to the home (FTTH) demonstration network has covered 290,000 users, bringing economic and social benefits. In the NGB project, standards such as "Transmission and Media Access Control Technical Specification of NGN Broadband Access System HINOC" have been issued. Other outputs of the project such as the architecture have been applied to the national cable TV network interconnection platform (Broadcasting 136 Project), which also produced enormous economic and social benefits.

For integrating the mobile Internet and broadband network, the network integration architecture and future evolution plans have been proposed, taking into account the demands of development and operation in 5–10 years. The Network Intelligence Capabilities Enhancement (NICE) standard was submitted to and approved by ITU. Core devices including distributed storage and smart cache have been developed.

For integrating space networks and terrestrial networks, the integrated architecture of Internet, mobile communication networks, and space networks were proposed. Key technologies are under development, and a demonstration environment has been established. The demonstration of the implementation plan of the major project has been launched, to construct the space-terrestrial integrated information network.

4.3 Research and Development of Optical Communications

China has the top rank of industrial manufacturing capability on optical communications. The independent products occupy 2/5 of the global optical network, 3/4 of the global optical access, and 1/2 of the global fiber-optic cable. However, there is still a lack of high-end core technology and a lack of design and manufacturing capabilities of high-end chips and precision instruments. Such instruments and chips are still depending on import. There are a few reasons for such a situation. China starts late in related technologies, and research funding is insufficient. There is a lack of equipment as well as professionals. Recently, there is a trend of optoelectronic integration of high-end devices and chips, bundled with high-speed analog-to-digital Conversion (ADC), digital-to-analog conversion (DAC), and digital signal processor (DSP), resulting in a multidisciplinary integration.

China carried out research on new ultra-large capacity all-optical switching networks. The next-generation architecture of colorless and non-blocking optical switching with ultra-large capacity has been proposed, which can greatly improve the networking performance. A batch of core technologies has been adopted by international standardization organizations including IETF and ITU-T. During the "Twelfth Five-Year Plan" period, a number of projects were funded and made important progresses, including "fundamental research on ultra-high-speed, ultra-large capacity, and ultra-long-distance optical transmission," "development and system verification of high-speed, high-performance ADC/DAC chips in next-generation optical transmission systems," "research and development of ultra-high-speed and long-distance optical transmission system," "photonic integration technology and system application," "research on new ultra-large capacity all-optical network architecture and key technologies," and "development and demonstration of intercity trunk optical transmission equipment supporting standard single-mode optical fiber transmission of 100T and 1000 km."

4.4   Wireless Mobile Communications

China starts late in the field of wireless mobile communications. After hard working for nearly 30 years, China has made a leap forward. China occupies an important position in international standardization organizations, and the level of fundamental research has been greatly improved. During the "Twelfth Five-Year Plan" period, the 863 program launched the major project of 5G mobile communication system. The 5G development vision has been proposed by a union of enterprises, research institutions, and universities, and the key technical metrics have been validated. Most results were approved by ITU and made a global wide-ranging impact.

In particular, the 5G key technologies, system framework, frequency requirements, business applications, and scenarios were extensively investigated and analyzed, and a series of research reports were completed, which laid a solid foundation for 5G standardization. The key technologies and development directions of 5G mobile communication systems were clarified. A first phase testbed based on cloud platform has been designed and is under implementation. To best of our knowledge, the testbed is the largest and most advanced testbed around the world. The testbed can support up to 1024 antennas, with the capability of reconfiguration and deployment either in a centralized way or in a distributed way. The testbed can be used to verify key technologies such as M-MIMO and UDN. The testbed also supports SON and SDN for network intelligence. Based on measured data, a large-scale antenna application scenario and a three-dimensional large-scale channel model were established, which can accurately reflect the characteristics of China's deployment environment. The theoretical performance and energy efficiency of large-scale antenna system have been analyzed and verified. Core enterprises such as Huawei, ZTE, Datang, and China Mobile are leading the R&D of 5G key technologies. They have started tests and experiments of the key technologies.

At present, however, China's R&D is still in a state of tracking in the field of information communication technologies. The abilities of R&D and industrialization on core chips are far from international advanced level. Although in recent years, China made a progress in fundamental R&D abilities on optical communications, mobile communications, and core networking/switching chips, the high-end chip market is still occupied by the developed countries. Especially, China falls behind at R&D of silicon photonics (SiPh), which represents the future direction of communication technologies. Such a situation is disadvantageous to China in the next round of industry competition, as the gap with leading companies such as Intel and IBM may be widened by leading companies such as Intel and IBM.

At present, the information and communications are entering a new round of technology and industrial reform. The telecommunication provider-centric paradigm is facing profound changes. The OTT paradigm is becoming one of the main driving forces for industrial development gradually. The hardware platforms of network devices are becoming generalized, distributed, and virtualized. The era of a deeply integrated ICT is on the way. Traditional telecommunication industry chain will face a huge shock. The industry is at an important turning point of rebuilding the technology chain and the industrial chain. Future communication systems face the demands of massive connections, huge capacities, and wide applications. New breakthroughs are supposed to be made on key technologies, and the architecture is facing revolutionary changes potentially.

5 Future Directions and Suggestions

5.1 Integrated Network Architecture and Key Technologies

With the continuous expansion of the cyberspace, the trend of integration and penetration of various heterogeneous networks is clear, including Internet, mobile communication networks, broadcasting networks, space networks, ocean networks, industrial Internet, energy Internet, and Internet of things. However, current network systems use independent, dedicated network architectures and protocol specifications, instead of unified ones. The networks can only be interconnected and integrated by various protocol gateways. Due to the lack of overall design and guidance of network architectures, the integration faces many challenges, including scalability, performance, and security. We should grasp the important opportunity of the next-generation Internet renewal, to carry out research and construction of integrated network architecture, and lead the development of China's future network.

Network virtualization technology separates the network hardware infrastructure from the network functionality for applications. Deeply integrated networks can be based on IPv6 and support individual needs by network virtualization technology over the physical infrastructure. Based on IPv6, Internet and integrated networks will become ubiquitous infrastructure of the society. We should also carry out the measurement and analysis of the basic network behaviors. The monitorable, analyzable, and manageable network behaviors are necessary for integrated networks to operate with high stability, high efficiency, and high reliability. It is necessary to master the measurement and analysis technologies independently.

5.2   Hardware and Software Systems for High-Performance Routing and Switching

In the past few years, the annual growth rate of China's Internet backbone traffic and bandwidth demand has exceeded 200%. It is expected that in the next few years, the annual growth rate will still be as high as 100%. The IP backbone is facing equipment upgrade and capacity expansion. Current 100–400G platform routers in the backbone cannot satisfy the rapid growth of bandwidth in the future.

In order to meet the capacity expansion requirements and build long-term sustainable IP carrier networks, it is extremely urgent to develop core chipsets with 1Tbps capacity on a single slot and high-end routers with a total capacity of 10T or more, based on China's independent intellectual property rights. At the same time, data center networks are developing rapidly with emerging of cloud computing and big data. Ultra-large capacity switches are core devices of the data center networks carrying cloud services. The research and development of such switches, especial switch with a total capacity of 100T, are significant. High-performance network equipment is not only the core of national network infrastructure construction and development, but also the commanding height of the national information industry technologies. R&D of such equipment will lay a solid industrial foundation, for China's products with independent intellectual property rights to obtain an international leadership.

Routing control is the core element of Internet architecture, and it is also the freest part in the evolution of Internet architecture. The IPv6 address space has grown by orders of magnitude compared to IPv4. The ability to perform efficient, fast, on-demand, and trusted routing will have a significant impact on the efficiency and quality of service of Internet and mobile Internet transmission. Routing protocols are required to be more lightweight and more scalable when Internet is connected to other networks such as IoT and industrial Internet. When the cyberspace expands to space networks, the network infrastructure is also mobile. The routing protocol must adapt to the characteristics of time-varying topology and large spatiotemporal span, to better support the infrastructure mobility. In the cyberspace, the security and trust of routing is extremely important. It is necessary to solve the routing-level security problems fundamentally, such as source address spoofing and route hijacking attacks, and realize a secure and trusted routing mechanism. In addition, efficient and trusted addressing in the huge address space of IPv6 will also help improve routing efficiency.

5.3 Key Technologies and Protocols for Secure and Trusted Networks

IP-centric Internet technology has achieved great success worldwide, greatly promoting social development and becoming the fifth territory besides land, sea, air, and space. However, the intrinsic feature of the IP protocol, i.e., connectionless communication, makes it difficult to track and monitor users, which introduces huge security defects to Internet. In connectionless IP networks, key technologies and protocols for realizing global security and trust are urgent needs and a major challenge. Such technologies and protocols include a collection of hardware and software used to realize Internet applications. The hardware includes routers, bridges, switches, communication links, etc. The software includes algorithms, programs, and systems to ensure the Internet operation, such as routing algorithms, domain name systems, network protocols. The new generation of secure and trusted networks needs to be based on authenticity. We should start from the three perspectives of construction, management, and prevention and construct a linkage between them to achieve the overall network security.

Building trusted network service is a prerequisite for network security. Network trust needs authenticated service provided by end nodes and authenticated traffic provided by networks. Authenticity of network traffic source needs network core devices that are with a trusted mechanism and are independently developed. Authenticity of network traffic source also needs authenticated addresses of end nodes. In order to effectively monitor user behaviors and ensure the authenticity of end-host services, a user identity authentication mechanism is necessary. By monitoring user behaviors, the network conditions can be grasped, and the network security situation can be evaluated. Based on trusted network system consisting of both trusted network side and trusted user side, access authorization, attack identification, and traceability can be realized. However, intrusions cannot be stopped effectively just by the system. Therefore, it is also necessary to establish a multi-level defense and intrusion prevention system, including dynamic network virtualization services, adaptive control on routing policies and self-healing routing, and encrypted security channels. Such a system aims to guarantee all network services or key services, when the network is under attack, or some equipment is damaged, or authenticated services are lost.

5.4   Intelligent and Integrated Networks

As the integration of traditional applications and emerging applications accelerates, tremendous changes have occurred on business form, transmission and distribution, and process of receiving services. The media content format is improving from 4k to 8 k and 3D, with higher definitions and higher dimensions. The media presentation form is evolving from a traditional single-screen live to an interactive multi-screen live mode. The media transmission is changing from single mode of wired or wireless or satellite to an intelligent and collaborated mode. The terminal receiving changes from traditional single device-based independent mode to a multi-device cooperative mode for personalized contents at any time and place. The above trends propose new requirements to service transmission methods and intelligence of integrated networks. Higher requirements on the quality of application service are proposed to integrated networks, and requirements are stricter on manageability, controllability, and trust. Such stringent requirements require innovations on video media network technologies and integration of new-generation information technologies such as cloud computing, big data, SDN, and NFV. Breakthroughs are desired in intelligent cooperation of wired and wireless satellite transmission and networking.

Industrial Internet is a new type of network formed by the integration of new-generation information technology and industrial systems. Industrial Internet is the key infrastructure for informatization and intelligent of industries. Industrial Internet will construct an integrated network with distributed sensing, efficient integration, low latency, high reliability, and wide coverage, which connects machines, things, people, and information systems to enable scientific decision making and intelligent control. Industrial Internet will promote deep integration of manufacturing and Internet.

6   Conclusion

Internet has been under a fast development and evolution for more than half a century, and remarkable progresses have been achieved. In the current situation that multiple networks are intended to be interconnected and integrated, network architecture and key technologies are in a critical period of revolution. This chapter reviews the situation and features of current Internet development and analyzes the network architectures and key technologies from both academic and practical perspectives. The chapter further presents a few suggestions on future directions for Internet in China, including integration, high performance, and intelligence.

Jianping Wu is Member of Chinese Academy of Engineering and Professor of Department of Computer Science at Tsinghua University. He is serving Chairman of the Department of Computer Science and Dean of Institute for Network Sciences and Cyberspace at Tsinghua University. He is also serving Directors of Network Center and Technical Board of China Education and Research Network (CERNET) and Director of the National Engineering Laboratory for Next Generation Internet, Member of Advisory Committee of National Information Infrastructure for Secretariat of State Council of China, and Vice President of Internet Society of China (ISC). He was also Chairman of Asia Pacific Advanced Network (2007–2011). He is the IEEE Fellow. He has been devoted himself to the research of computer network including technology research, network engineering, and cultivating talents for many years. And he has led his team to do an in-depth study of network design, network engineering, network core equipment development and network architecture, etc. He was the leader for establishing the China Education and Research Network (CERNET) in 1995 and establishing the first nationwide pure IPv6 network in 2005. He proposed the idea of Source Address Validation Architecture (SAVA) and the 4over6 transition technology from IPv4 to IPv6. He has published more than 300 academic papers, edited and co-edited eight books, and supervised more than 100 postgraduates. As the first author, he developed four IETF RFCs in IETF. He is inventor or co-inventor on 20 patents. He received Jonathan B. Postel Service Award of ISOC in 2010 and was inducted into Internet Hall of Fame in 2017.

Cyber-Infrastructure Requirements and Current Status of FAST and Other Astronomical Key Science Projects

Chenzhou Cui, Yihan Tao, Ce Yu, Changhua Li, Jian Xiao,
Youling Yue, Long Xu, Chao Wu, Feng Wang, Ge Zhang, Boliang He,
Dongwei Fan, Shanshan Li, Linying Mi, Yue Chen, Yunfei Xu
and Jun Han

Abstract With the development of modern science and technology, astronomy has entered the era of data-intensive scientific discovery. Cyber-infrastructure is becoming a crucial part of both key national science projects and daily research in astronomy. In this chapter, cyber-infrastructure requirements and implementation status of five cutting-edge astronomical projects are described, namely the five-hundred-meter aperture spherical radio telescope (FAST), the Large Sky Area Multi-Object Fiber Spectroscopic Telescope (LAMOST), the MingantU SpEctral Radioheliograph (MUSER), the Chinese–French SVOM mission, and the Square Kilometer Array (SKA). Furthermore, as a common cyber-infrastructure for astronomy research, a brief overview of Virtual Observatory including its research and development history and current applications is given. Several suggestions are listed in the end of the paper for further development of scientific research cyber-infrastructure.

Keywords Infrastructure · Astronomy · Astroinformatics · Virtual Observatory

C. Cui (✉) · Y. Tao · C. Li · Y. Yue · L. Xu · C. Wu · B. He · D. Fan · S. Li · L. Mi
Y. Chen · Y. Xu · J. Han
National Astronomical Observatories, Chinese Academy of Sciences, Beijing, China

C. Yu · J. Xiao
Tianjin University, Tianjin, China

F. Wang
Guangzhou University, Guangzhou, China

G. Zhang
Alibaba Cloud Computing Ltd., Hangzhou, China

© Publishing House of Electronics Industry 2020
China's e-Science Blue Book 2018,
https://doi.org/10.1007/978-981-13-9390-7_3

1 Introduction

Infrastructure refers to the facilities that serving social production and people's lives, which is crucial for ensuring the society and economy of a country or an area to function [1]. Infrastructure construction has the characteristics of long cycle, large investment, public welfare, and universal benefit.

E-science is the process of making full use of information technology and infrastructure, promoting the exchange, collection and sharing of scientific and technological resources, transforming scientific research institutes and activities, and promoting the transformation of scientific and technological development. E-science infrastructure includes all kinds of information infrastructure needed in the process of transferring traditional scientific research to e-science, such as high-speed scientific research network, supercomputing, and data application environments [2]. Infrastructure is the general material conditions for survival and development of society, and information infrastructure is a necessity for the smooth development of modern scientific research.

With the development of science and technology, astronomical research has entered the era of data-intensive scientific discovery. Either large research infrastructure or general research projects cannot undergo without the support of information infrastructure. In this chapter, we described the requirements and construction of information infrastructure for these key astronomical science projects. Five representative astronomical major scientific infrastructures, namely the five-hundred-meter aperture spherical radio telescope (FAST), the Large Sky Area Multi-Object Fiber Spectroscopic Telescope (LAMOST), the MingantU SpEctral Radioheliograph (MUSER), the Space multiband Variable Object Monitor (SVOM), and the Square Kilometer Array (SKA), are introduced as examples. We also presented the construction process and application of Chinese Virtual Observatory (China-VO) as a public platform supporting astronomical e-science in China. Finally, suggestions are given on the development of astronomical e-science infrastructure in China.

2 The Five-Hundred-Meter Aperture Spherical Radio Telescope (FAST)

The five-hundred-meter aperture spherical radio telescope (FAST) project as shown in Fig. 1, led by the National Astronomical Observatories, Chinese Academy of Sciences (CAS), is a Major National Science and Technology Infrastructure Program of "the 11th Five-Year Plan" in China [3]. It is the largest single-dish and the most sensitive radio telescope in China, and China has independent intellectual property rights. FAST was built in Dawodang depression, Jinke Village, Kedu Town, Pingtang County, the Buyei and Miao Autonomous Prefecture of Guizhou Province. The site of FAST makes use of the natural depression in Guizhou Karst

Fig. 1 Five-hundred-meter aperture spherical radio telescope known as "Tianyan" (lit. "the Eye of Heaven")

area. FAST was launched on September 25, 2016, and is now in the stage of testing and commissioning.

The FAST project consists of active reflector surface systems, feed support systems, measurement and control systems, receivers and terminals, and observation bases. At present, the construction of infrastructure and hardware has been finished. In order to achieve the best performance for observation, each subsystem and the underlying driver software are undergoing tuning phase. This phase is expected to be completed within 2 years. At the same time, another crucial task is to develop the FAST software system, which will comprehensively integrate FAST into the e-science network of the Chinese Academy of Sciences.

FAST supports six observation modes, each of which requires precise coordination of the subsystems. Based on the data acquisition rate, bandwidth and number of channels, the generation rate of observation data is about 1 TB/h, and the archived data volume is about 3 PB/year. The computing capability required for data processing is about 1P flops.

Compared with real-time streaming data processing, large-scale astronomical data archiving involves more software and hardware technologies, including data center hardware systems, network topology, data layout, storage management, backup and disaster recovery strategies, energy-saving strategies, cross-matching, and even need to consider how to support subsequent data analysis computing.

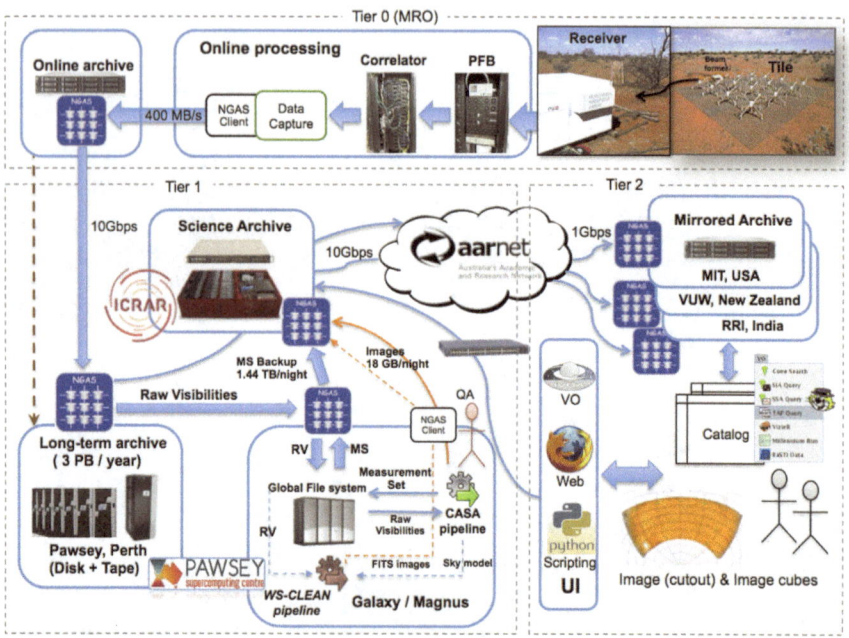

Fig. 2 Data archiving process for MWA

At present, the open-source software Next Generation Archive System (NGAS), which is launched by the European Southern Observatory (ESO) in 2004, is the most commonly used data archiving system in astronomy. Figure 2 shows the improved NGAS and its application in the Murchison Widefield Array (MWA). As a widely used astronomical data archiving system, the most outstanding feature of NGAS is its plug-in architecture, which allows it to be flexibly customized and expanded as needed while providing basic services. Based on the experience of NGAS, the data archiving system of large telescopes like FAST needs to establish not only a hardware environment, but more importantly a data flow specification and a scalable software system necessary to manage large amounts of data accumulated day-by-day.

In recent years, in the pre-research project of SKA data archiving system, researcher and developer were continuously optimizing NGAS. FAST has planned to use NGAS as the supporting software for its archiving system. While learning from NGAS, China-VO also explores the use of cloud storage technology to improve the performance and scalability of the data archiving systems.

The development of the FAST data archiving system is carried out in two phases. Phase one is building hardware environment for distributed and scalable elastic data center, which has been completed already. Besides the local storage and computing facilities at the FAST observatory, the data center and supercomputing platform of FAST are mainly distributed in Guizhou University, Guizhou Normal

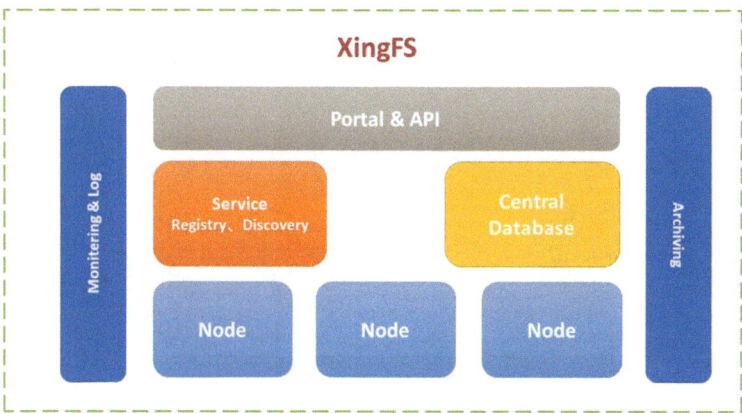

Fig. 3 Architecture of astronomical data management system "Zhuque"

University, and Qiannan Supercomputing Center. The Early Science Data Center of FAST, located in Guizhou Normal University, was established in September 2014 and aiming at storing and preliminary screening raw data. The current storage capacity is 2 PB, and the planned storage capacity is 5 PB, with 20 computing nodes. The National Astronomical Observatories, CAS, and Guizhou University Joint Astronomical Research Center will also provide storage and computing resources for FAST data and develop radio astronomy techniques and methods that are closely related to FAST. The Qiannan Supercomputing Center, jointly built by the High-Performance Computer Research Center of the Institute of Computing Technology, Chinese Academy of Sciences, Dawning Information Industry Co., Ltd (Sugon), and Guizhou Tianyan Group, deployed "Skyeye-1" supercomputing system to serve the data storage, computing, analysis, transmission, and other requirements of FAST project and also provide powerful computing services for Qiannan, Guizhou, and even Southwest China.

In the second phase, a distributed data management system similar to ESO NGAS will be built upon the hardware environment. The China-VO and the FAST Science Department are actively conquering the key technologies for FAST data processing and designed an astronomical data management system prototype named "Zhuque" based on Facebook Haystack and the OSS system of Alibaba Cloud. The system mainly composes of four parts, namely Jing—Data Center Management, Gui—Monitoring Subsystem, Xing—Distributed File System, and Yi—Data Transfer Subsystem. Figure 3 shows the architecture of the entire archiving system.

The project team sought support from various parties and began survey on data centers from the beginning of FAST construction. In 2012, a 5-node cluster was built at the Headquarters of National Astronomical Observatories, CAS, accumulated experience in operation and maintenance. In January 2016, the Guizhou Normal University node of the FAST Early Science Data Center (Fig. 4) has been built cooperatively by National Astronomical Observatories, CAS, and Guizhou

Fig. 4 FAST early scientific data center in Guiyang

Normal University. In June, the Dawodang node was built at the FAST site. Both nodes contain parallel storage systems and blade servers, with a total storage capacity of ~3 PB and a peak computing of 50 TFLOPS. In 2018, these data center nodes are expected to connect by high-speed optical cables.

The construction of distributed data centers employed various hardware combinations and network solutions and accumulated experience for the FAST data center. While the construction of the FAST data center has made great progress, it still faces technical challenges on operation and further development. In particular, without enough space and funding, the first phase of the data center was built to meet only the requirements of early scientific observations. With increasing demand from new scientific goals, the original design of data center can no longer meet the needs of high-speed data storage and high-performance computing. It needs to be optimized for new requirements and considers the expansion of the data center scale in future design.

In the future, the FAST data center will consist of at least three main nodes: the FAST site Dawodang, Guiyang, and Beijing. At the same time, the data center will connect to other nodes in cooperate research institutes and in Guizhou Big Data Center for computing. The FAST Guiyang data center is the central node of the FAST data center, providing long-term storage for FAST data and basic computing capacity.

In early scientific phase, FAST adopts the observation mode of drift scanning. The 19-beam receiver has a bandwidth of 400 MHz. Sampling at 1 GHz, the raw data generation rate is up to 38 GB/s. Such a high data acquisition speed is a technical challenge for both transmission and storage. The data signals at the FAST Dawodang site are mainly transferred through three components: the receiver, online processing, and on-site storage. The receiver converts the collected analog signals into digital signals and transmits to the online processor at a rate of up to 38 GB/s. The online processor includes FPGA-supported polyphase filter bank and GPU computing units. After the raw data being processed online, the data will be compressed to ~ 6 GB/s, temporarily stored at the site, and then quickly transferred to Guiyang.

The main task of the data center in Guiyang is data storage. The data will be transmitted from the Dawodang site to the Guiyang data center through a high-speed optical fiber network. Data storage is a hierarchical system, composed of memory, solid-state drives, RAID, and tape library. The hierarchical storage system automatically controls the movement of data and provides API for users to access the data stored on disk and tape and optimize data movement and hierarchical storage.

Computing service will be provided by a cluster consisting of 200 nodes, which is placed in the Guiyang data center. Part of the mirrored file inventory is placed in the Headquarters of the National Astronomical Observatories, CAS in Beijing. Data will be transferred from remote archives in Guiyang to the mirror archives in Beijing via private networks or public network-based VPN.

In the next 5 years, FAST is expected to generate more than 20 PB scientific data, or even more as the electronic technology develops continuously. The first phase of the Guiyang node of FAST data center will provide 20 PB storage, meeting the requirements for about 3-year operation, and keep the potential for growth. The data will be preserved for long term. In the next 5 years, the amount of FAST data could reach 30–50 PB.

Due to the limitations of natural conditions and radio-quiet requirements, large data centers cannot be built at FAST site. Storage and processing of FAST data will be mainly carried out in Guiyang. Simultaneous observations of pulsars, fast radio bursts, neutral hydrogen, and other scientific targets require the raw data to be transmitted to Guiyang at a rate of 6–8 GB/s, which needs a dedicated network of 100GbE. Dedicated networks have not been applied in the field of astronomy before. In the future, SKA will need more bandwidth. The dedicated network can solve the data transmission bottleneck in the next 5–10 years and also provide the research foundation for China to compete for SKA work package and drive the development the big data industry in Guizhou.

FAST is a single-dish telescope which has lower requirements on data transmission than array telescopes, for example, the SKA. As network technology develops, in the future, 400GbE and 1TbE network will be employed, so that the sampled data can be transmitted to Guiyang in real time. Also, with the development of GPU technology, the baseband data will be processed in real time. The development of technology will break the bottleneck of data transmission and computing and implement algorithms which were difficult to implement 10 years ago, such as coherent dispersion for pulsars and radio bursts search and open up new research directions in astronomy.

3 The Large Sky Area Multi-Object Fiber Spectroscopic Telescope (LAMOST)

The Guoshoujing telescope (LAMOST—The Large Sky Area Multi-Object Fiber Spectroscopic Telescope) is a special quasi-meridian reflecting Schmidt telescope which has both large fields of view and large aperture [4, 5]. LAMOST optical system consists of a reflecting Schmidt Ma at the northern end, a spherical primary mirror Mb at the southern end and a focal surface in between. Mb has a size of 6.67 m × 6.05 m, which consists of 37 hexagonal spherical sub-mirrors, each of them having a diagonal diameter of 1.1 m and a thickness of 75 mm. Ma is 5.72 m × 4.40 m, which consists of 24 hexagonal plane sub-mirrors, each of them having a diagonal diameter of 1.1 m and a thickness of 25 mm. Both the primary mirror and the focal surface are fixed on their ground bases, and the reflecting corrector tracks the motion of celestial objects. Hence, the celestial objects are observed around their meridian passages. The light collected is reflected from Ma to Mb, again reflected by Mb and forms image of the observed sky on the focal surface. The light of individual objects is fed into the front ends of optical fibers accurately positioned on the focal surface and then transferred into the spectrographs fixed in the room underneath, to be dispersed into spectra and recorded on the CCD detectors, respectively and simultaneously. The overall concept and key technical innovations make it a unique astronomical instrument in combining a large aperture with a wide field of view. The available large focal surface accommodates up to 4000 fibers, by which the collected light of distant and faint celestial objects down to 20.5 magnitudes is fed into the spectrographs, promising a very high spectrum acquiring rate of several ten thousands of spectra per night. LAMOST is located in Xinglong Station of National Astronomical Observatories, Chinese Academy of Sciences, on the south of the Yanshan main peak, Lianying Village, Xinglong County, Hebei Province (40°23′39″N, 117°34′30″E), 960 m above sea level.

As a national major scientific project, LAMOST started construction in September 2001 and was completed in October 2008 (Fig. 5). In April 2009, LAMOST passed the process appraisal organized by the Chinese Academy of Sciences and by experts review on construction, finance, devices, and archives. In June 4, 2009, LAMOST successfully passed the national acceptance of the National Development and Reform Commission was the completion of LAMOST made a breakthrough on the difficult problem that acquires both large telescope and large aperture of an astronomical telescope. It has become a large-field telescope which has the largest aperture currently in the world. It is a milestone in the development of optical telescopes in China, which has significantly improved China's capability of independent innovation in the field of multi-target fiber optical spectrum observation facilities.

On October 23, 2011, LAMOST officially launched the pilot survey and was ended on June 24, 2012. In September 2012, LAMOST officially entered the scientific survey phase. In June 2017, LAMOST successfully completed the 5-years

Fig. 5 LAMOST in October 2008 when construction is completed

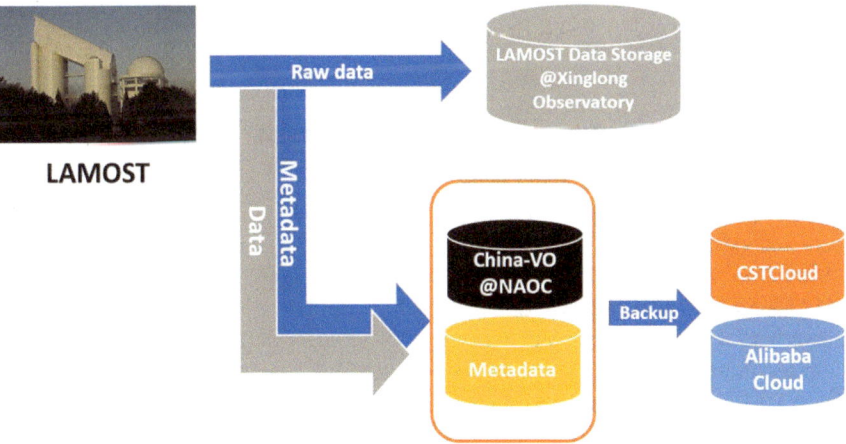

Fig. 6 LAMOST original observation data archiving process

first phase of the spectral survey mission. During this period, LAMOST conducted a total of 1668 days of scientific observations and accumulated 16.53 TB original observation data. The original observation data archiving process is shown in Fig. 6. First, a complete backup is saved at the National Observatory's Xinglong Observation Base. The next day, the original observation data and related metadata were transmitted to China-VO at the National astronomical Observatories

Headquarters through the fiber optic line connecting the National Astronomical Observatories Headquarters and the Xinglong Observatory Base to fulfill the storage of data and metadata. Furthermore, the data and metadata after storage will be uploaded to China Science and Technology Cloud and Alibaba Cloud for backup and remote disaster recovery.

By the end of the first phase of the spectral survey mission in June 2017, LAMOST had acquired approximately 9 million celestial spectral data, far exceeding the sum of the spectral numbers obtained by the Spectral Sky Survey project worldwide, including 7.25 million high-quality spectrum (signal-to-noise ratio greater than 10) and 4.92 million stellar parameters.

LAMOST is the first in the world to realize the large-scale galaxy spectral survey of the continuous coverage and statistical unbiased sky area and establish the world's largest inherited celestial spectrum database. Using these massive spectral data, astronomers have achieved a series of influential research results in important frontier areas such as the structure and evolution of the Milky Way and extragalactic astronomy.

In the past 6 years, the raw observation data accumulated in the entire first phase of LAMOST spectral survey was 16.53 TB. Why is it considered big data? Actually, because LAMOST is performing spectral surveys with an exposure time of at least 1.5 h per sky region. It is very different from photometry observations which can be done by a short exposure in a few seconds. Therefore, every spectrum is of great value, and its value of scientific output cannot be simply measured by the amount of data.

4   The MingantU SpEctral Radioheliograph (MUSER)

The MingantU SpEctral Radioheliograph (MUSER) is the most advanced solar radio imaging device developed by China (Fig. 7). It was approved in 2009 and accepted in July 2016. The initial investment for MUSER is 65.1 million RMB. It is one of the national major scientific research equipment development projects. The research of solar activity is a main direction of solar physics, and it is also one of the main research areas deployed in the discipline development and scientific frontier of China's Mid- and Long-term Science and Technology Development Plan.

MUSER aims at filling in the scientific gap of high-resolution radio imaging observations in the initial energy release area of solar burst [6, 7]. It will result in origin research outputs on the origin of intense solar activity, and its development pattern, therefore, lead to China's leadership in the field. It will also greatly enhance our ability to forecast solar activity, which provides a solid guarantee for aviation, aerospace, satellite communications, and defense construction.

The MUSER adopts aperture synthesis imaging technology, and its goal is to build a new generation solar-specific radio imaging observation equipment with high spatial resolution, high-frequency resolution, and high time resolution at the same time. The MUSER consists of two arrays of high and low frequencies, and a

Fig. 7 MingantU SpEctral Radioheliograph

total of 100 antennas are arranged on three spiral arms in the area about 10 square kilometers. All antennas are equipped with high-performance ultra-wideband dual-circular polarization feeds. The outdoor receiving unit uses high-precision thermostats. The solar radio signals received by the highly stable receiving system are transmitted via 100 optical cables which is 3.4 km long and in the unfrozen soil layer 2.5 m beneath the ground controlled by the central computer, to the central observation room for centralized analog receiving amplification and frequency conversion, large-scale high-speed digital correlation array processing, storage, calibration, and observation data processing in order to obtain real-time high-quality multi-dimensional image of the Sun.

The main features of the MUSER are high temporal, high spatial and high-frequency resolutions, and large dynamic range. In 0.4–15 GHz frequency band, it has a time resolution of hundreds of milliseconds, a spatial resolution up to 1.7 angular second and a frequency resolution of 25 MHz for continuous imaging observations of the Sun. It is far superior to other existing equipment around the world in terms of spatiotemporal and frequency resolutions. This makes the overall data processing very demanding. The details of MUSER are as follows:

- MUSER-I low-frequency array has a total of 40 antennas. The total number of baselines is (40 × 39/2). There are 64 frequency channels over 0.4–2.0 GHz. Each channel records 40 × 39/2 Fourier points within 3 ms. Assuming it observes 10 h per day, the highest time resolution is 3 ms, the data traffic is about 32 MB per second, 1.92 GB per minute, 1.2 TB per day, and about 36 TB per months.
- MUSER-II high-frequency array consists of 60 parabolic antennas and receiving equipment. The diameter of each antenna is 2 m. The astronomical image is reconstructed at 528 points of observation. It is assumed that everyday it will be observed for 10 h, and the amount of data acquired is approximately 3.3 TB. Therefore, when the MUSER system equipment enters normal observation and operation, the amount of observation data generated by MUSER every month will be close to 150 TB.

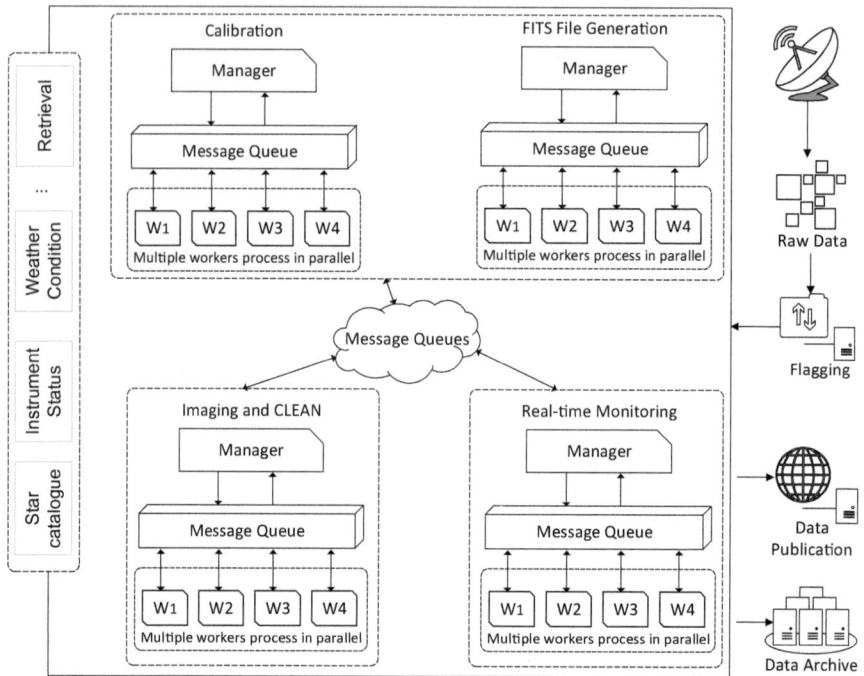

Fig. 8 Distributed data processing framework of MUSER

In addition, if MUSER's mapping calculation is performed in real time, it requires the images of 16 frequencies, from low to high, to be processed within 3 ms. The image size is 256 pixels at a low frequency around 400 MHz, but reaches 5120 pixels at high frequencies. At the same time, intensive calculations such as Fourier transform are frequently involved in image processing.

In order to meet requirements of high-speed data reception and processing for MUSER high-resolution observation, Fig. 8 shows the distributed data processing framework of MUSER. The data received by the MUSER antenna will first go through an algorithm for tagging the abnormal data and then enter the data processing system. The data processing system adopts multi-node distributed parallel computing. The workers are organized by masters to perform different tasks individually, including data preprocessing, data format conversion, image generation, and real-time monitoring. During the tasks, the system can communicate with the service window using ZeroMQ and obtain relevant information, such as data retrieval, instrument status, weather status, or star table calculation. The resulting image or FITS file will be published online.

MUSER passed the expert review on July 2016. By December 2016, the overall status of the Mingantu Sun Observation Base where MUSER is located is as follows:

- Internet access: China Mobile's 40 Mbps fiber optic line.
- Computing platform: 8 high-performance computing servers.
- Storage: 4 sets, for low frequency, high frequency, subsequent backup and real-time data processing, respectively. The current capacity is about 80 TB.
- Desktop computer: 20 computers.
- Application platform: OpenCluster (self-developed).
- Data publishing platform: self-developed, located at the National Astronomical Observatories Beijing Headquarters.
- Database software: MySQL and Redis.

The supporting capacity of current information infrastructure is still lacking behind the demand of MUSER's scientific observations and needs to be continuously expanded and upgraded in the future. MUSER's data storage and sharing are one of the major issues. The project team is actively planning the connection to Alibaba Cloud Node in North (Zhangbei, Hebei Province), hoping to achieve long-term preservation and sharing of data with the help of the private cloud of China-VO and Alibaba public cloud.

In the next few years, MUSER will be a world-leading solar radio telescope. It draws great attention from the international community of solar physics and is expected to obtain original results in exploring the origin, occurrence, and development of intense solar activities.

5   Space Multiband Variable Object Monitor (SVOM)

The Space multiband Variable Object Monitor (SVOM) (Fig. 9), planned to be launched in 2021, is a joint project between the Chinese and French governments. Its main scientific goal is multi-band detection of gamma-ray bursts in the universe [8, 9]. The onboard loads of SVOM satellite include wide field X-ray and gamma-ray telescope (ECLAIRs: French), Gamma-Ray Monitor (GRM: Chinese), Micro-channel X-ray Telescope (MXT: French), and Visible Telescope (VT: Chinese). And the ground telescope: Ground-based Wide-Angle Camera (GWAC: Chinese) and Ground-based Follow-up Telescopes independently built by China and France (CGFT: Chinese and FGFT: French).

The specially designed instrument for the SVOM project has higher sensibility and ability for detecting high-redshift gamma-ray bursts and neighboring dark/soft bursts than similar telescopes in the world. Also, the large-scale field-of-view optical telescope array designed by the Chinese team has benefited from its large field of view ($\sim$ 5000 square degrees) and high time sampling (15 s and one frame) design features, which will have great advantages in detecting gamma-ray bursts

Fig. 9 Space multiband Variable Object Monitor (SVOM)

and optical search for temporary sources of explosions in the universe. After completion of the project, it will bring great improvement to observational research on space astronomy and gamma-ray in China, and the observation and research of transient sources. At the same time, it will bring great improvement in information technology such as multi-band integrated tracking and measurement and big data processing.

The main scientific objectives of the SVOM project are multi-band observations of gamma-ray bursts and transient sources (gravitational wave electromagnetic counterparts, neutrino event electromagnetic counterparts, etc.) and multi-band joint measurements of the sky and the ground. This requires synchronous collaborative observations between telescopes around the world either on the ground or on the sky has reliable data transmission networks, as well as optimized transmission protocols and data formats. At the same time, it requires a simple, reliable, and uniform telescope remote control platform to carry out joint observation by each telescope observation network.

Ground-based Wide-Angle Cameras (GWACs) are the ground observation system of the SVOM project. It is designed to be consist of 36 wide-angle telescopes with an aperture of 18 cm and each equipped with a CCD camera with a resolution of 4 K × 4 K pixels. The camera array covers 5000 square degrees of the sky, and the time sampling rate is 15 s. In each observation night, continuous observations on fixed areas last 4–5 h. Due to its large sky area coverage and high time sampling rate, GWAC has unique advantages in discovery of the burst sources, short timescale transient sources, and study high time-resolution light curve.

The large field of view (5000 square degrees) and high time resolution (sampling every 15 s) of the GWAC observation device enabling the device to generate in every 15 s:a data stream of 0.66 gigapixel image data, 6×10^6 data records of the catalog (each record has about 20 columns of attribute parameters). The total amount of image data generated per observation night is 4.9 TB, and the catalog data is $\sim 1.0 \times 10^{10}$ records/night. The main scientific goal of GWAC is to find the source of transient sources and optical anomalies. The rarity of transient sources requires comprehensive monitoring and management of all observed data and real-time processing.

GWAC will generate raw data in TB everyday under full load operation. After data processing, a large number of advanced scientific products will be produced, including transient source samples, sky survey catalogs, various types of light-varying samples, small celestial brightness, and orbital parameters in solar system. In day-to-day operations, these data are geometrically increasing. All of these raw data, calibration data, advanced products constitute the following characteristics:

(1) **Large amount of data**: When running continuously for more than 3 years, the original data will reach PB. If other scientific products are included, the amount of data will increase a few times. Such large amount of data will be a great challenge for data management and distribution.

(2) **Various scientific data and product types**: For example, there are various types of star catalogs and various types of raw data and calibrated data.

(3) **Wide range of user needs**: Researchers interested in GWAC data may have different research interests on asteroids, binary or multi-star systems, supernovae, gamma-ray bursts, etc. Other astronomers might study other temporary sources, such as gravitational wave electromagnetic counterparts associated with black hole or neutron star fusion processes.

These features require the project team to consider the following two questions when publishing data: How to release the data? How can it better serve researchers in various fields?

Given the diversity of scientific products and data types, as well as the complexity of the needs of astronomers, data will be released in two ways:

(1) **Active release** such as the survey catalog, variable star light catalog and these data will mainly be published through public network data download and hard disk hard copy. However, both methods have their own risks. The former requires stability, speed, and security of the network and can provide multiple users with large data requests and downloads at the same time. The problem of the latter is mainly the time spent on the journey and data security carried by large hard drives.

(2) **Passive service publishing**: A data publishing system based on database management and web page request will be established for researchers interested in GWAC data in various fields of domestic and international astronomy to obtain specific types of data. This has the advantage of making good use of GWAC data and meeting the needs of different astronomers in various fields.

GWAC is a world-leading transient source surveys, featuring both large field of view and high time sampling (sampling at every 15 s with a total field of view of approximately 5000 square degrees). It challenges the existing technologies in several aspects of astronomical observation, for example, in big data management for time series, real-time fast data processing, rapid analysis, and mining of big data artificial intelligence. In response to the requirements and challenges of SVOM project on information technology, the Chinese team of SVOM project aimed at the international frontier and carried out research and program planning based on the actual situation in China.

(1) Integrated space-ground observation network. It requires real-time information transmission of transient source event and requires the establishment of an Internet-based messaging network with an alert trigger mechanism, i.e., when the SVOM satellite discovers a gamma-ray burst or GWAC and other ground-based telescopes discover transient sources, the source of the discovery information is transmitted to other telescopes in the network in a timely manner. Nowadays, the most popular messaging network is GCN. However, as the number of events increases, the instant event data increases, and the number of connected receivers also increases simultaneously. Considering security requirements such as encryption and password authentication, it is highly needed to develop a new message transmission system and specify data formats and protocols. The astronomy community is working on the pub/sub-based messaging mechanism and promotes the establishment of a VOEvent message data transmission standard format.

The SVOM Chinese team and French team collaborate on developing the SVOMNet transmission network. The real-time message transmission network mainly has the following candidate solutions: XMPP-based message mechanism implemented using Openfire, VTP VOEvent Transport Protocol (VTP) based on network event-driven engine-twisted, transmission scheme based on ZeroMQ communication protocol which is known as the "fastest message queue in history." Meanwhile, these candidate solutions are all been tested and evaluated. The main criteria of performance are the transmission speed of message data (the relationship between the number of users and the number of messages) and the reliability of message (i.e., the content is not lost, the correct delivery of message, the self-diagnosis function of the network interruption).

Research on automatic observing system is response to message data in real time. The workflow of data transmission mainly consists of receiving the alarm message transmitted in real time, the telescopes automatically respond to the alarm message, the telescopes automatically point to the observation target, automatically process the acquired data, automatically transmits the processing result back to the scientific center, observers can perform manual intervention through automated observing systems, make changes of observation strategies, and check status.

(2) Develop a space load simulation system. In response to the Chinese optical telescope load, in order to effectively verify the load design, develop data calibration strategies, and evaluate load detection capabilities, the Chinese team developed a space optical telescope image simulator. The simulator is a comprehensive integrated system involving CCD camera, noise simulation, optical imaging PSF contour, transmittance, response curve, spatial sky background, and observation target distribution. This is a basic work that has great significance for space astronomy in China, and the research goal is to develop a general basic platform.

(3) Fast and accurate real-time data processing capability. As required, the system should quickly identify and measure data attributes. The scientific goals of SVOM observations require fast real-time data processing, i.e., the ability to quickly identify gamma-ray bursts (transient sources). More specifically, the ability of real-time data processing reflected in fast cross-matching algorithm, multi-catalog fusion technology (i.e., support multi-catalog related query), hardware, and network resource stability.

Establish a catalog database for multiple survey catalog fusion to provide back-end cross-matching support for identification of gamma-ray burst. The main technical issues are the techniques of multiple catalogs fusion (including the consistency of data calibration), back-end database technology, cloud platform, or local quick data query support, carrying out research on fast cross-matching algorithm to support the rapid cross-matching of multiple catalogs.

(4) Capability of real-time big data processing. The self-developed GWAC system of the SVOM project will generate 40 4 K × 4 K images every 15 s. To avoid the accumulation of data streams, the images need to be processed within 15 s, including point source extraction, cross-matching, data written to database, searching for transient sources. Distributed parallel processing is adopted to effectively integrate hardware resources to achieve parallel processing. Research on accelerated parallel processing based on GPU technology, mainly includes rapid cross-matching, point source extraction, and image subtraction. Research on database acceleration based on PostgreSQL kernel technology.

(5) Artificial intelligence techniques for big data mining. The GWAC camera will produce $\sim 1.0 \times 10^{10}$ records per night. In order to dig out light-changing objects, such as gravitational micro-lens events, from large amount of data in time series catalog, we must rely on big data mining and artificial intelligence technology. In addition, the use of image methods to find transient sources also requires artificial intelligence techniques to screen transient sources from noises.

The China team of SVOM project uses one of the machine learning methods—random forest algorithm to effectively screen transient sources, and the autoregressive integrated moving average (ARIMA) model is used to analyze the optical data and automatically process the data stream. The advantages of this method are (i) adopting abnormal detection of dynamic stream data, and using a sliding window to determine in real time whether the sampling points within a fixed window width is time-varying, (ii) the statistics and predictions of historical data do not depend on the actual mathematical model parameters such as the optical variable period, which means no prior model is needed, (iii) manual intervention parameters are not required or minimized, and parameters can be adaptively adjusted during the calculation. Other machine learning methods are still under research and development.

(6) Big data management capabilities. High time sampling of the GWAC camera and its large field of view design make the time series catalog data grow by $\sim 1.0 \times 10^{10}$ records per night, and it is expected to reach 100 billion records within the 10 years project period. This poses a challenge to traditional database management platform, and a data management platform with big data storage capabilities is required. The SVOM Chinese team researches on the use of the column store database MonetDB as a database management platform and adopted the shared-nothing distributed storage solution. MonetDB has excellent performance in big data computing and storage. MonetDB has the advantage of bringing calculations near data, so that data can be processed quickly. Optimization of the data system provides an advantageous option for the big data management processing platform.

6 Square Kilometer Array (SKA)

The international big science projects Square Kilometer Array (SKA) is the world's largest aperture synthesis radio telescope proposed by the international astronomical community (Fig. 10). It has a receiving area of one square kilometer and a frequency coverage of 70 MHz–25 GHz [10]. The SKA consists of approximately 3000 15-meter-diameter paraboloid antennas, 250 sets of intermediate-frequency and low-frequency aperture arrays with a distribution range of more than 3000 km, forming a spiral-arm array telescope. SKA will be built in radio-quiet areas in Australia, South Africa, and other eight countries in southern Africa. Currently, SKA is in the pre-construction development stage (2012–2018), which is led by SKA Organization (SKAO), an independent legal entity, and consists of 10 member countries including UK, Australia, South Africa, China, Netherlands, Italy, New Zealand, Canada, Sweden, and India. The construction cost of SKA is about 8 billion euros, jointly funded by multi-government and national research institutions. It will be built in two phases: the first phase SKA1 (2019–2024) will build 10% of the SKA, and the second phase SKA2 (2025–2030) will construct the remaining part.

Fig. 10 SKA Telescopes

Since the launch of international cooperation in 1993, SKA has evolved in the past 20 years through the process of developing engineering concepts, site selection, refining scientific goal and key technology research and development and becomes a significant international big science and engineering project. SKA will provide great opportunities for humans to understand the universe, while enormous efforts will be made for the SKA project as it was for Hubble Space Telescope (HST) and the Large Hadron Collider (LHC). SKA project requires technological innovation in all aspects, such as low-cost antennas, large-capacity data transmission, high-speed parallel processing computers, and large-capacity data storage units.

The SKA1 mid-frequency array consists of 200 15-meter-diameter dish antennas distributed over 150 km². Each dish antenna collects the radio frequency signal into the feed, and then the signal is being amplified, filtered, mixed, synchronized, and converted into a digital signal by the ADC, and transmitted through optical fiber to the central data processing system (CSP). The data flow of each antenna is 90 Gb/s. The CSP system performs multiple correlation calculations (such as pulsar signal achromatic dispersion and timing processing) for each pair of antennas and each frequency channel and obtains "spatial coherence data." The input data to CSP is 28 Tb/s, and output data is 38 Tb/s. Spatial coherence data is transmitted over fiber to the scientific data processing system (SDP). The SDP needs to meet a computing requirement of 150 PFLOPS and a data storage requirement of 57 PB per day. The SDP performs iterations such as self-calibration and deconvolution to generate a radio source distribution image. The images of the radio sources that meet the scientific needs are stored in the storage system, and the data is analyzed by the astronomers to produce scientific research output.

The SKA data processing infrastructure has two parts: CSP and SDP. The CSP mainly performs beam synthesis, correlation calculation, and detection of rapid-changing radio signal sources (such as pulsars). The CSP is located near the antenna array where is sparsely populated and has limited electric power. It uses low-energy high-intense large-scale real-time hardware processing technology (FPGA/DSP) to complete task of single algorithm and huge data flow (correlation, folding, and shifting). The SDP element carries out image processing and scientific research tasks. The location of SDP is relatively convenient, and electric power is also more sufficient. The hardware platform of SDP will be composed of many computing islands, and each computing island is composed of many computing nodes. The computing islands are connected by shared file system and management network etc., but they are relatively independent. The number of computing nodes is not necessarily the same, and the configuration might also be different. The challenges for SDP are processing speed, massive parallelism, big data streams, large caches, and low power consumption. SKA regional center (SRC) is still in the stage of preliminary design, aiming to undertake SKA big data and provide scientific research platform for SKA users. SKA International Organization (SKAO) formed the SKA Regional Center Coordination Group to provide conceptual design and global coordination. China is actively planning the SKA China Regional Data Center.

The construction period of SKA1 is the whole year of 2019 and will be built in Australia and eight countries in South Africa/Africa. The SDP centers of the two site countries are located in Perth, Western Australia and Cape Town, South Africa, respectively. In order to timely transfer SKA's scientific data to China for domestic astronomers as well as provide services to international users in Asia Pacific, it is necessary to set up a 100 Gbps optical cable line from Perth, Western Australia and Cape Town, South Africa to the SKA regional data center in China, which has the capacity of 300–500 PB data transmission per year. Currently, such a big data transmission requirement has not yet found a feasible solution.

7 China-VO

As the amount of data in astronomy increases, scientific research collaboration is increasingly widespread, traditional research models must change to meet the requirement in data-intensive astronomy research. Around year 2000, astronomers realized the need to standardize the whole processes of astronomical data access and proposed a cross-disciplinary concept that involves astronomy, computer science, and information science—the Virtual Observatory (VO). In order to unite the efforts in Virtual Observatory from different countries, an international conference "Towards the International Virtual Observatory" was held in Germany on June 2002. The International Virtual Astronomical Observatory Alliance (IVOA) [11] was founded at the conference, and a number of working groups have formed to develop standards and specifications for data interoperability, and to enable data product generation, data publishing, data discovery, data access and retrieval under a standard VO framework. Astronomers only need to log on to the VO system to get rich data resources and powerful services, thus get rid of tedious tasks such as data collection and data processing and concentrate on their research problems.

The IVOA has now grown to 21 member projects. As an organization for the development of astronomical data interoperability standards and protocols, IVOA has successively set up working groups and interest groups including Applications, Data Access Layer, Data Model, Grid and Web services, Registry, Semantics, VOTable, VO query language, VOEvent, Data Curation & Preservation, Education, Knowledge Discovery in Databases, Operations, Theory, and Time Domain.

The massive data in astronomy is usually managed by a large data center or a small research team. Data and computing resources are provided to astronomers and other users through Internet, as a platform which forms the resource layer in the IVOA architecture. The consumers of data and computing resources, either individual astronomers, research teams, or computer systems, interact with resources via the user layer. In the architecture, VO is a middle layer that connects the resource layer and the user layer in a seamless and transparent manner. VO provides a set of technical frameworks for resource providers, allow resources to be shared in a way which users can easily find, access to and use. A series of protocols and specifications developed by IVOA provide guidance and constraints for the implementation of these features.

The emergence of VO solves the problem of inconsistency in access standards of different database systems and provides tools for cross-matching as well as analysis of image and spectral data. By using the tools, astronomers save time by avoiding repetitive work. Thanks to the efforts of VO teams around the world, some of the services described above have been implemented partially or completely under the framework of VO. For example, VOspec, Aladin, SPLAT, VOSesame, VOplot, TOPCAT, Iris are all very successful and useful VO applications. Nowadays, there are hundreds of astronomical data centers and astronomical projects around the world support VO standards. NASA and the European Space Agency (ESA) require their astronomical science projects to follow VO standards for their data release. The ESA has been greatly supporting VO development by requiring all of its astronomical databases to be compatible with VO standards, and actively participating in the development of VO applications and tools such as VOSpec, VOQuest, Registry, DALToolKit. Through participating in the European VO program and IVOA, the ESA VO project also plays an important role in the development of VO standards.

If the future astronomical data are all compatible with the VO standards, astronomers only need to master the VO tools to access to all the astronomical data resources for scientific research. VO is an important public information supporting platform for global astronomical research.

In 2002, led by National Astronomical Observatories, Chinese Academy of Sciences, the Chinese astronomical community proposed the idea of China-VO [12]. The same year, China-VO became a member of IVOA. Research areas of China-VO mainly include the development of China-VO platform, unified access to domestic and oversea astronomical research resources, VO-supported projects and observation facilities, VO-based astronomical research demonstrations, and

VO-based astronomical science education. China-VO is proposed as an applied research project since the very beginning, aiming at bridging astronomy and information technology, and enabling the application of advanced information technology to astronomical research.

In 2014, China-VO completed the development of a cloud platform for astronomy based on CloudStack, and deployed several cloud nodes in China. Since the second half of 2016, incorporated with Alibaba Cloud, China-VO begins to explore a new hybrid architecture of "public cloud + private cloud." Jointly developed by various astronomical observatories of Chinese Academy of Sciences and other partners, China-VO makes full use of the existing network, storage, computing, and other cyber-infrastructure, as well as advanced information technology and research outcomes in VO, provides services for observation time application for domestic telescopes, data sharing and use. Scientific data, cloud computing, software and tools needed for astronomical observation and scientific research activities are put together to form a physical distributed and logically unified e-science platform. The service provided by the platform covers the entire process of scientific research, from telescope time application to scientific paper writing, combines information infrastructure and resources with astronomical research activities. The platform improves the operational level of astronomical observation equipment and promotes sharing of equipment and scientific data. China-VO is a data-driven research and information environment, based on standard, complete, quality-assured metadata, and scientific data systems, and it serves astronomers and other scientific users through software, tools, and services with interoperability and provides an information resource platform for the whole life cycle of scientific research (Fig. 11).

The scientific research activities of observational astronomers mainly include telescope time application, scientific observation, data processing, and results from publication. Astronomers submit telescope time application to the telescope time allocation committee in the form of observation proposals and obtain the required raw observational data through scientific observations and then analyze and process the raw data to obtain data products using software tools; finally, the scientific research results are published and shared in the form of scientific research papers, and the data products are also shared according to the conventions in astronomy. Every step of the astronomer's scientific research activity, the input and output of each step are recorded and preserved in the VO environment, establishing the database for observation proposal, raw data database, software tool, data product, and thesis result, etc. Detailed log and metadata are also retained in the VO system. The VO system provides a comprehensive scientific research process management which is easily traceable through data generation, data acquisition, data processing, and results sharing.

The data is stored in physically distributed astronomical observatories. Each with a local data center maintains observational data of the telescopes. Among them, the National Astronomical Observatories, as the main node of the China-VO, maintaining not only its own data, but also some other data from other stations, as well as data mirroring of many large overseas survey projects and database systems.

Fig. 11 China-VO Portal

China-VO platform is a hybrid of three service modes of cloud computing, i.e., Software as a Service (SaaS), Platform as a Service (PaaS), and Infrastructure as a service (IaaS). The function structure diagram is shown in Fig. 12. In general, the system is based on China Science and Technology Cloud and Alibaba Cloud. Each observatory serves as resource nodes. Leveraging virtualization and cloud computing technology, distributed storage, data and computing resources are managed uniformly by the intermediate platform and provided to the upper-level integrated service platforms. User system and authorization service, resource monitoring and fault tolerant services are control centers of the system. They collect user operation logs and runtime information of each service platform, coordinate the use of system resources, and ensure the stability of system operation.

As shown in Fig. 13, The data archiving system is a hub connecting telescopes and service platforms. Data from each observatory is backed up in a shared cloud storage system on a regular basis. Each service platform can quickly get required data from the massive data through an efficient data retrieval interface. Results achieved on the service platforms will also optimize the telescope observation plans and management decisions. Data archiving is an important part in the research workflow, linking services platforms, and the base cloud nodes. Firstly, by referring to the relevant standards of IVOA, the standardization of metadata for various types of observation data was completed, and the archiving standard of domestic astronomical data was established. Secondly, considering the network bandwidth resources of each

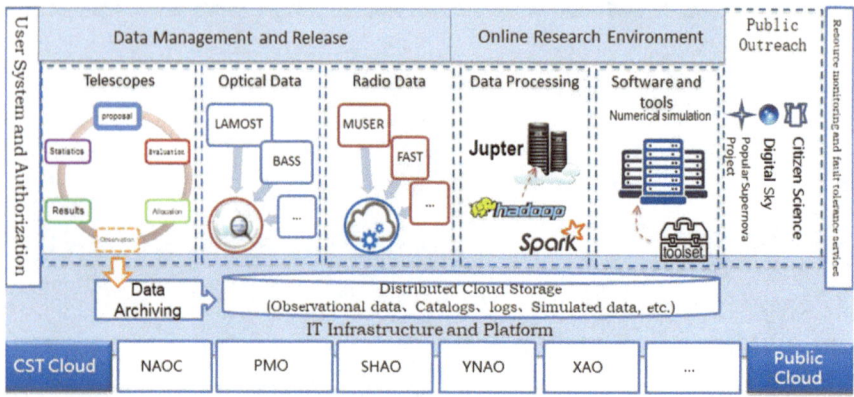

Fig. 12 Function structure of China-VO

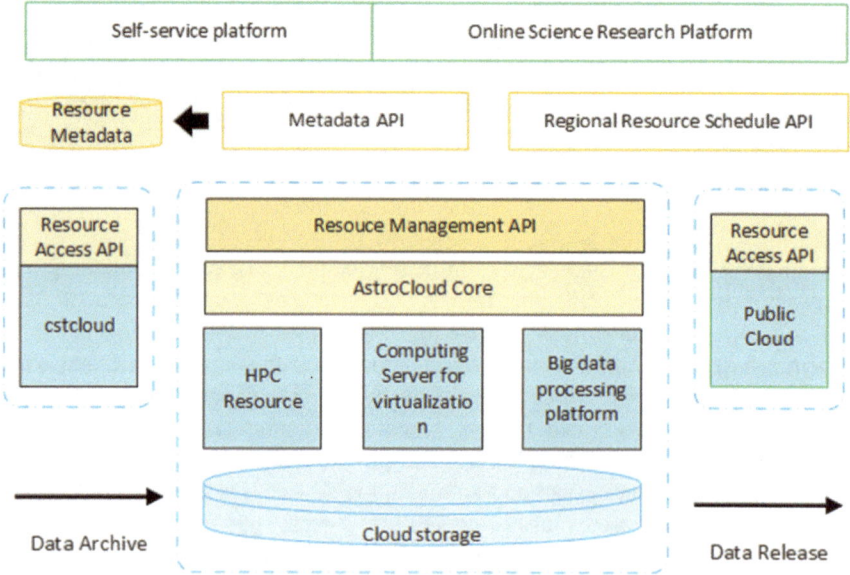

Fig. 13 Overall framework of China-VO Platform

observatory and the basic platform, appropriate remote synchronization strategy and backup strategy have been designed to ensure data security. Finally, for data that is difficult to move on a large scale in a short period, a direct access interface is provided at the data node to link to the service platforms in an in situ manner.

An efficient data retrieval engine based on cloud storage is the key to stable operation of the service platforms. The China-VO retrieval system overcomes the bottleneck of traditional database in managing massive scientific data and meets the needs of data analysis and processing by establishing a distributed metadata index, optimizing the data layout according to location on the sky, and making full use of object storage and caching technology to quickly locate the data and eliminating the dependence on traditional relational database.

The integrated service platforms are built on the base platform. At the service level, service platforms are independent of each other focusing on different functions. However, they share the same underlying storage, data and computing resources. Telescopes and public services focus on processes and logic, which have a lower demand for resources, while astronomical data mining and numerical simulation calculations on optical data, radio data, and time-domain data introduce more requirements on storage and computing resources. Besides providing the required software resources, the system also considers the distribution of data and computing nodes, as well as the network bandwidth, and coordinates the scheduling of various resources to improve the overall utilization and ensure the demand for resources.

The base platform is in the bottom layer of the entire system, responsible for integrating distributed storage and computing resources, and providing storage and computing services for upper-layer service platforms through virtualization. The base platform integrates storage and computing resources via establishing cloud nodes by regions and expanding based on existing techniques.

Since its launch in May 2014, the China-VO Service Portal has developed 7 cloud nodes in National Astronomical Observatories (Beijing), Purple Mountain Observatory (Nanjing), Shanghai Observatory (Shanghai), Yunnan Observatory (Kunming), Xinjiang Observatory (Urumqi), Nanjing University (Nanjing), and Alibaba Cloud (Hangzhou), providing data management and sharing services for more than 10 sets of observation equipment such as Xinglong LAMOST, 13.7 m-millimeter-wave telescope, 1.56 m Sheshan telescope, 2.4 m Lijiang telescope and 25 m Nanshan telescope, and support time application for Lijiang 2.4 m telescope, Xinglong 2.16 m telescope and other telescopes, providing virtual machines and other cloud services for more than 500 users in Chinese Academy of Sciences and various observatories. China-VO Portal has also facilitated the discovery of 13 supernovas and new stars based on domestic amateur astronomical observation data, and it has more than 20,000 registered users. In 2015, China-VO was awarded "the Top Ten Excellent Applications of E-science in Chinese Academy of Sciences." In the era of "Internet +" and big data, China-VO is playing an increasingly important role in supporting the development of astronomy.

8 Summary and Recommendations

Cyber-infrastructure has become an integrated part of astronomical research. This chapter introduces the demand and current state of the major astronomical science and technology projects for cyber-infrastructure, with examples of five representative astronomical big science and technology projects, namely FAST, LAMOST, MUSER, SVOM/GWAC, and SKA. This chapter also demonstrates the construction and application of China-VO as the public supporting platform for astronomical e-science and research.

After more than 20 years of development, the astronomy community in China has accumulated some experience in developing the infrastructure for e-science, in terms of scientific research networks, data storage capabilities, high-performance computing, and data application environments. The development of e-science infrastructure has met the basic needs of current scientific research. However, there is still a considerable gap in the demand for cyber-infrastructure from the astronomical observing facilities that are completed recently or still under construction. In order to give full play to the role of a new generation of major astronomical science and technology infrastructure, it is necessary to plan and develop e-science infrastructure as early as possible. Proposed recommendations are as follows:

Increasing the construction of cyber-infrastructure and supporting for maintenance and management and providing more bandwidth, storage, and computing capabilities. Strengthening the overall construction of astronomical e-scientific platform, efficiently managing shared astronomical data, hardware and software resources and providing a strong support throughout the whole life cycle of astronomical research.

Promoting balanced development of astronomical e-science infrastructure. Overcome the bottleneck and fill the shortboard. On the one hand, we should highly value the construction of high-speed scientific research networks in observing stations and remote areas, so that the massive data of new observation sites and equipment can be smoothly archived into data centers and share with the public. On the other hand, we need to facilitate the construction of astronomical e-science infrastructure and teams in domestic universities, so that the e-science forces of the Chinese Academy of Sciences and of universities can be balanced and promoted. In addition, the China-VO, as an effective astronomical e-science system, should be extended from the Chinese Academy of Sciences to universities and other research institutes and serve as a uniform coordination mechanism for the entire astronomy community in China.

The interdisciplinary intersection of astronomy and information technology has become an irreversible trend of the times. Since 2006, the National Natural Science Foundation of China and the Chinese Academy of Sciences have jointly established astronomical fund which listed "astronomical big data storage, computing, sharing and Virtual Observaory technology" as one of the five research areas, supporting the development of VO and astroinformatics. The guidance of the 2003 National Natural Science Foundation of China describes the research area more explicitly as "application-based research to address issues such as data, calculations, and information extraction for major astronomical projects, including, storage and sharing of massive astronomical data, data mining, high-performance computing and Virtual Observatory technology." In 2008, the National Astronomical Observatories, CAS, and Tianjin University jointly established an astronomical information technology laboratory and upgraded to the Astronomical Information Technology Joint Research Center in 2015. In 2011, the National Astronomical Observatories, CAS, and Kunming University of Science and Technology Astronomical Information Technology Joint Laboratory were established. In 2011, "Astronomical Information Technology" was included in the 2011 Master and

Doctoral Admissions Directory of National Astronomical Observatories as a research direction of "Astronomy Technology and Methods." Astronomical information technology has also been added to the 2018 Ph.D. Admissions Directory of Tianjin University as a research direction of in the computer science and technology major. In 2016, Astroinformatics was considered a sub-discipline of astronomy the third edition of the Encyclopedia of China (Astronomical Volume). At the meantime, the Chinese Astronomical Society is planning to establish an astroinformatics committee to better promote the interdisciplinary development of astronomy and information technology.

Acknowledgements This work is supported by National Natural Science Foundation of China (NSFC) (11573019, 11803055), the Joint Research Fund in Astronomy (U1531246, U1731125, U1731243) under cooperative agreement between the NSFC and Chinese Academy of Sciences (CAS), the 13th Five-year Informatization Plan of Chinese Academy of Sciences (No. XXH13503-03-107). We would like to thank the National R&D Infrastructure and Facility Development Program of China, "Earth System Science Data Sharing Platform" and "Fundamental Science Data Sharing Platform" (DKA2017-12-02-07). Data resources are supported by China National Astronomical Data Center (NADC) and Chinese Virtual Observatory (China-VO). This work is supported by Astronomical Big Data Joint Research Center, co-founded by National Astronomical Observatories, Chinese Academy of Sciences and Alibaba Cloud.

References

1. Jiang Mianheng. Develop information technology vigorously for scientific research and serve national scientific and technological innovation. China's e-Science Blue Book 2011, Beijing: Science Press, 2011, 1(1):1–2.
2. Nan, Kai. The Cyberinfrastructure for Science and Innovation of Chinese Academy of Sciences and Its Applications. Bulletin of Chinese Academy of Sciences, 2013. Vol 28 (4): 476–481.
3. Nan, Rendong; Li, Di; Jin, Chengjin, et al. The Five-Hundred Aperture Spherical Radio Telescope (fast) Project. International Journal of Modern Physics D, Volume 20, Issue 06, pp. 989–1024 (2011). https://doi.org/10.1142/s0218271811019335.
4. Chu, Y. Q. The Large Sky Area Multi-object Fibre Spectroscopy Telescope (LAMOST) Project. Proceedings of the 21st Century Chinese Astronomy Conference, held August 1–4, 1996. Edited by K. S. Cheng and K. L. Chan. Published by World Scientific Publishing Co. Pte. Ltd., P. O. Box 128, Farrer Road, Singapore 912805, ISBN 981-02-3226-8, p. 155.
5. Cui, Xiangqun; Su, Ding-qiang; Wang, Ya-nan. Progress in the LAMOST optical system. Proc. SPIE Vol. 4003, p. 347–354, Optical Design, Materials, Fabrication, and Maintenance, Philippe Dierickx; https://doi.org/10.1117/12.391524.
6. Yan Yihua, Zhang Jian, Chen Zhijun, et al. Progress on Chinese Solar Radioheliograph in cm-dm Wavebands. [J]. PNAOC, 2006, 3(2): 91–8.
7. Yan Y, Zhang J, Wang W, et al. The Chinese Spectral Radioheliograph—CSRH [J]. Earth Moon & Planets, 2009, 104(1–4): 97–100.
8. Cordier B., Wei J., Atteia J.-L., et al. The SVOM gamma-ray burst mission, 2015, published by PoS, proceedings of the conference Swift: 10 Years of Discovery, 2–5 December 2014, La Sapienza University, Rome, Italy. eprint arXiv:1512.03323.
9. Wei J., Cordier B., Antier S., The Deep and Transient Universe in the SVOM Era: New Challenges and Opportunities-Scientific prospects of the SVOM mission, 2016, Report on the

Scientific prospects of the SVOM mission. Proceedings of the Workshop held from 11th to 15th April 2016 at Les Houches School of Physics, France. arXiv:1610.06892.

10. Dewdney, P. E.; Hall, P. J.; Schilizzi, R. T. et al. The Square Kilometre Array. 2009. Proceedings of the IEEE, Vol. 97, Issue 8, p. 1482–1496. 10.1109/JPROC.2009.2021005.

11. Quinn, P.J., Barnes, D.G., Csabai, et al., 2004. The International Virtual Observatory Alliance: recent technical developments and the road ahead. in: Oschmann, J. M., Glasgow, Jr. (Eds.), Ground-based Telescopes, Proceedings of the SPIE. 5493, 137–145.

12. Cui, C.Z., Zhao, Y.H. Worldwide R&D of Virtual Observatory. [C].In: Jin, W.J., Platais, I., Perryman, M.A.C. (Eds.), A Giant Step: from Milli- to Micro-arcsecond Astrometry, Proceedings of the International Astronomical Union (2007) Symposium S248. 2008, 3, 563–564.

Chenzhou Cui Head of Information and Computing Center of National Astronomical Observatories, Chinese Academy of Sciences, Vice President of the International Virtual Observatory Alliance (IVOA), Director of the National Astronomical Observatories, and Alibaba Cloud Astronomical Big Data Research Center engaged in research of Virtual Observatory and astroinformatics and has published more than 70 academic papers. He has been the principal investigator for more than 20 scientific research projects and established strong cooperative relations with over ten world-leading scientific research institutions such as Johns Hopkins University, California Institute of Technology, and Microsoft Research. He has awarded Beijing Science and Technology Star Program, Innovative Talents Program of Chinese Academy of Sciences, and Outstanding Contribution Award of the China Astronomical Society.

e-Science Applications of China's Space Science Satellite Missions

Ziming Zou, Senlin Xiong and Xiaoyan Hu

Abstract The fact that satellite planning in Strategic Pilot Projects in space science had opened a new vigorous era in China's space science was firstly depicted, and then the demands, chances, and challenges of space science e-Science that meet the condition of this new era were analyzed. The "Domain Cloud of Space Science and Technology" (DCSST) project supported by "e-Science projects" in the "12th Five-Year Plan" was taken as a typical example to introduce the space science e-Science practice and progress. Its framework, construction process, and system achievements—the STAR-Network was subsequently introduced, and STAR-Network functions and abilities were described in detail. Furthermore, its application results and serving effects in satellite mission collaborative demonstration, operation, and scientific research assistance were also elaborated. Finally, the future goals were determined, focusing on improving the e-Science core element levels and abilities, and devoting to the construction of a deeply fused and extensive cloud e-Science platform.

Keywords Space science · Satellites mission · e-Science · Domain cloud of space science and technology · STAR-Network

Z. Zou (✉) · S. Xiong · X. Hu
National Space Science Center, Chinese Academy of Sciences, Beijing, China
e-mail: mzou@nssc.ac.cn

S. Xiong
e-mail: xsl@nssc.ac.cn

X. Hu
e-mail: huxiaoyan@nssc.ac.cn

© Publishing House of Electronics Industry 2020
China's e-Science Blue Book 2018,
https://doi.org/10.1007/978-981-13-9390-7_4

63

1   e-Science Opportunities and Challenges in the Vigorous Era of Space Science

1.1   Space Science Satellite Missions Are Flourishing

Based on the spacecraft platform, space science mainly studies the physical and chemical phenomena and the laws that occur in Earth, solar-terrestrial space, interplanetary space, and the whole cosmic space. It involves both macroscopic and microscopic. And it focuses on solving the origin and evolution of the universe and life, and the relationship between the solar system and mankind, and other important basic issues. Satellite mission planning and implementation had become an important and effective way to understand the frontier problems of space science and satellite missions are one of the main methods to obtain first-hand exploration science data and experimental data. Furthermore, it also had become a competitive highland between countries worldwide that own powerful abilities in space science and technology.

The successful launch of the "Dongfanghong-1" satellite in 1970 brought China's "satellite dream" into reality, while the launches of "TC-1" and "TC-2" satellites in December 2003 and July 2004 made the world understand China's space science in another viewpoint. Significant discoveries and achievements had been made through TC-1 and TC-2 scientific data analysis, such as research progresses in magnetosphere sub-storm onset, magnetopause reconnection, plasma sheet flapping, and radiation belt and ring current exploration.

Meanwhile, the "Double Star Project" opened the prelude to China's space science exploration 13 years ago, and the "Space Science Strategic Pioneer Science and Technology Project" (Strategic Pioneer Project, SPP, for short) is presently opening a new vigorous development era for space science in China. With the support of the SPP (Phase I), China has successively launched four advanced science satellites, namely Dark Matter Particle Explorer ("Wukong"), Shijian-10 Returnable Scientific Experiment Satellite ("SJ-10"), Quantum Experiment Scientific Satellite ("Mozi"), and Hard X-ray Modulation Telescope ("Huiyan"). The successful implementation of these four science satellite missions promoted a series of major breakthroughs in both space science research and key technologies. Furthermore, these results also attracted the extensive attention of government leaders and scholars, as well as general public. It also has greatly enhanced China's space science international influence. President Xi gave a high appraisal of the space science satellite programs in the 2016 and 2017 New Year's greetings.

At present, the SPP for space science (Phase II) is progressing in an orderly manner. It is expected to launch the Solar Wind Magnetosphere Ionosphere Link Explorer (SMILE), the Advanced Space-based Solar Observatory (ASO-S), the Einstein Probe (EP), the Gravitational wave high-energy Electromagnetic Counterpart All-sky Monitor (GECAM), the Space-based multi-band astronomical Variable Objects Monitor (SVOM) from 2020 to 2023. Meanwhile, the long-term development blueprint of space science in China has been described as follows: The planning and implementation of a series space satellite programs would allow for original discoveries and breakthroughs in the field of the mechanism of universe formation and evolution, the cognition of extragalactic planets and extraterrestrial life, the process of solar system formation and evolution, and solar activity and the Earth's space environment response. The "Research Report on Space Science Planning 2016–2030" [1] has incorporated the "Black Hole Probe," the "Astronomical Signal Pulse," and the "Astronomical Portrait." There are 23 plans in the space science development plan for 2030. It is also expected that 6–8 new satellite missions would be selected, established, and developed in each Five-Year Plan, and all selected missions would be launched and finish its orbit test in the next Five-Year Plan. Thanks to the government's great attention and support, space science and satellite mission development have entered a vigorous golden period.

1.2 e-Science Requirements in All Life-Cycle Activities of Satellite Missions

e-Science is actually the informatization of scientific research activities, which is characterized by taking full use of network information infrastructure and technology to promote science and technology resource exchange, collection and sharing, finally transforming the manner and mode of activities of science research organizations [2]. Vice Premier Liu Yandong once pointed out that, e-Science is the only way to realize the modernization of scientific research, and the durable innovation and development of space science need the support of e-Science. Space science is an experimental subject, while satellite platforms are mobile laboratories which own powerful ability in scientific data acquiring. And the scientific data is the driving force of space science development. Therefore, to a large extent, the e-Science of space science is the informatization of all life-cycle activities in satellite mission implementation.

Complete satellite mission implementation processes usually consist of four key stages: pre-research period, project demonstration and engineering development period, orbit operation period, and results output and evaluation period.

In the first stage, key technologies are mainly tackled, followed by scientific goals advancement and accessibility, together with analysis and demonstration of rational and feasible engineering tasks. Therefore, engineers and researchers often need simulation tools or models to develop conceptual designs, task simulations, and comprehensive evaluations in this process.

In the mission development phase, the satellite system, scientific application system (SAS), TT&C system, carrier system, launching site system, and ground support system are mainly developed. In addition, ground simulation experiments and docking tests between different systems are carried out to check system function and system interface correctness, during which IT infrastructure, such as communication network, computing resources, and storage resources, are needed.

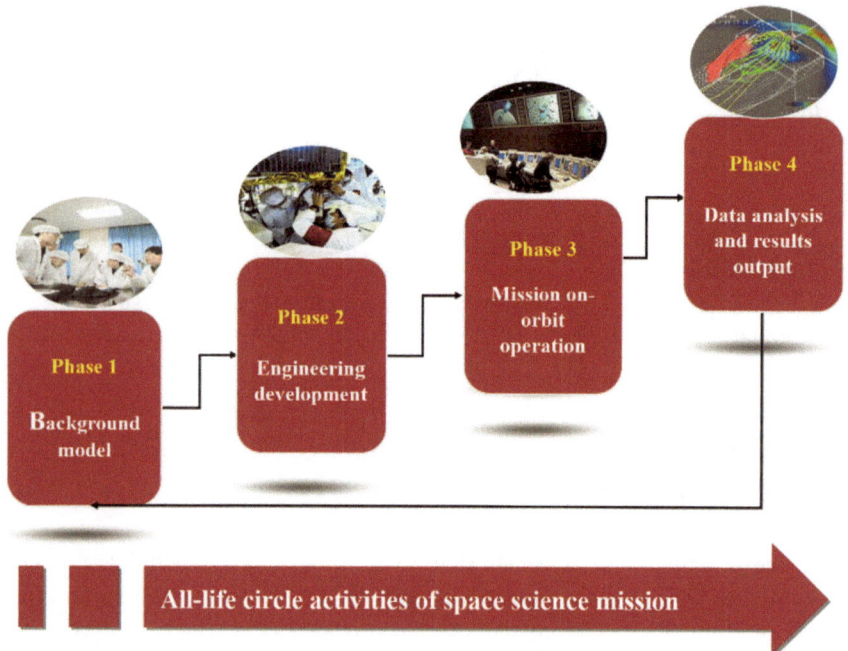

Fig. 1 All life-cycle periods of space science satellite missions

During the on-orbit operation phase, the ordinary control of satellites and data transportation are implemented, including downlink data processing, distribution and management, telemetry command generation and uplink, and payload status monitoring. In addition, the above-mentioned data services need business software support, such as command generation, data processing, data transfer.

In the data output and evaluation stage, observation data is deeply processed and analyzed to achieve mission scientific goals, in which mature algorithms and software libraries are often used. Furthermore, other mission scientific data and simulation model results are also comprehensively studied and analyzed.

In summary, the implementation process of the satellite mission program is extensive, complex, and challenging, which generally require scientists and engineers with great wisdom and advanced technology to tackle the problems that might be met. Furthermore, there are also strong demands for e-Science resources, such as knowledge, science data, simulation tools, physical models, storage environment, computing resource, communication network, and other general IT infrastructures. Meanwhile, scientists and engineers need to work together for knowledge, information and document exchange, software sharing, interface coordination, data flow, and infrastructure interconnection. That is, the implementation of satellite mission programs cannot move a step without the support from knowledge, data, software, tools, and other infrastructures (Fig. 1).

1.3 Opportunities and Challenges that Coexist for Space Science e-Science

The vigorous progress of space science creates an opportunity for e-Science, while scientific data, models, simulation analysis tools and IT infrastructure resources provide basic conditions for the fast development of space science e-Science. With the support from the "Informatizition Project in Chinese Academy of Science," the National Space Science Center (NSSC), the Chinese Academy of Science constructed the data application environment of the Space Science Virtual Observatory (VSSO), which integrates space science data with satellite programs, such as "Double Star," "Fengyun Project," and "Lunar Exploration Project," as well as the ground-based data from the "Meridian Project" [3]. Meanwhile, with the support of the "National Natural Science Foundation of China (NSFC)" and the innovative projects of CAS, NSSC has accumulated a number of space physical models and space weather prediction models. NSSC has also constructed a series of mission demonstration software libraries, such as interactive conceptual design software, payloads field of view analysis software, thermal flux analysis software, data transmission analysis software, and simulation visualization software, with the help of the "Space Science Background Model Project" of SPP. Furthermore, after years of continuous investments and construction, CAS established sets of e-Science infrastructures, including high-speed network and supercomputing resources, which can effectively support the construction of the scientific research information system [4].

Although the development opportunities and conditions of e-Science in space science are satisfactory, there were still some blocks and insufficiencies that would stop this development. This was due to the following:

(1) The space science research community did not own enough cognition, lacked enthusiasm on e-Science, and there was not stable e-Science community in space science, since most of the scientists stayed in the stage of "fighting for themselves and self-sufficiency."

(2) Most of the existing models, tools and software were kept or only shared in small-scale groups and group members, and online sharing or large-scale publicity degree was not satisfactory.

(3) The connections between e-Science elements were not smooth, such as the connections between data models and tools, and the connections between data, models, tools and cloud computing, and cloud storage were not seamless, which to some extent resulted in inconvenience in tool and software usage. Furthermore, this caused too much time to be spent in model calculation and the deficiency for complex serviceability of scientific data.

(4) A comprehensive information service platform that can satisfy whole life-cycle activities in mission implementation has not been established at present. Hence, it was difficult to meet e-Science needs in an era where space science mission projects are vigorously developed.

2 Focusing on National Key Special Project Requirements, the "Domain Cloud of Space Science and Technology" Project Was Constructed Following These Trends

The construction of the "Domain Cloud of Space Science and Technology" (DCSST) project started at the end of 2013, with the support of the e-Science of CAS in the "12th Five-Year Plan," undertaken by the NSSC. Its construction was under the guidance of the Committee of e-Science of Space Science. Since this project was oriented to e-Science needs in space satellite missions' full life-cycle activities and the space science innovation research process, it was aimed to establish a comprehensive information platform that should integrate IT infrastructure resources and discipline knowledge resources, supporting the implementation of SPP and other satellite missions, and finally becoming a core research supporting platform for the Space Science Innovation Research Institute, which could promotes the space science and technology output and assists in the progress of space science (Figs. 2, 3).

The thought for the DCSST construction was "Resources comes from research communities also serving for communities," which uses IT infrastructure resources provided by CAS and NSSC, and knowledge resources fully originating from NSSC. Through the transformation, upgrade, and integration of existing knowledge (data, model, software, and tools) in the research community, platforms were attempted to be established, including one portal and four sub-platforms. These could provide mission collaborative demonstration service, mission operation

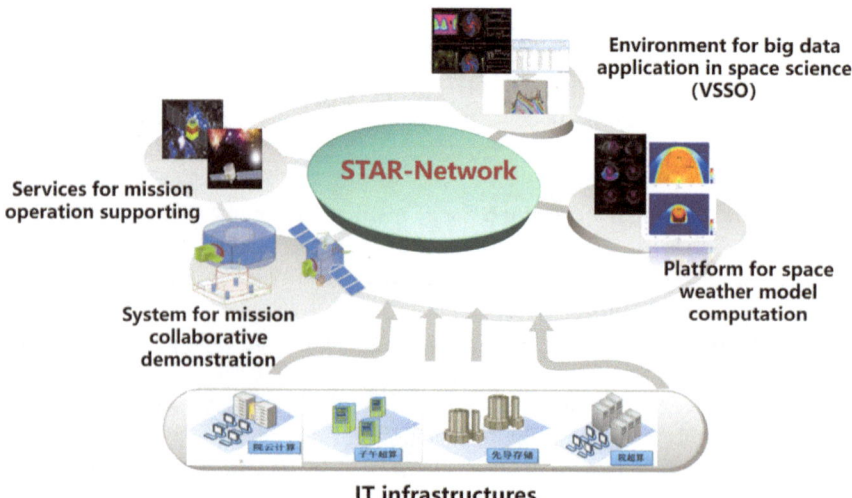

Fig. 2 Structure of DCSST

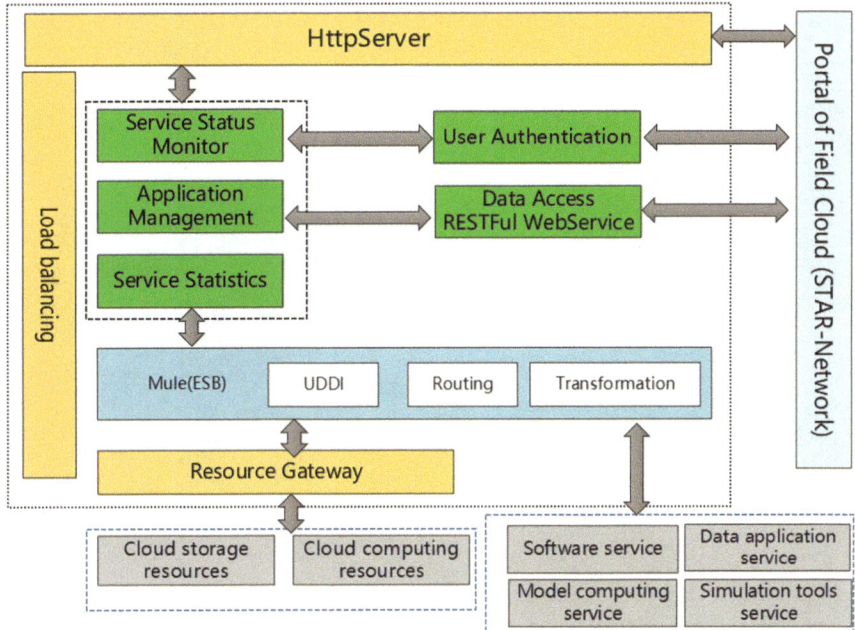

Fig. 3 Framework of DCSST

support service, space science data application service, and space weather calculation and analysis service through the DCSST portal—one-stop online service.

The project team chose the Service Oriented Architecture (SOA) as the main scheme, for its coarse-grained, open and loosely coupled service structure made the resources integration process more flexible and elastic. In addition, all kinds of resources were independent to each other, which could be extended and upgraded independently merely through application integration. Furthermore, unified integration standards and open interface protocols enable system function to be dynamically multiplicable, which also meets the renewal and expansion needs of future knowledge resources and ensures the vitality of the sustainable development of DCSST.

Using this technical framework, and through the joint efforts of the project team, the integration and transformation of infrastructure resources from NSSC and CAS were completed, and the storage and computing resource pools that satisfy this running platform were also established. Meanwhile, mission task collaborative analysis tools, space weather calculation models, and the space science data application environment were basically transformed and integrated. A prototype system for DCSST called "Solar-terrestrial Astronomy Research-Network, STAR-Network" was constructed, which began to provide an online trial service in May 2014.

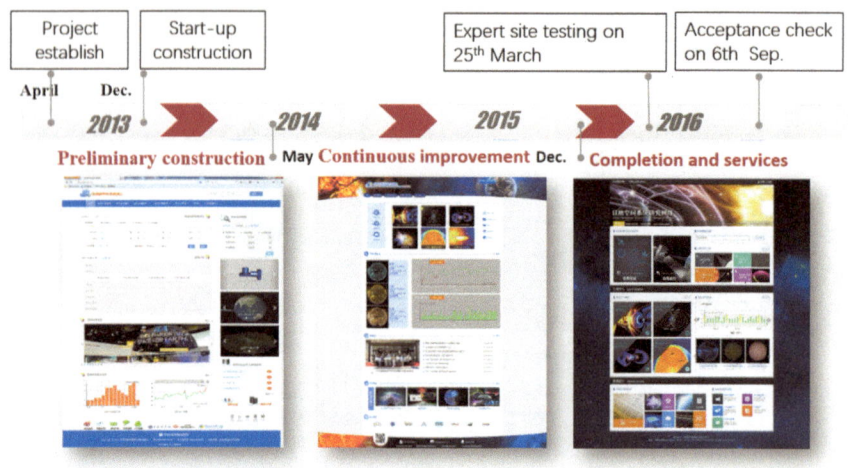

Fig. 4 Construction progress of DCSST

After nearly one year of operation trial and continuous improvement, the system functioned more perfectly and the performance was stable. Furthermore, this successfully passed the expert scene test. In December 2015, the satellite mission operation support service was launched. To date, all functions designed in this project were provided online, and the system remains officially open at present (Fig. 4).

3 Construction Results of the DCSST Project

The DCSST project team built a "One-stop" platform system running in a cloud environment (STAR-Network), which had packaged mission simulation tool sets, on-orbit control systems packages, space weather calculation and analysis models, and big data analysis tool sets, and its successful establishment was mainly through seamless interconnections between resources of scientific data, simulation modes, analysis tools, and IT infrastructures. Hence, it has capabilities that serve for satellite missions in all life-cycle activity processes, for big data retrieval, visualization, and application, and for physics process simulations and key parameters analysis that occurs in solar-terrestrial systems. In addition, it provides IT services to the space science community (Fig. 5).

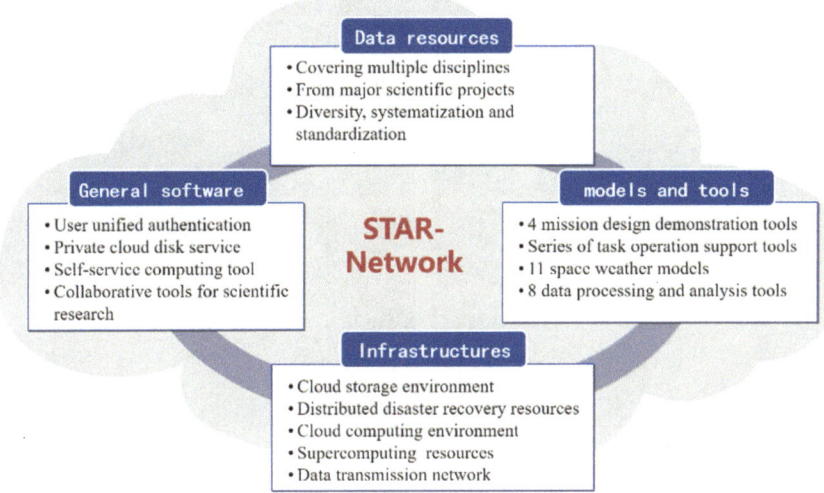

Fig. 5 Resources integrated by DCSST

3.1 Ubiquitous Serving Capability Covers for All Life-Cycle Activities in Satellite Mission Implementation

In the phase of preliminary planning and collaborative demonstration, the mission collaborative demonstration system in the STAR-Network can offer simulation tools to the scientist team, supporting their point to mission scientific objectives and carrying out coordinated analysis and comprehensive planning in the fields of orbit and attitude design, space configuration, payload layout, payload detection area, and data transmission feasibility analysis. The capabilities of these simulation tools are summarized below:

(1) The interactive orbit design tool provides functions for fast orbit design, target orbit modification, and satellite orbit characteristic analysis;

(2) The satellite structures and payloads layout design tools support the rapid installation and layout of satellite platforms and its payloads, displaying satellite platform architecture, payload geometry, and installation location in detail;

(3) The payloads field analysis tools providing payload observation area simulation and analysis by giving the effective observation arc based on the user input information of orbit, payload field, and the definition of the observation area;

(4) The downlink data transmission analysis tool provides the satisfaction analysis of downlink data transmission, which is the basis of the scientific data transmission feasibility analysis (Fig. 6).

During the satellite mission orbit operation period, the mission operation support services of the STAR-Network provide services on mission workflow and data flow aspects, which ensures the stability and safety of satellite missions in its operation

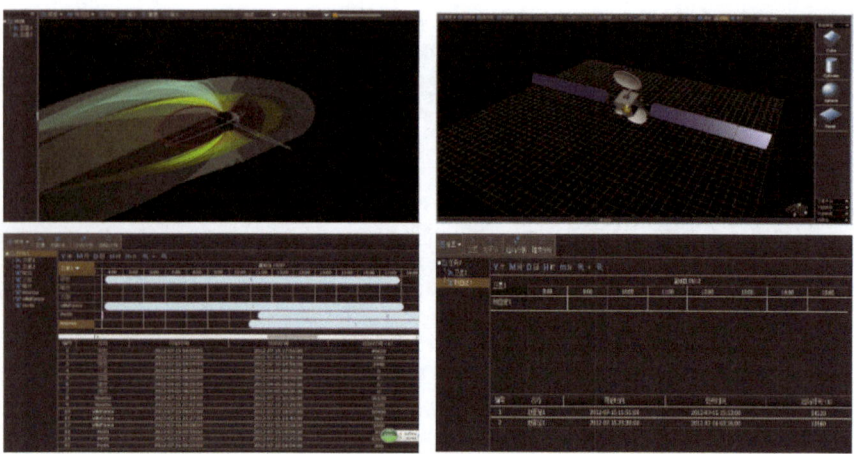

Fig. 6 Simulation tools for mission demonstration

and realizes the observation of scientific targets. The mission operation supporting services are as follows:

(1) *Full scan of the mission situation*

This service mainly faces users from satellite system, TT&C system and scientific application system (SAS) teams. It offers overviews of the status of the satellite platform operation, science plan execution, satellite-to-ground data transmission, and real-time space environment. Basically, it can meet the needs of mission teams, in order to master the whole mission status in real time. Moreover, it provides a channel for the public understanding of the overall state of the science satellites mission (Fig. 7).

(2) *Scientific exploration plan generation*

This service includes functions for scientific exploration plan analysis and generation, satellite orbit and attitude calculation, and station-passing prediction, and all of which are used for the rapid and accurate production of scientific exploration plans or experiment schedules.

(3) *Payload status monitoring*

By processing platform engineer parameters in real time and obtaining statistical results based on special rules, the mission operation platform can support scientists and payload teams when monitoring and evaluating the health status of payloads and the execution status of the scientific plan.

(4) *Online publication for quick-look data*

For SAS and payload teams, the mission operation platform provides a quick-look scientific data browse and download service, which supports the overview and fast-sight of data in real time, and helps in rapid discoveries, to a certain extent.

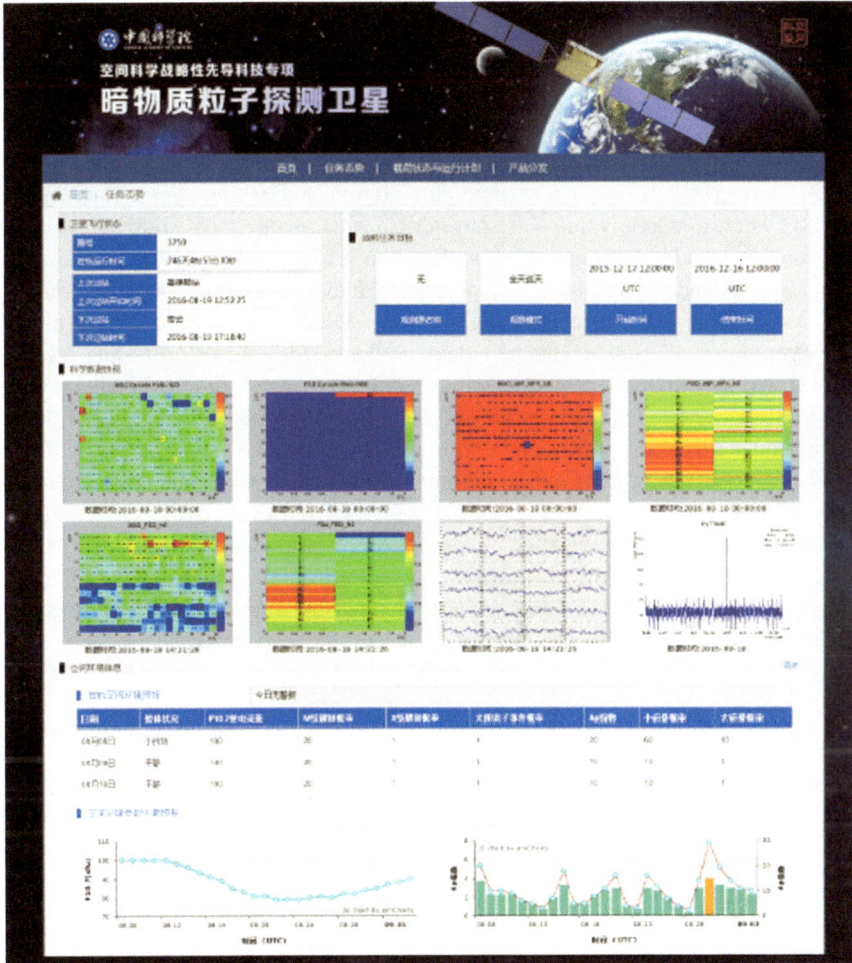

Fig. 7 Webpage of the mission's comprehensive situation

(5) *Real-time data product distribution*

The mission operation platform distributes satellite edited level data products (preliminary processed products from received raw data) and auxiliary data products (ephemeris data, orbital elements data, satellite-to-ground time difference data and science exploration plan), which are the main input data for data calibration, deep analysis and research, to SAS, payload teams and the satellite system in quasi-real time.

3.2　Capability for Big Data Application in Data Retrieval and Visualization

The Space Science Virtual Observatory (VSSO) [3] is a data publication and serving system that works in a cloud environment, and this was integrated by the STAR-Network. It is a space science data archive publishing system that covers space physics, astronomy and solar physics subjects, mainly providing data sharing and analysis service for users worldwide, aiming to promote science data sharing from space science programs, and supporting the maximization of scientific output.

Data products from major program, such as "Double Star," "Meridian Project," "Fengyun Project," and exchanged data products with the World Data System (WDS), are continuously provided online services through VSSO, while the release of "Wukong" scientific data is on the way. However, the scientific data of "Huiyan" is still under the protection period and has not been released at present.

VSSO offers services of online retrieval, browsing, downloading and subscription for space science data, supporting keyword retrieval, combined condition retrieval, and subject navigation. Meanwhile, it provides online data analysis functions, such as coordinate system conversion, data smoothing and filtering, space weather event correlation analysis, single-point time data visualization, and time-series data visualization. Basically, it has the capability for massive data analysis based on cloud computing technology and large-scale computing platforms (Fig. 8).

3.3　Capability for Physics Process Simulation and Key Parameters Analysis Occurring in the Solar-Terrestrial System

It is a constant pattern in space science research that scientists combine science data and model calculation results for joint analysis and discussion to obtain meaningful scientific results. In particular, space physics research models generally consist of numerical simulation models and numerical prediction models. The former have been used to simulate space weather burst processes that occur in solar-terrestrial space, complex processes, and laws of energy coupling and wave-particle interaction in various regions, which belong to natural science research, while the latter are often used for early warning and prediction for important parameters in space environments, which bias toward practical application.

The space weather model computing and analysis platform have integrated eight domestic models that independently developed, and five international models are generally used. This can provide a three-dimensional visualization for the model calculation results based on the server, as well as for the results saved in personal cloud storage (Fig. 9). The 13 models group can effectively support the simulation of physical processes or the prediction of key space environmental parameters that

Fig. 8 Home page of the VSSO Web site

occur in space, such as solar coronal, interplanetary, magnetosphere, ionosphere, and upper and middle atmospheres.

(1) The "L1-magnetospheric–ionospheric causal chain model" provides the results of solar wind density, thermal pressure, the three components of velocity, and the magnetic field at each grid point of computational space, which could support researches in many aspects, such as interplanetary shock wave, bow shock wave, magnetopause configuration, plasma and magnetic field parameters in the magnetic sheath, ionosphere convective structure and electric field characteristics, and magnetosphere K-H instability.

(2) The "Earth's bow shock wave and magnetopause topography prediction model" is an analytical model, which adopts upper solar wind parameters

that originate from the "L1-magnetospheric–ionospheric causal chain model" as input to simulate magnetopause topography in a complete equatorial and meridian plane.

(3)   The "Solar high-energy particle propagation model" takes solar burst data as input, simulates, and calculates the interplanetary propagation process of solar high-energy particles, which realizes the prediction of arrival time and particle anisotropy of various energetic particles in heliosphere space.

(4)   The "Geomagnetic storm forecasting model" uses data from the ground network of cosmic ray as input, adopts the wavelet analysis method to analyze the characteristics of the cosmic ray flux variation, and obtains the response of the amplitude and asymmetry of cosmic ray flux declination to the geomagnetic storm, thereby realizing the prediction of the geomagnetic storm one day ahead of schedule, and warning three hours ahead of schedule.

(5)   The "Near-space atmospheric forecasting model" is mainly based on 11-year atmospheric temperature data to forecast the atmospheric wind field, temperature, density, and pressure. It is often used for the study of the distribution of atmospheric temperature and wind field in the near space of middle and low latitudes, as well as variations with local time, seasons, and year.

(6)   The "Multi-station, mid-upper atmospheric climate model" uses the migratory diurnal and semidiurnal tidal data, including amplitude and phase. This originates from the mesospheric climate model (MMACM and GSWM00) and calculates atmospheric temperature, density, pressure, meridional and zonal winds over a global altitude range of 20–90 km.

(7)   Based on ionospheric parameters foF2 (critical frequency of F2 layer) and TEC (total electron content) from the ground station, the "Ionosphere prediction model" determines the accurate estimation of ionospheric foF2 and TEC over the monitoring stations one day ahead of schedule.

(8)   The "China low latitude ionosphere scintillation model" calculates probability of ionospheric GPS-L band scintillation index S4 greater than 0.1, 0.3, and 0.5, based on the date, t;ime, region code, solar radiation flux, and geomagnetic activity index.

(9)   The International Geomagnetic Reference Field (IGRF) is one of the most widely used empirical model for geomagnetic fields, which can be used for the calculation of dipole moment, dip angle, deflection angle, total strength and L value of the geomagnetic field.

(10)   T96 is a semiempirical external model for geomagnetic fields, which was established through the combination of huge observation data of the magnetic field and a certain physical theory. In addition, it also takes full account of solar wind conditions, the tilt of the geomagnetic axis, and current systems at the magnetopause. Therefore, it is widely used for geomagnetic field calculation in both the geomagnetic field static state and disturbed period.

(11)   The "Geomagnetic field cutoff rigidity calculation model" offers functions in tracking particles trajectories in the Earth's magnetic field and provides calculation service for the cutoff rigidity of cosmic ray particles with different incident energies and at different positions (different longitude, latitude, and

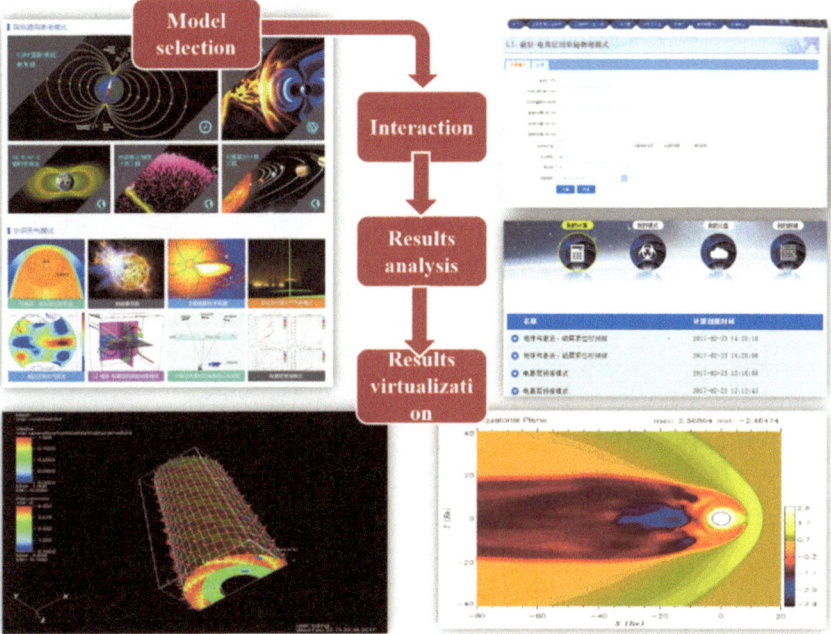

Fig. 9 Scenario of space physics model computation usage

height) in the magnetosphere. It can also support the visualization of particle trajectory in the geomagnetic field.

(12)   The "Planetary ephemeris calculation model" offers a calculation function for the three-dimensional position and velocity of sun, moon and nine planets in the solar system in the J2000 geocentric inertial system, based on planetary and lunar ephemeris data in the DE/LE series provided by the NASA Jet Propulsion Laboratory, which can serve for satellite ephemeris prediction.

(13)   AE-8 and AP-8 are classical models for radiation belts, which provide calculation functions for omnidirectional integral and differential fluxes of electrons and protons in solar maximum and minimum years. In addition, the energy spectrum range is 0.1–400 MeV for protons and 0.04–7 MeV for electrons. The simulation results are often used for energetic particle radiation dose assessment in satellite orbit (Fig. 9).

3.4   IT Services Capability

Through the integration and connection of super-computing, cloud computing and cloud storage from CAS and NSSC, DCSST established its own virtual IT resource environment for running the STAR-Network system, providing basic resource support for big data analysis in space science, space physical model calculation and simulation tool operation. In addition, the virtual IT resource pools provide cloud disk service for the space science research community, which allows users to apply and manage their personal cloud disks, create and manage personal virtual machines, and submit self-service mode calculations on personal virtual machines. Based on the distributed disaster environment of CAS, a disaster system particularly for space science satellite data was built to ensure the permanent security of data products. This special science and technology network line can directly serve the safe and high-speed transmission of science data in SPP. The integrated "Ducking" environment for collaboration research provides collaborative services, such as team-document library, science and technology e-mail, science letter APP and conference platforms.

4 Application Demonstration and Service Effectiveness

The construction of the DCSST project adopted the principle of "building while serving, while optimizing"; during which space science satellite programs, SPP, for instance, were mainly tracked and served for teams of DCSST. It had provided directional docking services for researches in the field of space science and then achieved remarkable service benefits and results; a group of stable subjects and community users have been condensed.

4.1 Demonstration Support for Background Missions in Space Science

STAR-Network provided an efficient and virtual collaboration platform service for ASO-S, EP, and SMILE projects in the preliminary planning and conceptual design periods. The design results from the present analysis tools lay a good foundation for further mission demonstrations, effectively support the official establishment of SMILE and ASO-S satellite programs, and the adoption of the EP satellite plan has been approved as a consequence.

Taking the Einstein Probe (EP) project for instance, its demonstration teams demonstrated and analyzed the mission's progressiveness and feasibility with the assistance of satellite orbit design tools, satellite structure, and payload design tools provided by the STAR-Network, those services mainly included point to orbit designation, attitude control, measurement and control, data transmission, thermal control, power supply, space environment, ground segment, and other aspects. The space science mission demonstration system tools play a key role in the approval of EP program by providing re-calculation and mission-level simulation analysis, which were the basis for the qualitative and quantitative evaluation of this satellite program.

4.2 Serving Effectiveness for On-Orbit Space Science Satellite Missions

STAR-Network also provided solid infrastructure support and system platform support services for the successful implementation of four science satellite projects launched by SPP in space science (Phase I), namely Dark Matter Particle Exploration satellite ("Wukong"), Returnable scientific experiment satellite ("SJ-10"), Quantum Science Experiment Satellite ("Mozi"), and Hard X-ray Modulation Telescope ("Huiyan"). In the phase of mission development and joint system debugging, STAR-Network provided a high-speed data transmission network, and cloud computing and cloud storage services to different teams. Furthermore, during the phases of the on-orbit test and long operation of "Wukong," "SJ-10," "Mozi," and "Huiyan" satellites, STAR-Network mainly provided space science mission operation platform services, which strongly supported the mission's operation and tasks execution, such as downlink data processing and quasi-real-time distribution, and ensured the implementation of regular and opportunity science exploration and experiments (Fig. 10).

For the "Wukong" mission, the satellite mission operation support service effectively guaranteed the integrity and correctness of science data, ensured science data does not lose even a source package, and provided quasi-real-time edited and auxiliary data products distribution service to the scientific application system. Up to October 16, 2017 (also the time for the following statistical results), 43 TB of data had been accumulatively delivered to the scientific application system, which was the basic input for subsequent data calibration, deep processing, and research. "Wukong" has finished three observations in all-sky scans, and detected a total of more than three billion high-energy particles. These important observations and analysis results were published in NATURE magazine.

During the 21-day experiment period of the SJ-10 satellite mission, the recycle module ran for 12 days, and the orbital module actually ran for 21 days. The

Fig. 10 Schematic diagram of the mission support service for space science satellites

mission operation platform of STAR-Network provided quick-look data service, as well as edited and auxiliary data product (exceeded 8.8 TB) distribution service, in quasi-real time to SAS and the payloads system, and those measures satisfied the requirements of SAS with rapid data products processing, analysis, and research. In addition, the mission operation platform effectively coordinated with SAS for the flexible deployment and launch of onboard experiment tasks. SJ-10 has carried out a total of 28 scientific experiments in the microgravity environment at an orbital altitude of 250 km [5], and 15 original experimental results were achieved, such as the first mammal embryonic development and the first observation of coal combustion in the microgravity environment.

Meanwhile, the mission operation platform has supported satellite-Earth quantum scientific experiments for 884 times and provided more than 1.08 TB of data product distribution service to the SAS of "Mozi" to support experiment analysis. Relevant experiments have yielded significant original results: the first time of Satellite-to-ground quantum key distribution, and the long-distance free-space quantum key distribution in daylight toward inter-satellite communication, and satellite-based entanglement distribution over 1200 km. These results have been published in top international academic journals of Science and Nature [6–8] (Fig. 11).

The Hard X-ray Modulation Telescope needs a timing switch off to avoid damage to the payload caused by high-dose particle radiation in the South Atlantic Anomalous area (SAA). Therefore, based on the calculation results of the AP-8/AE-8 model, the mission operation platform provided SAA space environment prediction service to SAS, which determines the time on when the high-energy telescope should be switched off in its operation orbit. This measure greatly ensures the safe and stable operation of the satellite. Furthermore, after the occurrence of the gravitational wave event of two neutron star combinations [9], the mission

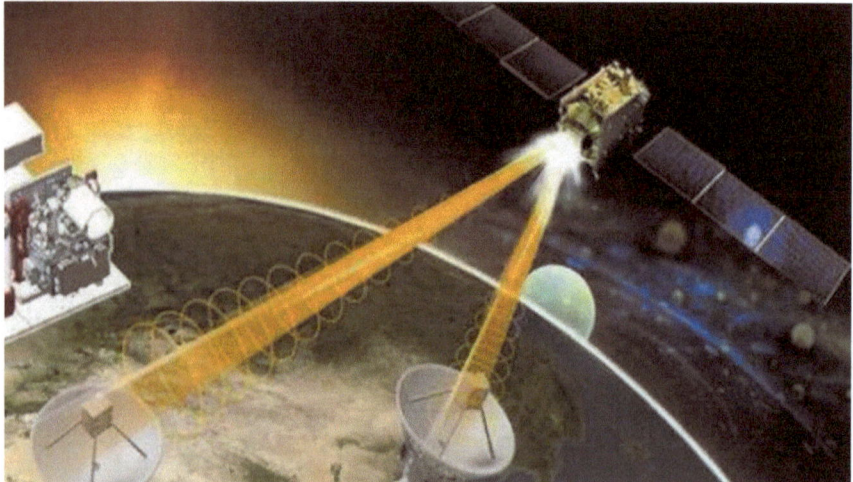

Fig. 11 An illustration of the quantum communication experiment of "Mozi"

operation platform responded to the SAS's emergency opportunities observation in real time, performed the observation planning and data distribution at a minute level. In the whole period of the gravitational wave event, the total volume of the distribution data product was 1059, with a volume of 23.17 GB. Those measures effectively served for "Huiyan" telescope's joint observation on gravitational wave events and the rapid analysis of its observation data.

4.3 Effectiveness of Services in the Field Space Science Research

Focusing on the study of basic processes and laws, the causal chain model, and the prediction method of the space weather in the solar-terrestrial system, DCSST has supported more than 40 scientific research projects, such as 973,863 projects and the National Outstanding Youth Science Foundation, by providing scientific data services and models computing services, standard specification services, and information infrastructure services through the space science big data application environment and space weather calculation platform.

In the subject of "Physical mechanism of short-term variation of tidal waves in the middle and upper atmospheres," the space science big data application environment of the STAR-Network provided OH airglow imager data from six ground stations of the Meridian Project for the project team, with the data time coverage from January 2012 to December 2013, as well as the wind and temperature data service originating from the Meteor Radar installed in the Fuke and Qujing station, and its visualization service. Using the above scientific data, the project team carried out researches on the propagation characteristics of high-frequency gravity waves in the middle–low troposphere and made remarkable achievements in the influence of high-frequency gravity waves on atmospheric circulation [10, 11].

The space weather model computing and analysis platform of STAR-Network had provided models computation services for the teams on the subjects of "Simulation of the basis, processes and laws of space weather," namely "Global numerical simulation of the time-varying solar wind–magnetosphere–ionospheric system," "The modulation of galactic cosmic rays by solar activity," and "Research on Solar high-energy particle diffusion coefficient." These models included the "L1-magnetospheric–ionospheric causal chain model," "Earth's bow shock wave and magnetopause topography prediction model," and the "Solar high-energy particle propagation model." In addition, high-performance computing resource service and multi-satellite joint data service were also provided to the subject teams. The present simulation tools and services effectively supported the research work of the project team, key progresses have been acquired in the diffusion coefficient of particles in the slowly varying high-energy particle event, as well as the processes of solar high-energy particle propagation, research on complex chain time-varying processes, and the coupling relationship of space weather occurring in the solar-terrestrial system [12–14].

5 Summary

Facing the demands of e-Science in space science in the new era, the "Domain Cloud of Space Science and Technology" project was proposed, which had integrated mission collaborative demonstration and analysis service, mission operation support service, space science big data application environmental services, and space weather models computation service in the close cooperation with various departments of NSSC. And all of these services are provided to users through the STAR-Network portal. A large number of steady user teams in the research community have been gathered. The STAR-Network also provides a solid backing guarantee for a stable mission operation, and major research results output by providing a stable infrastructure and business software support to the four satellites launched in SPP in space science (Phase I). Furthermore, it has effectively assisted the conceptual designation and technology selection of the space background model in SPP (Phase II) by offering simulation and analysis services in the period of mission planning and demonstration. Meanwhile, the STAR-Network has promoted the output of scientific research in the basic phenomena, processes and physical laws of space weather by providing services of space science data, model computing, and high-performance computing resources.

In the future, by following the long-term goal of the application development of space science big data, and focusing on the four core connotation of e-Science: cloud infrastructure, subject area data resources, big data application support platform, and typical application demonstration. The focuses will be given to the following:

(1) Powerful infrastructure environment construction, which includes the establishment of a flexible computing support platform, a software-defined fusion storage system, and a high-speed and stable international communication network.
(2) Special and deep fusion data production in both South American Anomaly (SAA) analysis and particle radiation environment estimation in geostationary orbit, by handling and processing multi-satellite observations, and developing conversion algorithms in pitch angle directional fluxes to omnidirectional fluxes.
(3) The development of tool sets with independent intellectual property rights. One of these is spacecraft payload simulators for the assessment of both electromagnetic radiation and energetic particles radiation in space, as well as for the simulation of payload responses to radiation. Parallel-coordinate data visualization is another tool to be developed for the visualization of the fine structure of regional parameters, electromagnetic field intensity, and streamline shapes in solar-terrestrial space.
(4) Digital space engine technologies' development, which consists of technologies in the optimization of sequential computing operators that work in the framework of spatiotemporal for data organization, mainly to support spatiotemporal data retrieval, association mining and feature extraction. Meanwhile, it includes technologies for spatiotemporal data organization models construction with the functions of topology establishment, mapping, and time coding in this framework.

(5) Core business software developing. As to mission operations, software mainly targeted for missions' process reconfiguration and results' presentation, as well as for the cooperative response of opportunity events, supporting for system test and verification. As for science research, the software will focus on the quick recognition of celestial objects based on machine learning technology, as well as on the multi-dimensional reconstruction and analysis for key physical fields.

The above methods may be employed to create a ubiquitous, integrated, and intelligent digital space, and also they could accurately serve for future space science missions and substantially promote knowledge discovery and technological innovation in space science.

References

1. Wu Ji. Research report on Space Science Planning in 2016–2030. Science Press. 2016. (in Chinese)
2. Chen MQ, Wu LH. Review and Perspective on Chinese Academy of Sciences' e-Science. Bull. Chin. Acad. Sci. 4, 2013. (in Chinese)
3. Zou ZZ, et al. e-Science practice and results in space science under an era of big data. Big data, 2, 06, 2016. (in Chinese)
4. Nan K. The Cyber infrastructure for Science and Innovation of Chinese Academy of Sciences and Its Applications. Chin. Acad. Sci. 4, 2013. (in Chinese)
5. Kang Q, Hu WR. Microgravity Experimental Satellites——SJ10. Chin. Acad. Sci. 31, 05, 2016. (in Chinese)
6. Liao S K, Cai W Q, Liu W Y, et al. Satellite-to-ground quantum key distribution. Nature, 549, 7670, 2017.
7. Liao S K, Yong H L, Liu C, et al. Long-distance free-space quantum key distribution in daylight towards inter-satellite communication. Nature Photonics, 2017.
8. Yin J, Cao Y, Li Y H, et al. Satellite-based entanglement distribution over 1200 kilometers. Science, 356, 6463, 2017.
9. Castelvecchi D. Rumours swell over new kind of gravitational-wave sighting. Nature, 2017.
10. Wang C M, et al. Gravity wave characteristics from multi-station observation with OH all-sky airglow imagers over mid-latitude regions of China. Chinese Journal of Geophysics, 59, 05, 2016. (in Chinese)
11. Wang C M, et al. Statistical characteristics analysis of atmospheric gravity waves with OH all-sky airglow at low-latitude region of China. Chinese Journal of Geophysics, 57, 11, 2014. (in Chinese)
12. Qin G, Shalchi A. Perpendicular diffusion of energetic particles: Numerical test of the theorem on reduced dimensionality. Physics of Plasmas, 22, 1, 2015.
13. Wang Y, Qin G. Estimation of the release time of solar energetic particles near the Sun. Astrophysical Journal, 799, 1, 2015.
14. Wang J Y, Wang C, Huang Z H, et al. Effects of the interplanetary magnetic field on the twisting of the magnetotail: Global MHD results. Journal of Geophysical Research-Space Physics, 119, 3, 2014.

Ziming Zou Deputy Director of the National Space Science Center, Chinese Academy of Sciences (CAS), is the chief designer of the Ground Support System in the CAS Strategic Priority Program on Space Science and the Director of the Chinese Space Science Data Center in World Data System. As a Member of the Information Technology Committee in the Chinese Geophysical Society and a Member of the Science and Technology Committee in the CAS Space Environment Prediction Center, he is mainly engaged in space environment information system integration and space science satellite data processing system construction and also develops research on space information, such as solar-terrestrial information organization, retrieval, representation, visual analysis, and interoperability. Till now, he has published a book (co-authored) and more than 30 papers, gained two first prize of the Military Science and Technology Progress Award, acquired the CAS Manned Space Program important contribution award and received the honorary title of "the excellent worker attending the CAS Manned Space Program."

Application of the Antarctic Survey Telescopes AST3 in the Detection of Electromagnetic Counterparts of Gravitational Waves

Lifan Wang

Abstract Chinese Antarctic Survey Telescopes AST3 consist of three large fields of view optical/infrared survey telescopes placed at the highest point in the Antarctic Dome A. Chinese Antarctic astronomy collaboration team led by Purple Mountain Observatory, and Chinese Center for Antarctic Astronomy has developed independently hardware and software system of AST3 telescope and realized the continuous and highly reliable remote automatic observation on the extreme cold condition. AST3 is the first unique full automatic Survey Telescope in the Antarctic. Benefit from the excellent observational conditions in Dome A, AST3 team have carried out a series of time-domain astronomical research using AST3 telescope, including searching for gravitational wave sources, fast radio bursts, supernovae, exoplanets. Recently, the discovery of gravitational waves and their electromagnetic counterparts have initiated the era of multi-messenger astronomy. In the context of global collaboration and data sharing, fast response, multi-messenger and multi-band global joint observation, AST3 team have made a series of achievements in collaboration and independent innovation, and we succeeded in detecting the first electromagnetic counterpart of the binary neutron star merger event GW170817.

Keywords Antarctic · Survey telescope · Transient · Gravitation wave

1 Background

Dome A (latitude 80°22′02″S, longitude 77°32′21″E, elevation 4093 m) is the highest point on the Antarctic Plateau. The Antarctic Survey Telescopes, AST3, are the second-generation Chinese telescopes on Dome A [1–4]. They are designed and manufactured by Purple Mountain Observatory of CAS, Nanjing institute of optical technology, National astronomical observatory of China, Tsinghua University, Nanjing University, Peking Normal University. AST3 project consists of three

L. Wang (✉)
Purple Mountain Observatory, Chinese Academy of Sciences, Nanjing, China
e-mail: lifanwang@gmail.com

© Publishing House of Electronics Industry 2020
China's e-Science Blue Book 2018,
https://doi.org/10.1007/978-981-13-9390-7_5

catadioptric telescopes with 680 mm primary mirror (500 mm entrance diameter), 3.73 f-ratio, and 2.92° × 2.92° field of view matched with different filter and 10 K × 10 K CCD. AST3 carries three different filters, and two filters were designed at the focal plane which can be remotely switched over. i-band filter was used for AST3-1, and g- and r-band filters were used for AST3-2. The designed image quality of AST3 is 80% energy encircled in 1 arcsecond and distortion in the whole field less than 1 pixel. Compared with conventional refracting and reflecting telescopes, AST3 has the advantages of large field of view, flat plane, distortion elimination, short lens barrel, and atmospheric dispersion correction. They are the first group of large field of view survey catadioptric telescopes in Antarctic that can be remotely controlled, automatically observed, automatically focused.

The main scientific goal of AST3 is to make use of the 2670 h of continuous Antarctic polar night for the observation of transients and variable sources, searching for gravitational wave optical counterparts [5, 6], fast radio burst optical counterparts, gamma-ray bursts, supernova, exoplanets, etc. The research includes key scientific issues such as dark energy, dark matter, different cosmological models, and modified gravity theory. Searching for transients is one of the main tasks of AST3. Most of the transient outbreaks occur on a very short timescale ranging from milliseconds to hours. Transient sources with light variation scale less than 1 day include fast radio bursts (FRBs), certain gravitational wave (GW) sources such as neutron star-neutron star mergers producing kilonova, supernova shock breakouts, gamma-ray bursts that include "bursty" or "dark" events, type-Ia and fast-rising supernovae, ICECUBE high-energy neutrinos, flare stars, certain tidal disruption events, ultra-luminous X-ray outbursts, flare stars, and classical novae (Fig. 1). However, to date, there has been no systematic observational exploration of transients on millisecond to hours timescales, largely as a result of the challenges of data acquisition and the necessary real-time analysis and identification for rapid follow-up spectroscopy and imaging. For example, in the first event of gravity wave GW 150914 [6], the localization of the event remains relatively poor (600 deg^2). To improve the localization accuracy and the probability of finding the optical counterparts of gravitational waves and other fast transients, global cooperation is necessary. Using the global information and communication network, a number of gravitational wave detectors and telescopes in various wavebands around the world are combined for collaborative observation. Modern astronomy has evolved into an era of rapid sharing of information and data and global collaboration.

In Galileo's day, the only way to observe the universe was visible light, and the observation was not only limited by weather conditions, but also by the carrier of visible light. With the development of modern science, the methods for people to understand the universe are gradually increasing, and X-ray and radio waves are used to explore the universe. X-rays, visible light, and radio waves are all electromagnetic waves, but with different wavelengths. Studying the properties of the same astronomical phenomenon in different bands, we can get a more profound understanding of it, and full-band astronomy was born on this basis. The discovery of gravitational waves gives us another tool to probe the universe. Gravitational

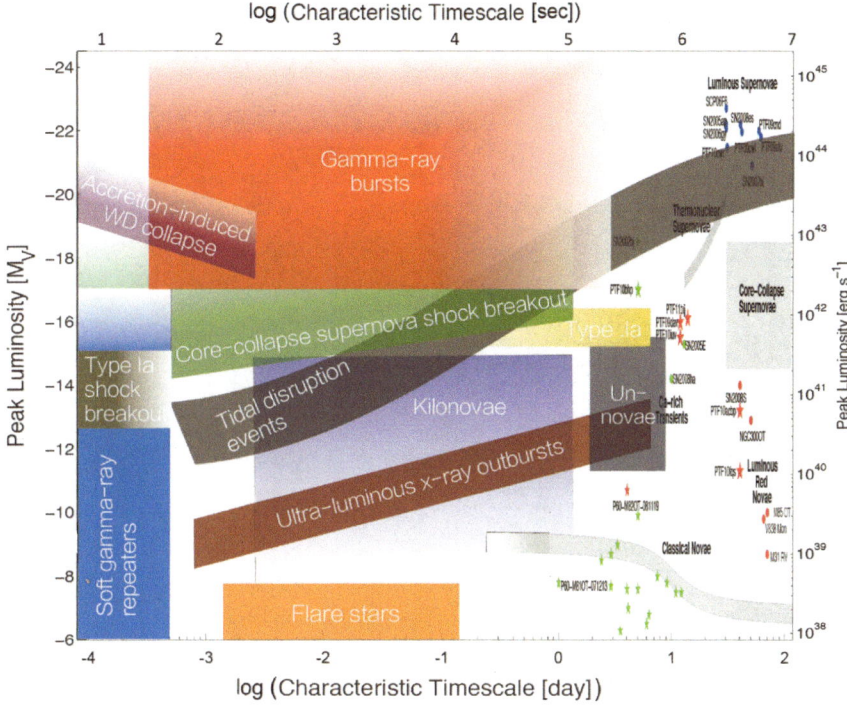

Fig. 1 Relationship between peak luminosity and timescale of transient sources with different physical mechanisms [7]

waves are the fluctuations of space and time, which are completely different from electromagnetic waves in nature. Gravitational waves and electromagnetic waves act as different "messengers" that can tell us different aspects of the same astronomical event. The gravitational wave event GW170817, from the binary neutron star (BNS) merger, was the perfect example of a global collaboration that ushered in a new era of "multi-messenger astronomy" [8]. At 20:41 Beijing time on August 18, 2017, Laser Interferometer Gravitational-Wave Observatory (LIGO/VIRGO) detected the new gravitational wave event GW170817 [5], which lasted about 100 s. This is the first time that humans have directly detected the gravitational wave event caused by the binary neutron star merger. 1.7 s later, NASA's Fermi gamma-ray satellite and European INTEGRAL satellite detected very weak short gamma bursts, named as GRB170817A [8], and this is the first time that electromagnetic and gravitational signals have been unquestionably linked. Fermi's gamma-ray detector automatically sent an alert to the GCN system as it detected the gamma-ray signal GRB170817A. Since the automatic data analysis of LIGO took about 6 min, the information was released a little later than Fermi satellite. As soon as the LIGO team confirmed, it also issued an alert to the organization that signed the agreement. Since then, more than 50 astronomical instruments around the world

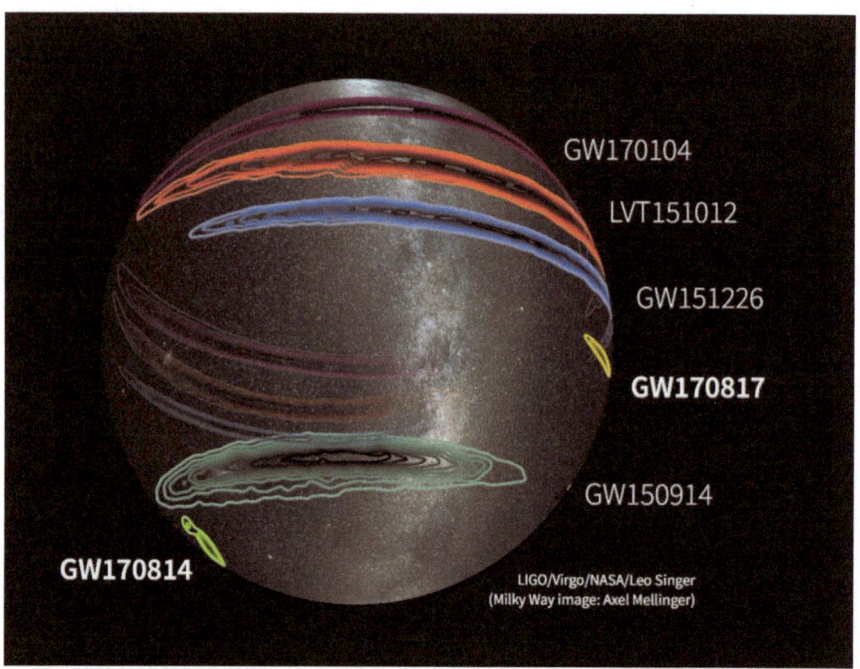

Fig. 2 Spatial localization of five gravitational wave events detected by LIGO. GW150914, GW151012, GW151226, and GW170104 were detected by LIGO alone. Since August 2017, the joint detection of VIRGO and LIGO has greatly improved the space positioning accuracy. GW170814 and GW170817 were detected by the joint detection of ligo-virgo. (https://www.ligo. caltech.edu)

have made detailed observations about GW170817, searching for its optical counterpart. The Antarctic survey telescopes AST3 cooperative team also made intensive observations of GW170817 using AST3-2 mounted at the Dome Antarctica [9]. The AST3 team are also a partner of the Deeper, Wider, Faster (DWF) program in Australia, and we observed the optical counterpart of GW170817 at optical to near-infrared band using almost all Australian telescopes and ESO's 8-m telescope. Finally, we successfully obtained a large amount of data of AT2017gfo/SSS17a [10], the optical counterpart of GW170817 in galaxy NGC4993. In this event of binary neutron star merger, the three detectors of LIGO and VIRGO were observed together (Fig. 2), resulting in a source positioning error of 28 degree. It is because of the great improvement of the spatial localization accuracy of gravitational waves that it is possible to find the optical counterpart.

Coordinated global observations require rapid response. The duration of transients such as gravitational waves is very short (from seconds to hours), and preliminary data processing must be carried out within a short time. Then, the candidate position information will be released to other observers through the global information system for follow-up observation. For example, once a satellite detects a gamma-ray signal, the GCN mail system of gamma-ray bursts will send the location information of gamma-ray bursts to the system as soon as possible, and other observers who subscribe to the mail system will receive the prompt and carry out following observation immediately. In the case of GW170817, Fermi satellite used GCN system to inform organizations around the world as quickly as possible, and many telescopes were then able to join in. LIGO/VIRGO has memorandum contracts with nearly 70 observatories around the world (nearly 10 in China), and if gravitational-wave signals are detected, they pass on information through their own channels.

Modern high-precision wide-field surveys generate enormous amounts of data, such as hundreds of gigabytes a day for the AST3 telescope. In the 2016 observation season, the data volume of AST3 reached 20 TB. If the local computer is unable to analyze the data in real time to identify the candidate of transient, the data needs to be transferred to other powerful computers for processing, and the data transmission process itself will cause the bottleneck of the whole transient search. Therefore, in the context of big data, the requirement of rapid response poses a huge challenge to the rapid processing of data.

2 Application of e-Science in AST3 Hardware Control and Observation Data Processing

2.1 The Specificity of AST3 Site Selection and Design Difficulties

AST3 is installed at Dome A, the highest point in the Antarctic interior, at 4093 m above sea level. Dome A has 2670 h of continuous dark night during the annual polar night, which is comparable to the conditions in space and has unique advantages for observing fast transients and variable sources and obtaining their continuous and complete light curves [11]. Dome A is one of the lowest temperature regions on earth, and the lowest temperature measured is −82.5 °C. It is extremely cold and dry here, water vapor content is less than any other sites in the world. The artificial light source here has the least interference, less dust in the atmosphere, high atmospheric transmittance, average wind speed less than 2 m/s, less atmospheric turbulence, with extremely high seeing; infrared background radiation is more than 10 times lower than the usual site, which makes the high sensitivity of infrared observation possible [12, 13] (Fig. 3).

The AST3 is controlled remotely through iridium's satellite network and operates unattended throughout the year in extreme cold temperatures not experienced at other sites. This puts forward extremely high requirements for reliability in extremely cold conditions, which is a great challenge to China's telescope manufacturing industry. The AST3 team solved a number of hardware problems. We have completed the design of a large field of view and short barrel distortion optical system for the AST3 telescope [14].

Fig. 3 Dome A after the 32nd Antarctic scientific expedition

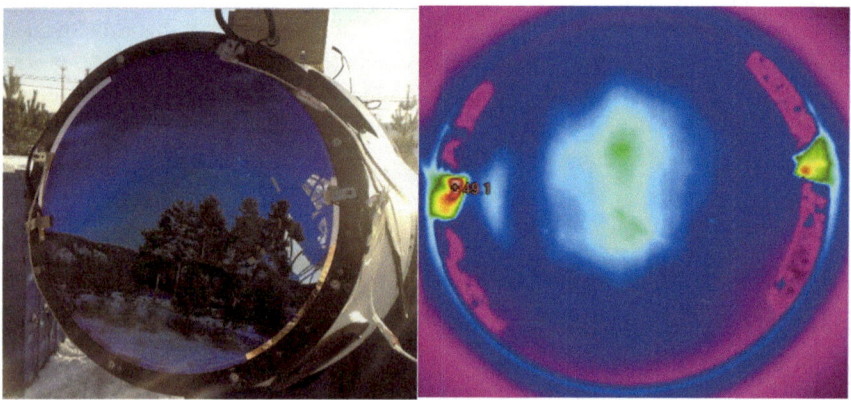

Fig. 4 ITO coating of aspheric surface correction mirror

Though Antarctic is very cold and dry, the surface of telescope is easily to get frosted and iced; to solve this problem, we developed an adaptive indium tin oxide (ITO) coating with adaptive power input to keep the temperature difference within 2 °C which is expected to have much less effect on the final image quality (Fig. 4). We also added an active blower device to the outer mirror of the telescope and a hot air device to the inner of the mirror barrel as a backup plan of ITO mirror heating film to prevent frosting of the mirror in extreme cold environments.

Through the temperature control design, we have solved the problem that the high-precision photoelectric encoder cannot be applied in the low-temperature environment. This design cannot only be used in AST3, but also has done technical verification for the 2.5-m KDUST telescope [15, 16] to be built in the 12th five-year plan. We have developed an optical interval automatic temperature compensation mechanism and a precise focusing mechanism. In addition, a flat reflector is added in front of the focal surface correction mirror group (Fig. 5), which solves the problem of seeing and CCD maintenance difficulty caused by CCD heating. In terms of software, we have developed local control software and remote control software for AST3. The control software is mainly used for telescope debugging,

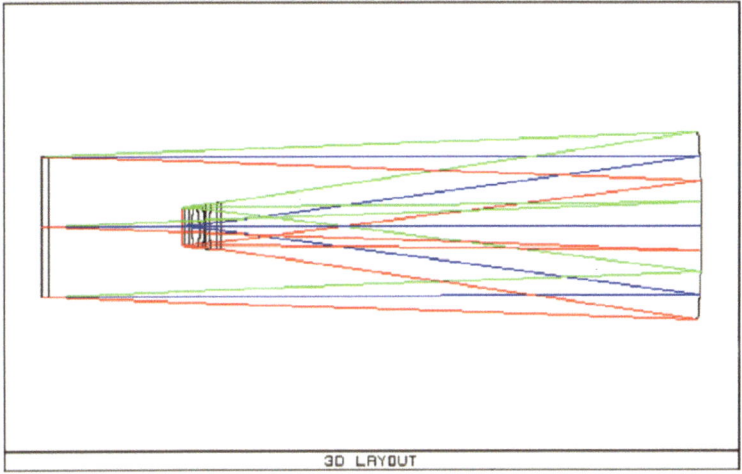

Fig. 5 Optical layout of AST3

survey observation, remote command analysis and transmission, error handling, etc. The debugging part is to control and debug the driving system, temperature control system, and focusing system of the telescope through software. The observation part allows the telescope to operate automatically according to the set target. This is a comprehensive and flexible multi level instruction system [17] (Fig. 6).

2.2 CCD Camera of AST3

In mmag (millimagnitude, 1‰) high-precision photometry, the error mainly comes from the stability of the instrument, so it is necessary to understand the performance of the instrument in detail, especially the detector. AST3 is equipped with 10 k × 10 k frame transfer CCD camera of STA (Fig. 7), and the AST3 team have tested its performance in detail to determine more accurately the observation exposure parameters and subsequent image processing. The test includes linearity, gain, full well electron number, readout noise, undercurrent, and charge transfer efficiency. [18, 19]. The gain is 1.64 e/ADU, which can be obtained from the reciprocal of the slope of the photon transfer curve (Fig. 8). We can see from the growth curve of the signal with time (Fig. 9) that the linearity within 60,000 ADU is better than 1%, and the electrons in the full well is about 1×10^5. According to the statistics of the overscan area, the readout noise is 4e for slow reading (40 s) and 11e for fast reading (2.5 s). CCD dark current is 0.09 e/pix/sec at −80 °C. Using the Extended Pixel Edge Response (EPER) method, that is, the number of residual electrons in the first few columns of the scanned area, we can obtain that the charge transfer efficiency is 0.999999.

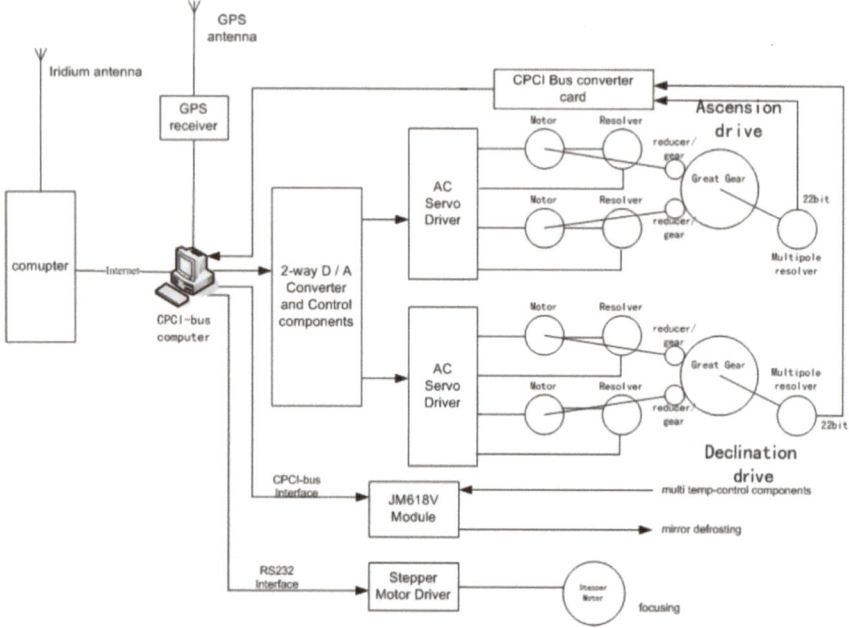

Fig. 6 Schematic description of the control system of AST3

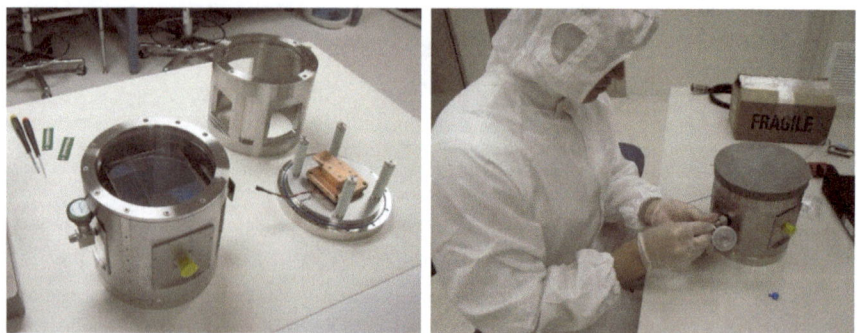

Fig. 7 CCD camera of AST3

In the test of AST3 CCD, AST3 team found that the photon transfer curve showed more and more obvious nonlinearity with the increase of signal intensity, and this nonlinearity could be better fitted with the quadratic curve (Fig. 10). The gain can be calculated more accurately by using the linear term coefficients of the conic than by directly using the linear fitting of the low signal region, with a difference of about 10%.

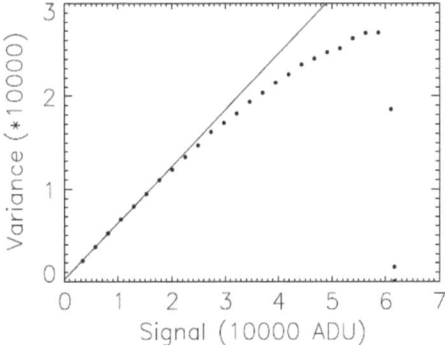

Fig. 8 CCD photon transfer curve of AST3

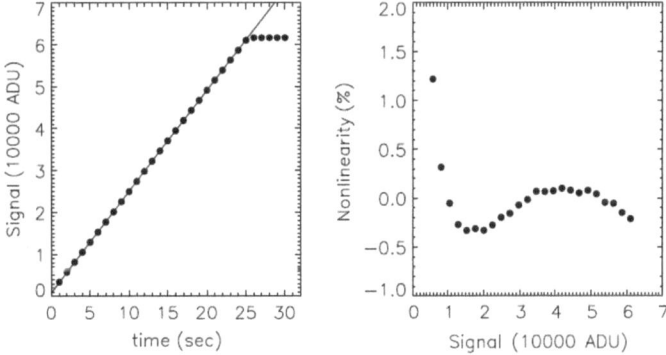

Fig. 9 Linearity test results of CCD quantum response of AST3

This is the charge sharing caused by the Coulomb force that accumulates electrons in the pixel and repels subsequent electrons. We put forward the concept of charge-sharing point diffusion function [19], described the device effect directly with the point diffusion function PSF commonly used in observation images, developed a method to calculate charge-sharing point diffusion function from flat-field images, and predicted the influence of charge-sharing point diffusion function on star image and photometry. As can be seen from Fig. 10, the stronger the signal is, the more serious the charge diffusion is, and the vertical direction (readout direction) is stronger than the horizontal direction. As a result, the star image becomes darker in the center and brighter around, and the vertical direction is more obvious. This effect makes stars of different brightness and has different point diffusion functions, which not only affects photometry, but also introduces systematic bias for high-precision shape measurement, such as the measurement of galaxy morphological changes in weak gravitational lens.

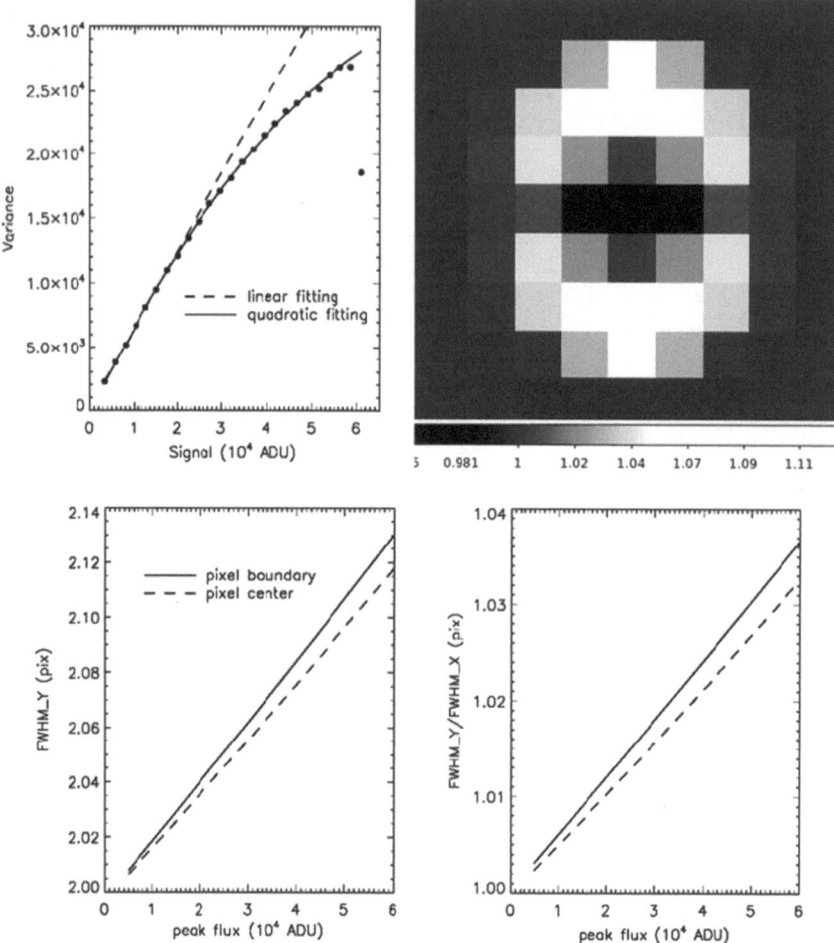

Fig. 10 Top left: The nonlinearity of photon transfer curve increases with the signal; Upper right: The ratio of the flux of the star image before and after the charge-sharing point diffusion function is taken into account. The center is weakened, the edge is strengthened, and the Y-direction is lengthened; Bottom left: effect of charge-sharing point diffusion function on FWHM of star image with different peak flow; Bottom right: influence of charge-sharing point diffusion function on ellipticity of star images with different peak flow

2.3 Operation and Control System of AST3

AST3's operational control system [20, 21] is different from those at ordinary observatory sites, where the infrastructure is mature. AST3 faces a series of special difficulties that are not found at other sites, such as the extremely low-temperature and low-pressure environment in the Antarctic. The AST3 work unattended throughout the polar night, which requires extremely high reliability of the

Fig. 11 Left: First generation operation control system of AST3. Right: Second generation operation control system of AST3

equipment, but the power consumption budget of the whole operation control system, is only 200 watts (Fig. 11).

2.3.1 Hardware of Operation and Control System

The first-generation AST3 operational control system contains three subsystems: master control system, real-time data processing system, and data storage system (Fig. 12). In order to prevent single point failure to cause the shutdown of the entire operation and control system, each subsystem has redundant backup machine. The main control system sends control commands to the telescope, collects and distributes CCD images, and allocates power. The processing system processes the observation data sent from the main control in real time. Storage systems manage hard disks and store data. Each switch of the power supply distribution unit on the main control is controlled by a parallel circuit of two chips, which can further improve the stability of the system.

The storage system is equipped with PCIe Raid card and port multiplexer, which enables a set of storage system to connect 40 1 TB hard disks. The total capacity of

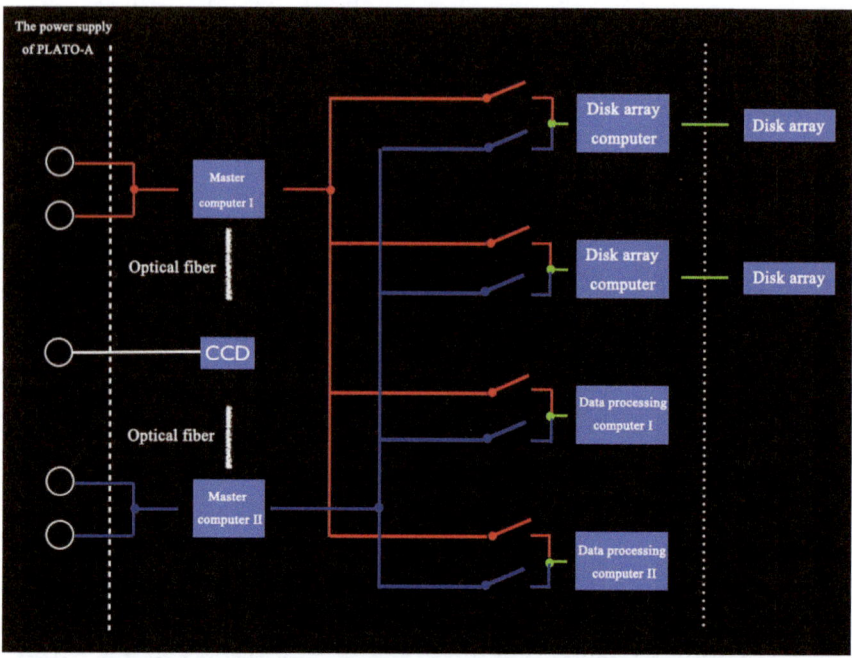

Fig. 12 Schematic diagram of the first-generation operational control system

40 TB meets the requirements of AST3 for one year. We have developed new technology that enables the storage system to work with only one hard disk open at a time. This means that the actual power consumption of the storage system is only 30 W, less than 1/20 of the available commercial storage servers of the same capacity.

After the AST3 operation and control system was put into use, the AST3 team further optimized the operation and control system based on the feedback obtained from the equipment operation and developed the second generation of AST3 operation and control system (Fig. 13). In the second generation of operation and control system, the subsystem of power distribution unit (PDU) is added to separate the functions of power switch and temperature acquisition in the main control system, which makes the operation and control system more modular. The power supply distribution unit can control 32 switches, which is twice of the first-generation operational control system. The power distribution unit can control 32 switches, twice as many as the first-generation operation control system, and it can connect at most 16 temperature sensors, four more than its predecessors. At the same time, we unified the power input and output and temperature acquisition interfaces of the operation and control system, which facilitated the installation work of the scientific research team on site and greatly reduced the possibility of connection errors. The 1 TB hard disk used in the first-generation operational control system was replaced with 8 TB helium-filled disk with larger capacity and

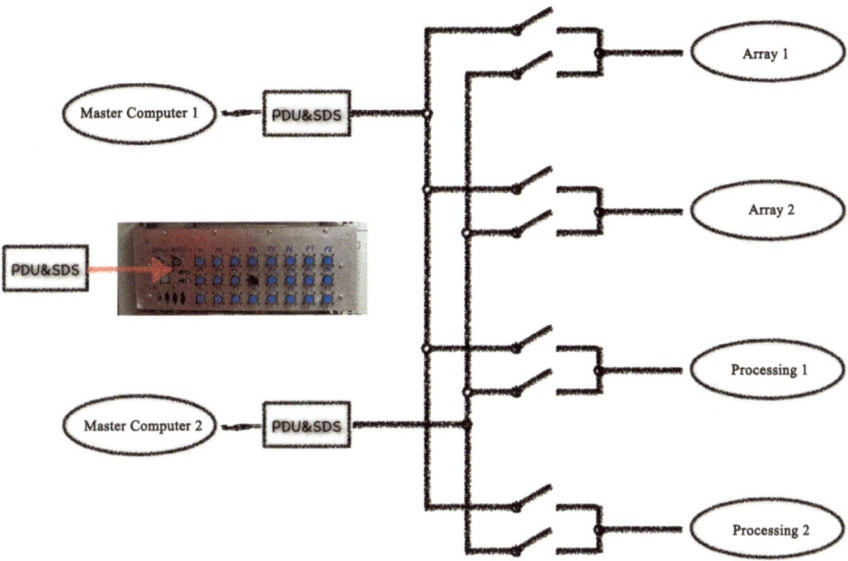

Fig. 13 Schematic diagram of the second-generation operation and control system

especially superior performance at low temperature and low pressure, which not only met the requirements of exoplanet search for hard disk capacity in this project, but also further improved the reliability of the operational control system.

In February 2017, a staff member found that the data of the real-time display data of the CCD on the Web site run by AST3 was not updated in real time. He logged in remotely and found that the main control system lost connection with the CCD. At the same time, he also found that before the CCD connection is lost, the temperature of the chipset on the CCD image acquisition card suddenly rises above 90 °C. After discussion, it was determined that the CCD image acquisition card was damaged (cosmic ray, aging, etc.), so it was decided to switch to the standby main control system. After switching, the CCD restores the connection and starts working properly. Using two computers to control, a CCD camera is a technique originally developed by the AST3 team. This is one of the various redundant designs we have done to solve the reliability problem in unattended conditions. This example verifies the high reliability of the redundant design of AST3 operational control system under unattended conditions.

2.3.2 Software of Operation Control System

The software of operation and control system includes three parts: daemon, basic command, and sky survey script. Daemons have been running in the background since the computer was started. They directly manipulate the hardware of the AST3 operating system. Each daemon implements a function, such as CCD image acquisition, sensor data acquisition, power switch control, data receiving and storage, and system log recording. The basic commands are the client programs for the daemon, the telescope control program, and the CCD control program (provided by CCD manufacturer STA). The user performs an operation in the operational control system through a basic command. The sky survey script integrates basic commands to realize the full set of functions of automatic operation of the telescope without human intervention. These scripts control AST3 to obtain the coordinates of the observation sky area, telescope pointing and tracking, CCD exposure and acquisition of images, image distribution processing and storage, etc.

We also added SMS alert in the automatic observation program. When the automatic program detects the error that the telescope cannot be automatically recovered by the program, it will send the alarm information back to the domestic server through the data transmission system developed by the AST3 team, and trigger the server to send the alarm information to the person in charge of the relevant equipment via SMS for remote manual intervention.

The expandability is fully considered in the design of the operation and control system software. After the AST3 operation and control system hardware optimization and upgrade, the AST3 team only need to modify a small amount of code and daemon configuration files, so that the operation and control system software can be used in the new hardware system, greatly improving the work efficiency.

2.4   Survey Strategy of AST3

In order to improve the efficiency of the telescope, save observation time and realize automatic sky survey, the AST3 team developed a special survey strategy program for AST3 (Fig. 14). The first version of the survey strategy program can select the best observation target from the sky area of the survey target according to the current time, the current position of the telescope, the zenith distance of the target, the distance between the target and the moon and other parameters. The automatic survey script calls the survey strategy program to obtain the target sky area to be observed next time.

High-precision photometry requires excellent flat-field images, so the AST3 team further optimized the survey strategy program and added the automatic twilight flat-field observation. When the altitude of the sun is within the range set by the program, the survey program will select the appropriate flat-field observation coordinates and exposure time. By calling the survey strategy program, the automatic survey script program will shoot the flat field before and after the beginning of each day's survey, for the use of subsequent off-line high-precision photometry.

The strategy takes into account the needs of different scientific objectives. For example, we set different observation modes for supernova survey and exoplanet search. With the survey program, we can automatically open and close the telescope, survey mode selection, survey area selection, etc. At the same time, the survey strategy also sets a flexible special observation mode, which can be inserted

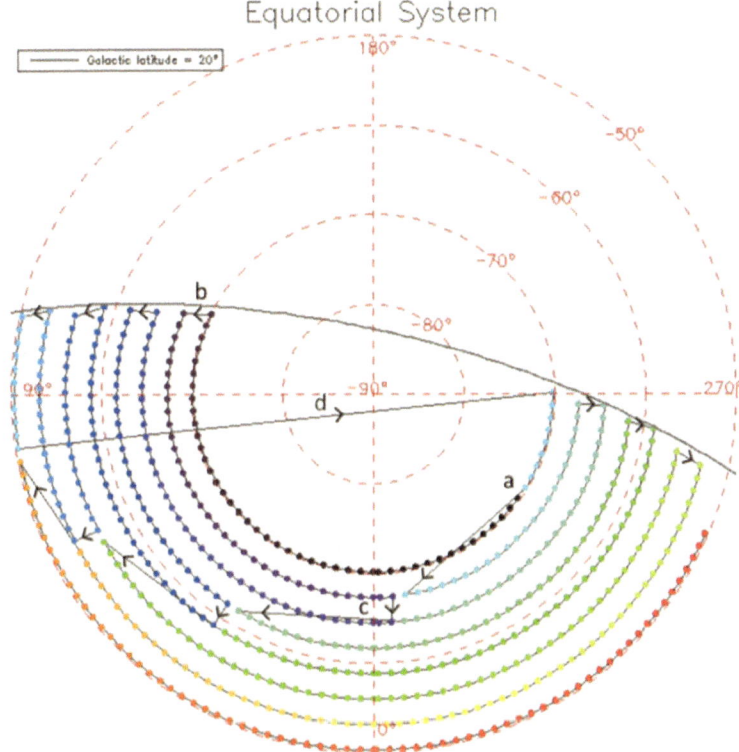

Fig. 14 Simulation diagram of sky survey strategy

at any time to observe celestial bodies with special requirements (such as GRB and supernova) and automatically return after the completion of observation to continue the unfinished sky survey observation. Sky survey strategy is a key component of automatic sky survey.

2.5 AST3 Data Reduction

2.5.1 Data Reduction at Antarctica

In order to discover transient candidates as soon as possible, real-time image processing is of crucial importance in a transient survey. Considering the dramatically high expense on data transfer from Antarctica via the Iridium satellites, it is hence necessary to run data reduction at Antarctica for the observed images with huge file size. We developed a robust pipeline for the low-power and high-performance computer working for AST3.

- **Overscan Correction** measures the median value for each one of the 16 overscan regions and then removes these offsets from corresponding pixels.
- **Image Correction (dark current and flat field)** The dark current will be not corrected at Antarctica to shorten the computation time and save electrical power, while twilight flat fields are prepared in March and April before the polar nights. The downloaded APSS and PPMX catalogs are employed for flux and position calibrations, respectively, where APSS catalog covers photometry on g- , r- , i-bands and PPMX has high position accuracy about 0.04 arcsec such that the standard catalogs are sufficient for AST3.
- **World Coordinate System (WCS)** Initially estimate a rough WCS in according to the pointing information from the FITS header, and then a basic photometry catalog can be generated by the software Source Extractor. The final accurate of WCS is calculated using Scamp with the photometry catalog as input.
- **Flux Calibration** APASS photometry catalog on i-band is used for flux calibration. The program is based on IDL.

Since there is a gradual shift in Antarctica from polar day to polar night, the sky brightness is not likely to be very dark at the beginning of the transient survey; therefore, the images took at that time can only produce a relatively shallow template. We designed a dynamical automatic update module for template making. As survey observations continue, updated templates will be produced with increased detection limit by stacking new observed images.

- **Image Subtraction** The preliminary operation for this process is image registration, reference, and science images will be aligned by wcsremap. The software Hotpants is employed for image subtraction, and then all transient candidates with brightness variation are expected to appear on the difference image. These candidate snapshots are then transferred back and posted on our Web site.

Empirically, artifacts will dominate the set of all positive detections. We designed specific exclusion criterion and ranked all remaining candidates. Priority is assigned for these candidates by checking the existence of nearby galaxies and visible counterpart on the corresponding template. The ones with high priorities are transferred back as soon as possible and posted on the top of our Web site. The real-time Web site also provides useful links for each candidate, including NED, DSS image, MPS (Fig. 15).

2.5.2 Real-Time Monitoring of the AST3 Sky Survey

High-precision photometry requires assistance with information from the environment and state of the device during exposing. After finished the automatic survey, we carried out real-time monitoring for Antarctica. The AST3 team developed the second-generation Kunlun automatic weather station KLAWS-2G and the Kunlun Cloud Aurora Monitor KLCAM independently (Fig. 16). KLAWS-2G is a multi-layer weather station with temperature sensors and wind speed/direction

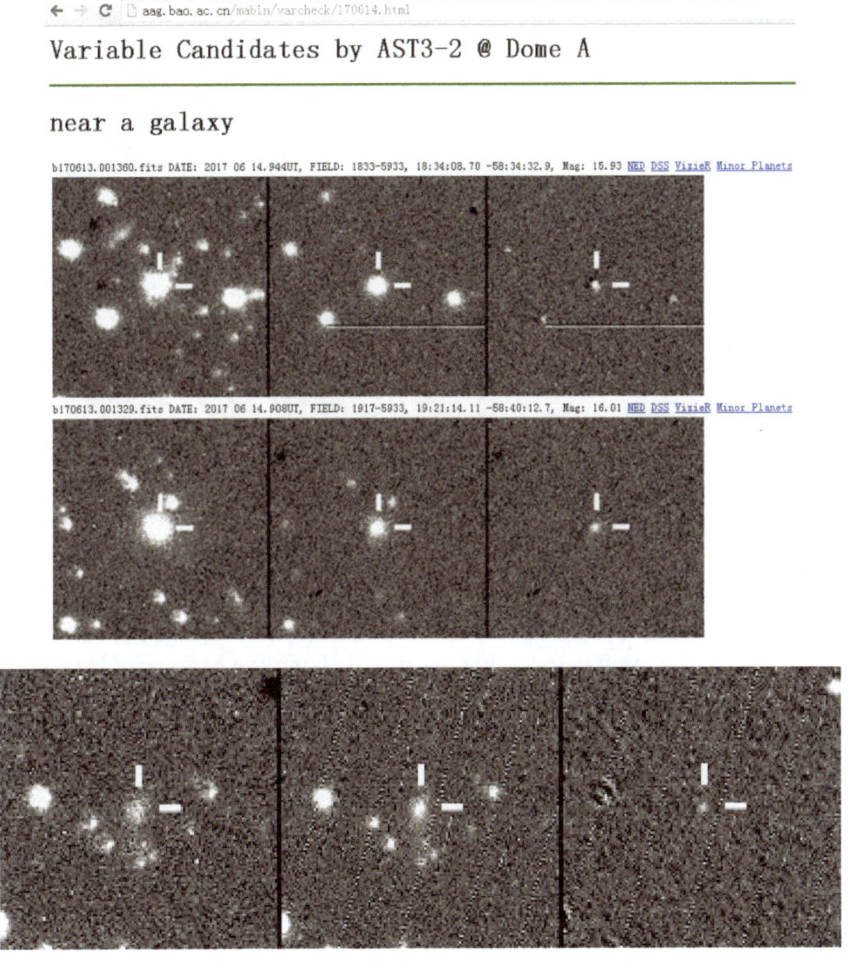

Fig. 15 Above: AST3 variable source candidate page. Below: supernova SN 2017fbq (AST-17a) discovered by AST3-2 in 2017

indicators on each layer. Data is collected every 10 s and passed back to the server every 15 min. KLAWS-2G was successfully installed at Kunlun Station in January 2015 and continued running until September 2016. KLAWS-2G is still working today after simple maintenance by the expedition team in 2017. KLCAM uses a commercial camera equipped with a fisheye lens to monitor the cloud and Aurora at Kunlun Station by taking photographs every half hour. Restricted by communication costs, KLCAM returns photographs at local midnight and noon.

To make it easier for researchers to understand current and retrospective instruments status' and weather information in real time, AST3 team developed a real-time Web site to display the instruments status and weather data (Fig. 17). The Web site not only shows on-site monitoring but also on all instruments and

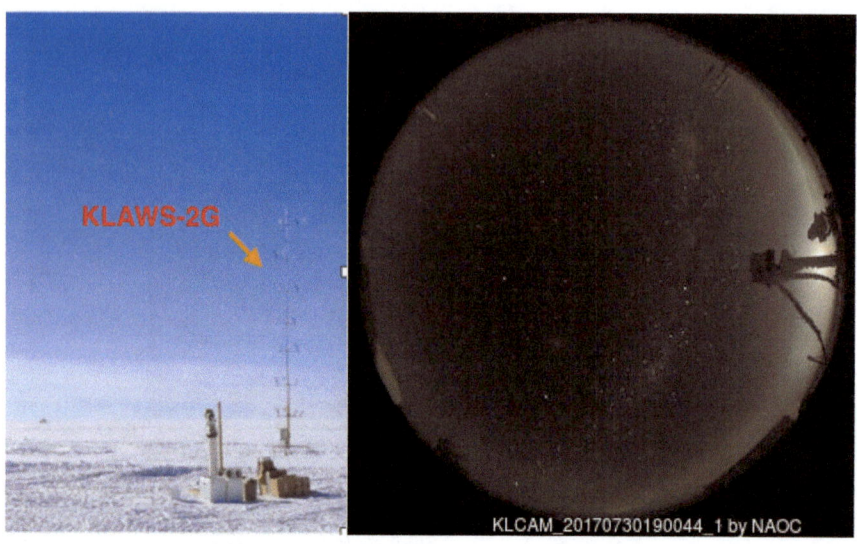

Fig. 16 Left: Kunlun automatic weather station KLAWS-2G; Right: photograph taken by Kunlun Cloud Aurora Monitor KLCAM

equipment maintained by AST3. Users can visit http://aag.bao.ac.cn/klaws to get the latest data or the data of some given time range interested. These operational and running proofed that real-time monitoring Web site provided data not only helps researchers to evaluate data quality, improve photometry accuracy, but also guaranteed the safe running of AST3 telescope greatly, such as stopping observing when the wind is too fast to protect the telescope.

2.5.3   High Accuracy Photometry Based on AST3-1 2012's Data

AST3-1 Antarctic data was carried back by scientific exploration team at 2013. Afterward, the AST3 team managed to increase the threshold magnitude by 1 mag and increase the bright star photometric accuracy by 2–3 mmag.

The high accuracy photometry for AST3-1 data includes the cross-talk effect induced in CCD readout process, CCD's overscan bias, dark current, flat field. Our AST3 team emphasize on large FoV flat-field correction and dark current correction.

As the Kunlun Station is fully automatic, it is impossible to maintain a dome for flat-field correction. Moreover, the 2012 mission focusing on dense star-field and the lack in pre-observation also barricade the acquisition of super-sky for flat-field correction. Consequently, we adopt twilight for flat-field correction. Considering the twilight's inhomogeneity (about 1–10% luminosity gradient per degree, depending on the solar-telescope angle) which will be considerable in our AST3's relatively large FoV ($3 \times 1.5°$), we stack plenty of twilight pictures to acquire an initial flat-field graph, then divide every individual twilight graph with the initial

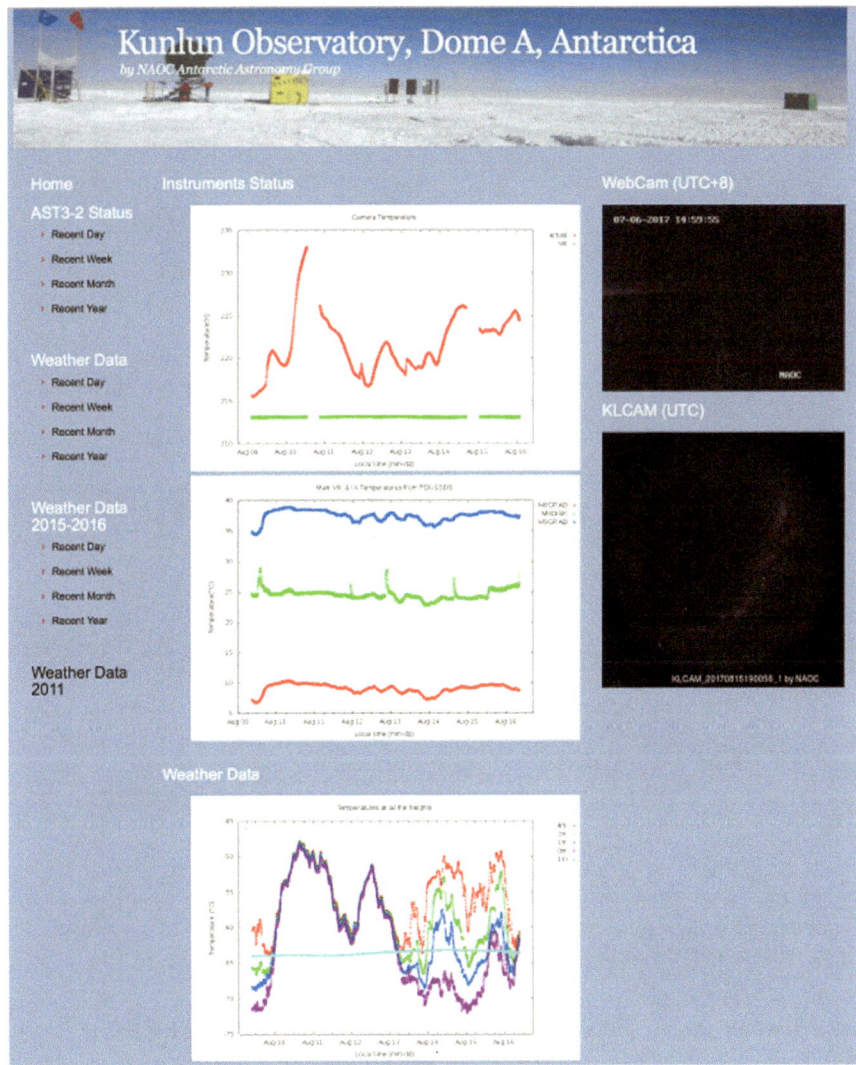

Fig. 17 Screenshot of real-time instrument status and site monitoring

flat-field graph in order to subtract the effects from vignetting and the channel's amplifications. In the next step, we fit the twilight gradient with a smooth function (Fig. 18) to finally get the flat-field graph. As is shown in Fig. 19, the twilight graph before correction is with a large root-mean-square near the boundary (0.3–0.5%) and a smaller RMS at the center (0.2%). In contrast, the twilight graph's RMS is uniformly decreased to 0.1% after flat-field correction.

Considering the relatively high temperature in the CCD causing large dark current, we tailored the dark current correction for our mission. Our CCD works in

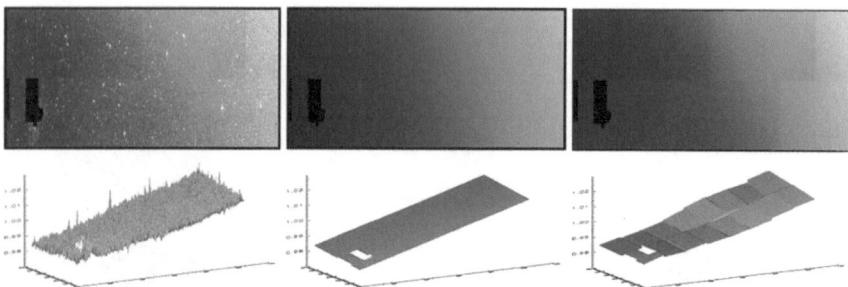

Fig. 18 Left: a twilight image divided by the initial flat field; The middle figure: fitting residual skylight gradient in the full field of view; Right figure: fitting residual skylight gradient into 16 readout channels. The figure above is a two-dimensional image with black areas as bad pixels. The picture below is a three-dimensional image

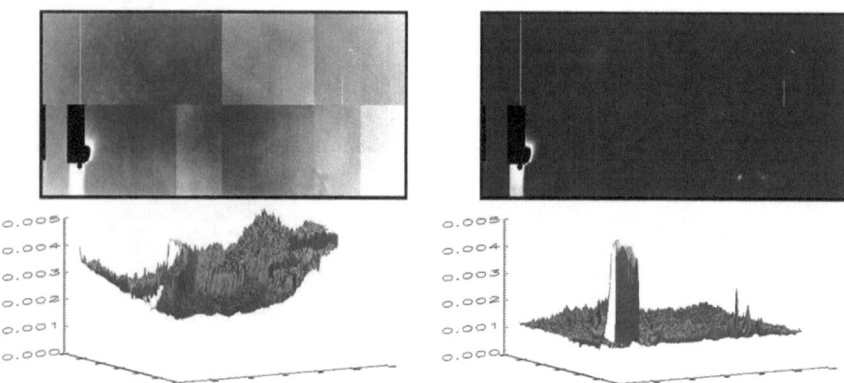

Fig. 19 Comparison of mean square values of twilight images stacking directly (left) and stacking after deducting skylight gradient (right). The figure above is a two-dimensional diagram, and the figure below is a three-dimensional diagram

frame transfer mode, and we cannot obtain dark current data on the scene, worse still, the dark current data obtained in the laboratory do not match the observational data (Fig. 20). To tackle with the problem, we developed an algorithm which can effectively correct the dark current effect by extracting the data from a pair of observational graphs. This method curtailed the background noise down to near photon noise (Fig. 21) helped us to increase the threshold magnitude and the accuracy. For a typical graph snatched when CCD temperature is −40 °C, the method can increase the threshold magnitude by 1 mag.

After the as-mentioned CCD cross-talk correction, overscan correction, dark current correction, and the flat-field correction, we applied aperture metering, coordination locating, photometric locating, etc. The ultimate limit magnitude of the i-band of AST3 reaches 19 mag, the luminous star photometric accuracy is 2–3 mmag, the absolute accuracy of the position measurement is about 0.1", and the

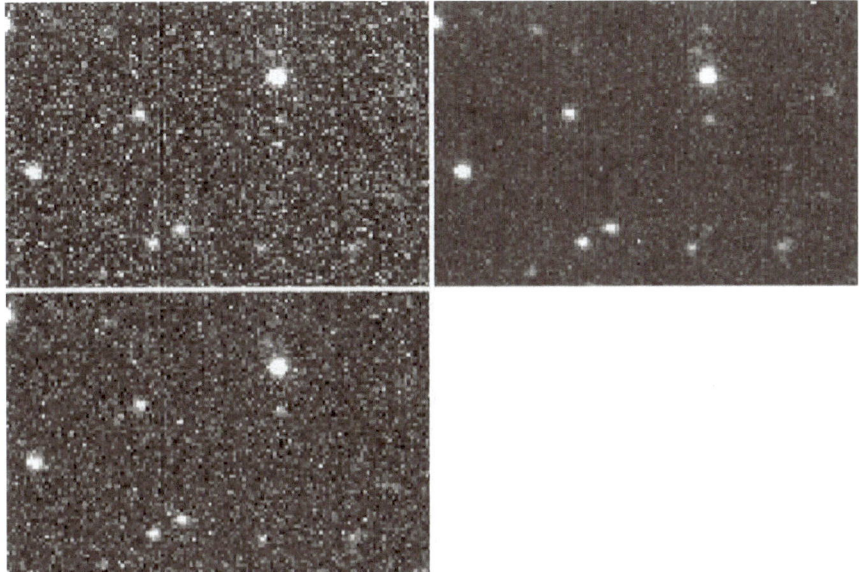

Fig. 20 Top left: AST3-1 original image; Bottom left: corrected by laboratory dark current; Upper right: After correcting by the dark current calculated from the observation image, the background noise is greatly reduced and the dark source is more obvious

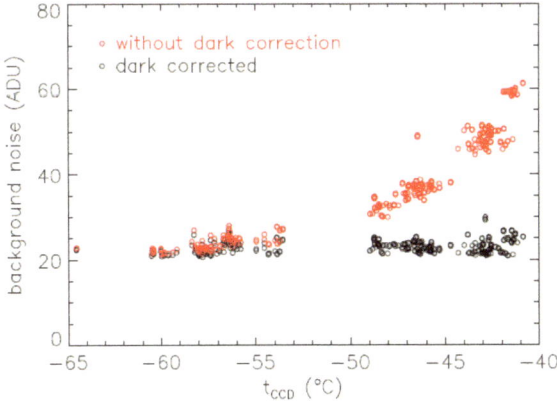

Fig. 21 For some images with similar background values, the background noise increases rapidly with the CCD temperature. After correcting the dark current calculated from the observation image, the background noise is reduced to the same level, close to the photon noise, which indicates that the calculation of dark current is very effective

relative accuracy of the bright star position can reach 0.04". All these results embody AST3's enormous potential in time-domain astronomy. The corrected graphs, star sheets, and the light curves during AST3 2012 mission are available on Chinese Virtual Observatory (http://explore.china-vo.org/).

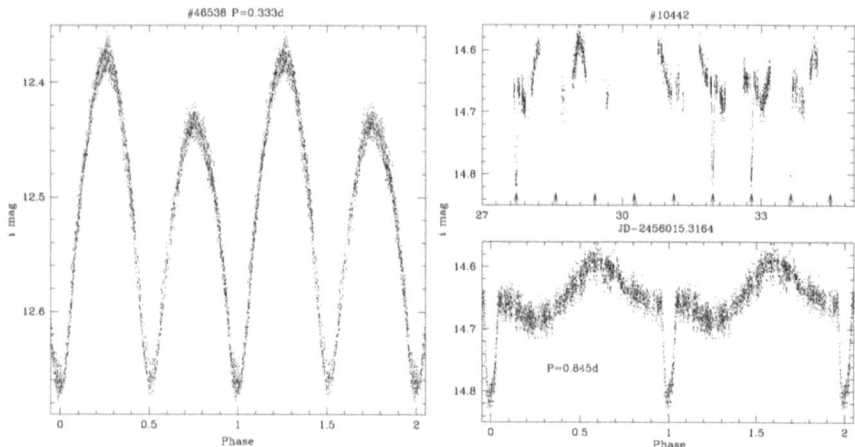

Fig. 22 Examples of light curves in ast3-1's 2012 data. Left: Phase diagram of the variable star with the strongest O'Connell effect in the sample; Right: Variable star with RS CVn-like light curves, the figure above is the light curve, and the figure below is the phase diagram

During the variable source search with other teams, we found 560 variable stars in a single sky region with several days' observational data. In the 339 of them which are newly detected stars, 85 eclipsing binaries (EW, EB, EA) and 27 pulsating variable stars (Delta Scuti, Gamma Doradus, Delta Cephei variable, RR Lyrae star) are identified. Particularly, we acquired a binary star system's light curve, which shows RS CVn-like features. We have applied for the following spectroscopic observation on Gemini South telescope for this target (Fig. 22).

2.5.4 Cooperative Developments on AST3 Controlling, Data Analysis, and Auxiliary Programs

Due to the satellite bandwidth and the transport expense, the observational data cannot send back in total, which enforces the station to analyze the data on the scene. As a result, AST3 team amalgamate with Tianjin University's computer science school for the development of AST3 domestic high-performance data analysis and storage program, as well as AST3 Antarctic data analysis and daemon system. The basic procession for simultaneous data analysis is designed. In these works, AST3 team contribute most on the astronomical data analysis program and present not only CCD data preprocessing program but also aperture photometry and other functions, while Tianjin University's team focus on the background daemon process (Fig. 23), general data analysis and database managing. In order to increase the system's stability and reliability, daemon process is with multiple protection modes, as shown in Fig. 24. Data analysis process is shown in Fig. 25. With these integrated designs, AST3 is capable for real-time data analysis.

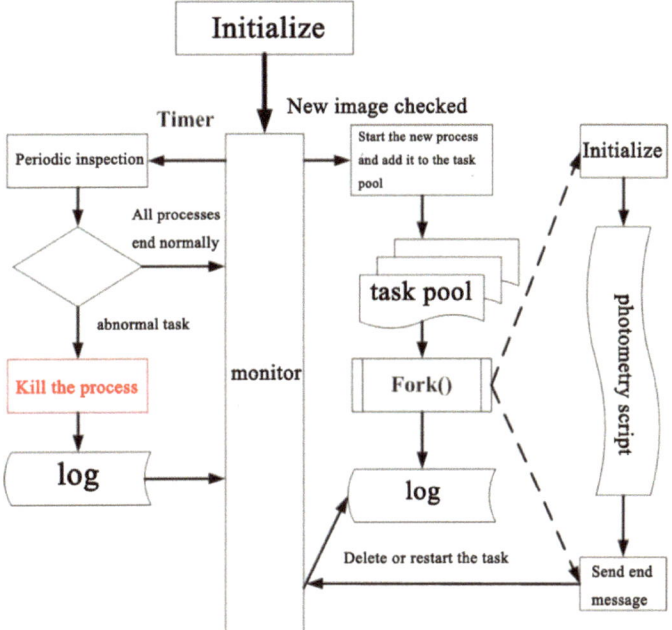

Fig. 23 Daemon logic diagram for data reduction

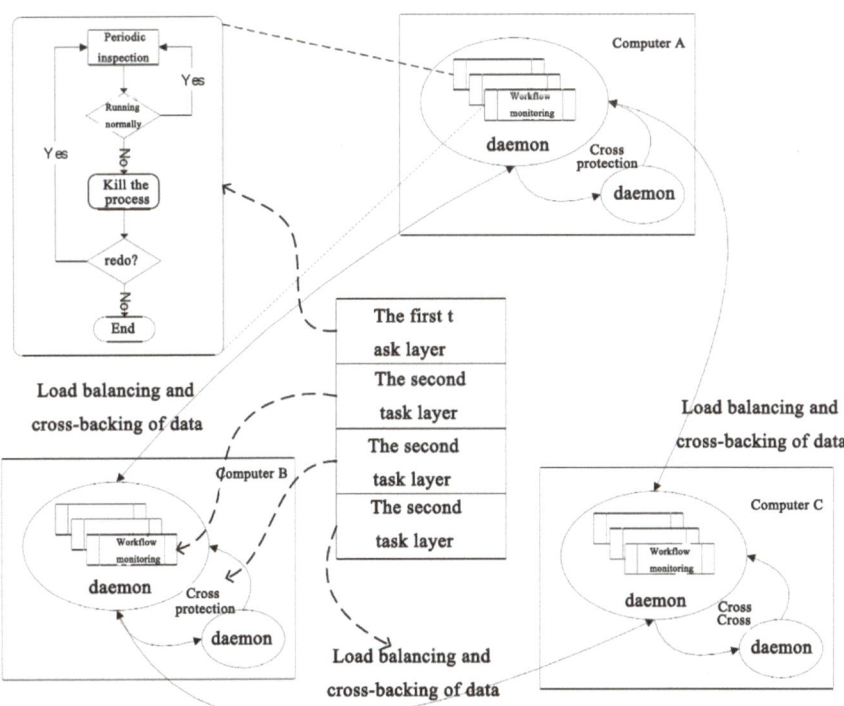

Fig. 24 Multi-level, multi-node protection for the data processing daemon

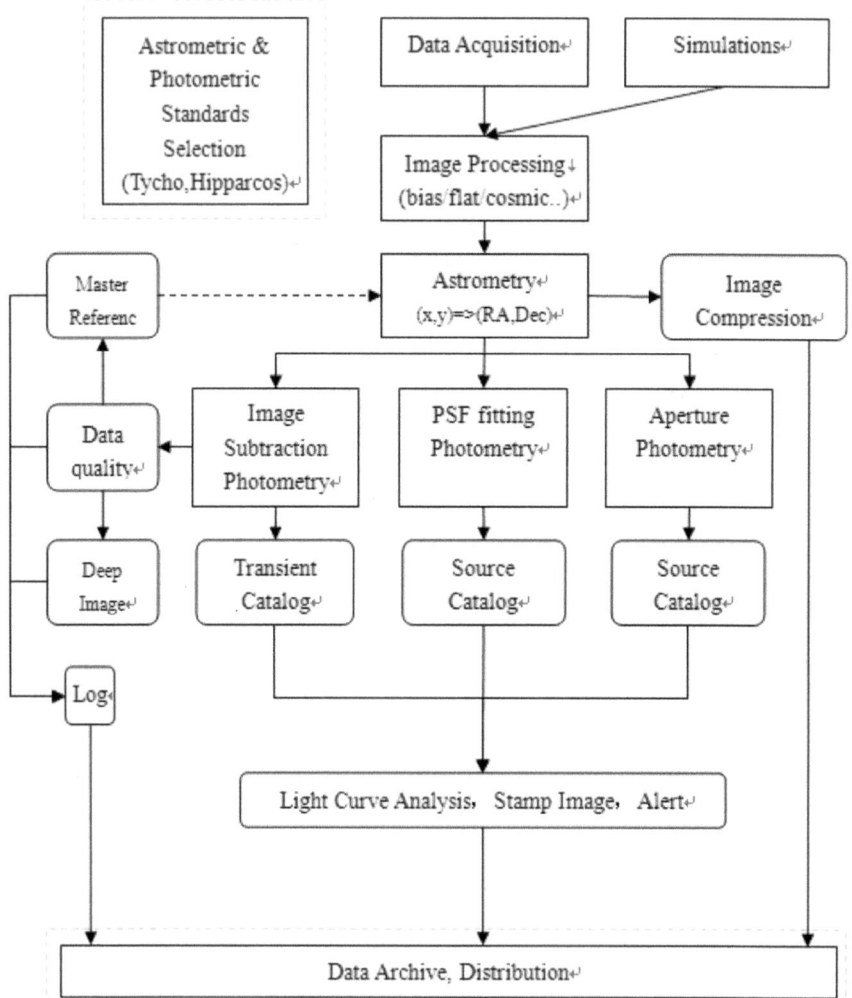

Fig. 25 Process of data reduction

The database system presented by AST3 team and Tianjin University is designed to store the star sheet data. With the help of non-relation type MongoDB database structure, inserting 500 k data will cost one-thirds of the time comparing to MySQL database (Fig. 26).

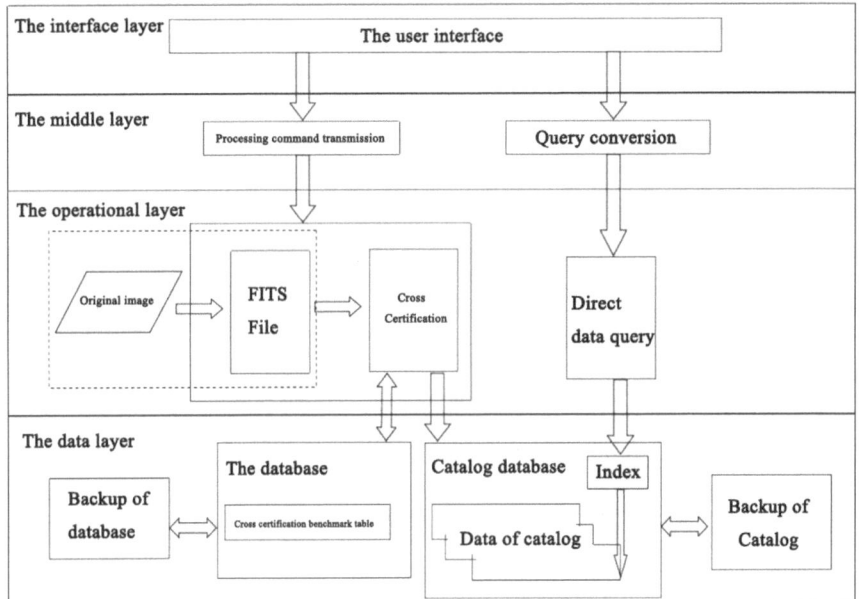

Fig. 26 AST3 catalog data storage architecture

3 Scientific Research Around Antarctic Survey Telescopes AST3

3.1 AST3 Detected the First Optical Counterpart of a Gravitational Wave Event

Advanced Laser Interferometer Gravitational-Wave Observatory (LIGO) and Advanced Virgo observatories made the discovery of gravitational waves from GW170817 [5] at 20:41:04 Beijing time on August 17, 2017. The gravitational waves from GW170817 were produced by the merging of two neutron stars in a binary system. In less than 2 s after the LIGO/Virgo discovery, both NASA's Fermi gamma-ray telescope and the European INTEGRAL satellite detected a very weak short-duration γ-ray burst (GRB 170817A) [8]. The association of a γ-ray burst with GW170817 triggered a massive number of follow-up observations by ground-based telescopes all over the world. It was quickly found that the gravitational waves of GW170817 originated from a galaxy called NGC 4993 about 130 million light-years from the Earth. This was the first time in human history that a gravitational waves event and its optical counterpart had been identified. It is a milestone achievement in astronomy.

AST3 is located at Dome A, the highest point of the Antarctic continent, with geographical latitude of −80°22′. Although GW170817 is located in the southern

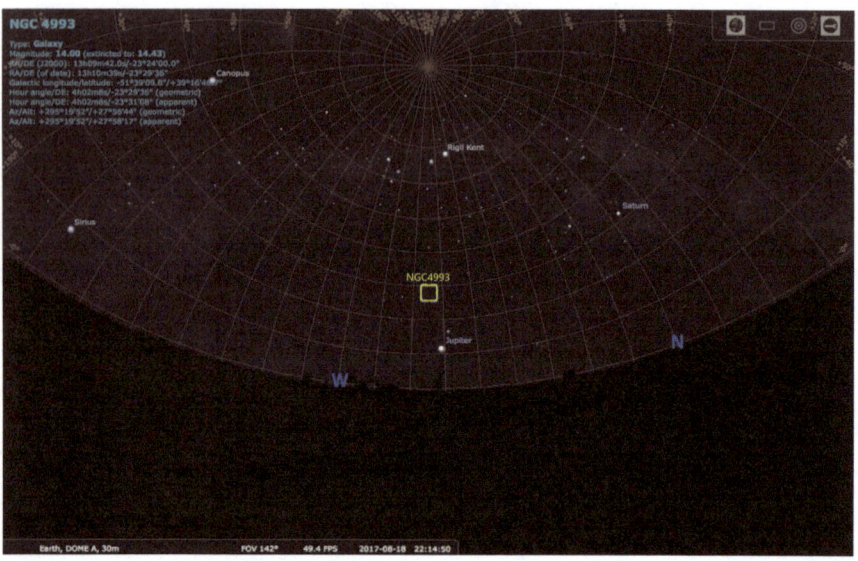

Fig. 27 Diagram of the horizon height (< 30°) of the object observed by AST3 during the 18 August observation window

celestial region, it is close to the horizon as it points at declination −23°22′53.350″. AST3 removed the original software limit and collected valid data from target for 8 days (18th, 20th, 21st, 23rd, 24th, 25th, 27th, and 28th) at a very low horizontal height (15°–30°), a total of 218 images. In order to calculate the optimal observation window of the event at the Antarctic, we need to analyze the altitude of the object and the sun and the moon as well as the Angle relationship between them in real time. Since this region is not an AST3 regular survey area and the polar night has passed, it is important to calculate the observation window time as soon as possible so that we can adjust our observation strategy in time. Based on the real-time information of the sun and the moon (such as the altitude of the earth and the coordinates on the celestial sphere) provided by the python astronomical expansion package astropy, we made a corresponding webpage for quickly determining the best observation time and adjusting the exposure time of the observation. After calculation and analysis, the Antarctic observation window appeared at about 21:00 PM on August 18, Beijing time (Fig. 27). Thanks to AST3's stable spindle and mechanical design, the telescope was able to observe the sky continuously for the next two and a half hours, collecting 21 long-exposure images and eventually detecting the first signal from the optical counterpart of the binary neutron star merger. To record as much as possible the evolution of the brightness over time, the observations continued until August 28.

The data analysis and processing are carried out simultaneously with the observation. Due to the unique geographical location of the Antarctica, the images taken by AST3 are transmitted back to China via iridium satellite. Iridium is slow and expensive to transmit, so the vast amounts of data stored in the Antarctica need to be analyzed immediately and then only useful data can be transmitted back through the satellite. In this observation of the optical counterpart of gravitational waves, we used the local computer in the Antarctic to run the image statistical analysis code, calculate the skylight background of each image, the full width half maximum (seeing) of the astrological image, the limiting magnitude and other statistical data, selecting the image with better observation quality. WCSTools' relevant tools (sky2xy, imcopy, etc.) were used to crop the image. Sky2xy could quickly convert the celestial coordinates of the target into the pixel coordinates in the image, and imcopy could be used to intercept the small range of images of the gravitational wave event occurrence area. The size of the CCD used in AST3 is 10 K × 10 K, while the size of the trimmed image is only about 1 K × 1 K, and the file size is reduced to about 1% of the original image, thus greatly reducing the time needed for data transmission through iridium satellite and accelerating the progress of subsequent data processing. In the final data obtained, the observation data on July 18, 23, and 24 were of the best quality. The exposure time was 5 min, and the extinction was close to 3 magnitude, which also provided enough depth for AST3 to have the opportunity to track and detect the early optical signals of the gravitational wave. This event occurred in the periphery of galaxy NGC 4993. It was difficult to distinguish the signal of the gravitational wave optical counterpart from the signal of the brighter galaxy itself in a single image, so we used the method of image subtraction. Before image subtraction, the image needs to be aligned, which requires re-sampling of the image based on the WCS information of the image. The most common difference is Lanczos-3, which computes each pixel individually. Even with the cropped image, it still takes millions of calculations. In addition, since images in the same region photographed at different times have different seeing and point diffusion functions, image subtraction itself requires the use of space-related kernel functions to convolution, which also involves a large number of calculations. To improve computational efficiency, we use Cython and OpenMP to speed up the original code. Cython lets you use a simple python-like syntax to get close to the speed of a C program by constraining data types. OpenMP, on the other hand, can simultaneously call multiple CPU cores of a computer for parallel computation. In this way, the overall speed is increased by about 10 times, which greatly improves the efficiency of image processing. The adjusted code successfully picked up the optical counterpart signal from the host galaxy NGC4993 (Fig. 28), and the celestial coordinates were exactly the same as the binary neutron star merger given by LIGO, with a deviation of less than an arc second. Since then, we have confirmed that AST3 has successfully detected the optical counterpart from the GW170817 gravitational wave event.

With subsequent photometry and kilonova model fitting [9] (Fig. 29), we found that this binary neutron star merger threw out more than 3000 earth-mass neutron-rich materials. The projectile has about 1% of the mass of the sun and is moving at about 30% of the speed of light. Some of these materials are synthesized by fast neutron capture, while some form super heavy elements heavier than iron. Together with data from other international telescopes, the AST3 observations provide important information about the physics of binary neutron star mergers.

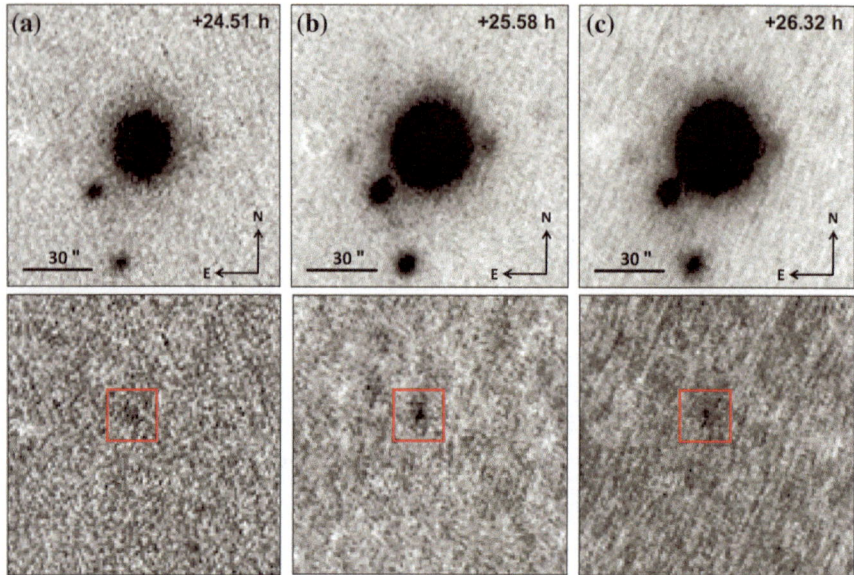

Fig. 28 Optical signals of gravitational waves detected by AST3-2 (in box)

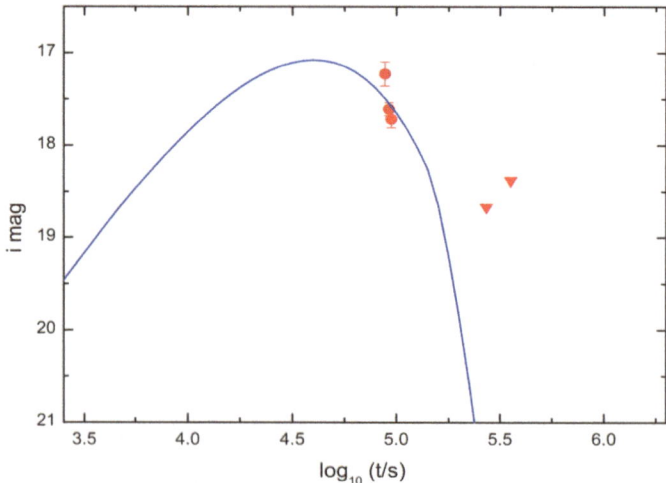

Fig. 29 Comparison of the time evolution of the GW170817 optical counterpart obtained from AST3-2 photometric data with the fitting results of the thousand stars semi-analytical model

3.2 Exoplanet Detection and Asteroseismology in Antarctica

The excellent astronomical observation conditions in the Antarctic and the continuous darkness lasting for more than 100 days also provide conditions far superior to other ground observation sites for the study of exoplanets and asteroseismology. The AST3 team carried out a series of related scientific studies using the excellent observation conditions in Antarctica. In order to search for transits of exoplanets, stellar light changes need to be measured with an accuracy of 0.1% (i.e., 1 mmag change). In order to increase the efficiency of exoplanet detection, it is necessary to use large-field telescopes as much as possible, so that tens of thousands of stars can be observed at the same time and the suspected transit signal can be detected automatically. Such automated, large-field, high-precision observation is almost the limit of surface photometric observation at present.

We have developed and improved a set of software system, which can automatically process massive star photometric images, generate high-precision light curve, and detect exoplanet transit signals. Our software can measure light with an accuracy of more than 1 mmag at the bright star end (Fig. 30) and is more than 30% efficient at detecting planets from super-earth to Neptune, and more than 90% efficient at detecting hot Jupiter. Its photometric accuracy is in a leading position among similar projects in the world (Table 1).

The AST3 team processed and analyzed the data of AST3 and Kepler by using the software developed by ourselves. In 2014, the first batch of exoplanet candidates was found in the South Pole. The first batch of six candidates was found with an occultation depth of 0.7–2.1% and a period of 1.4 to 10 days. It is the world's first candidate exoplanet in the Antarctic.

During the Antarctic polar night in 2016, AST3 carried out a month-long continuous observation on the key sky area to be covered by the TESS satellite

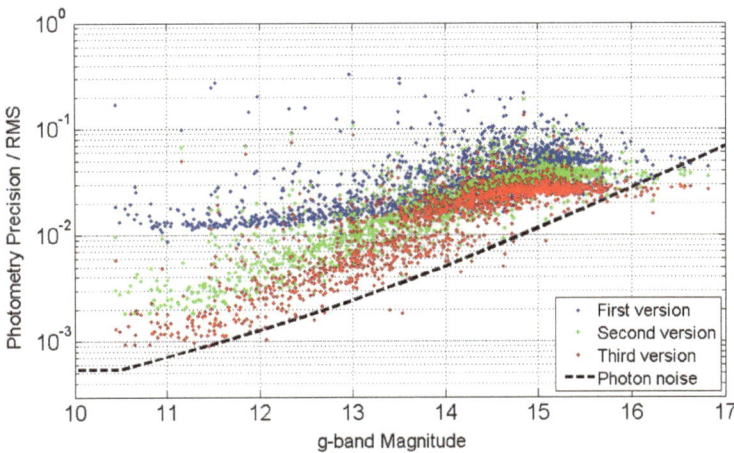

Fig. 30 Accuracy of our automated photometry program has reached 1 mmag, meeting the needs of exoplanet detection

Table 1 AST3 photometric accuracy compared with other similar projects in the world

Project		Optimum photometry accuracy (%)
Hungarian Automated Telescope Network (HATNetwork)	HAT-N	0.5
	HAT-S	0.1
Wide Angle Search for Planets (WASP)	WASP-S	0.5
	Super-WASP-N	0.5
XO		1
Kilodegree Extremely Little Telescope (KELT-North)		1
Transatlantic Exoplanet Survey (TrES)		0.2
AST3		0.1

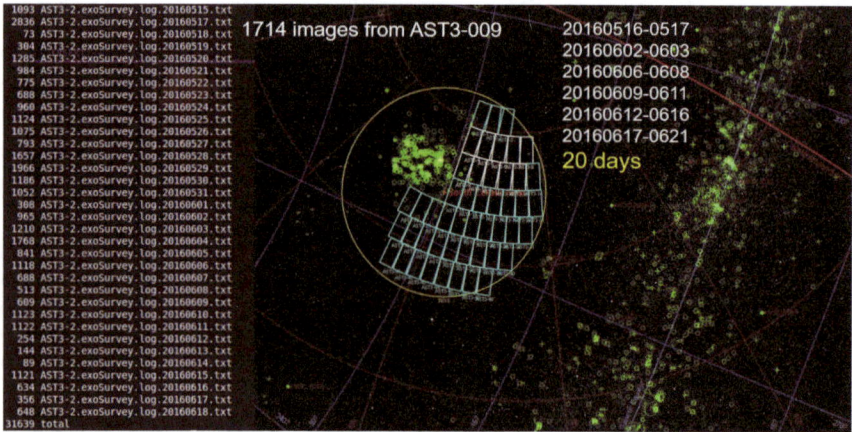

Fig. 31 In 2016, the AST3 made observations of TESS in the sky near the South Ecliptic Pole for more than 20 days

program. This is the first time that AST3 has used our self-developed automatic control software to automatically observe for a long time during the polar night, showing that our hardware and software system has reached the expected level. TESS is the next-generation exoplanet survey satellite, which will be launched around 2017–2018 to look for exoplanets. TESS will cover most of the sky for only one month because of the limited field of view, so it will cover mainly short-period planets. However, in the area near the north and south ecliptic poles, TESS will continue to observe for more than 12 months. Therefore, any planets and variable star targets in this part of the sky will become a research hotspot and be concerned by the whole world. Since AST3 is located in the Antarctic Dome A, it can cover TESS very well in the sky area near the ecliptic pole. The AST3 team selected 10 fields in the target region for observation (Fig. 31), obtained about 32,000 images, and discovered more than 50 extrasolar planet candidates. The smallest transit depth is only 0.5% (Fig. 32).

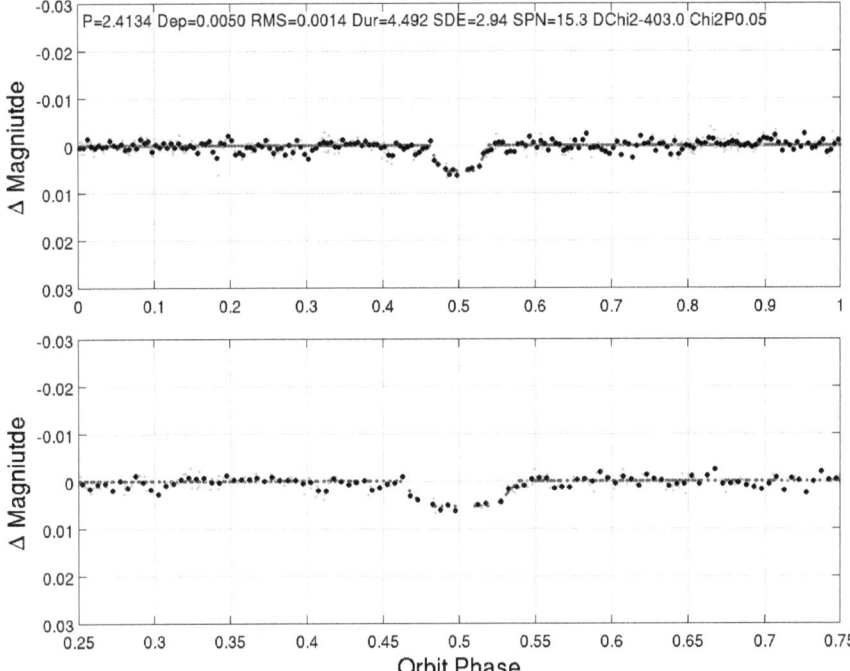

Fig. 32 AST3 found a candidate for an exoplanet with a magnitude of 0.005, but at AST3's level of photometry accuracy it was remarkable. This shows that AST3 is capable of searching for small exoplanets

This is the first time that China has discovered a large number of exoplanet candidates in Antarctica with its own telescope after the first batch of six candidates in Antarctica, which indicates that the software and hardware level of AST3 has been mature.

The AST3 team also performed asteroseismological studies using Antarctic data. The observation of Antarctic Dome A can provide long-time uninterrupted, massive high-precision, multi-band photometric data for the study of asteroseismology, while the polychromatic data is not available for the space telescope, which can carry out pattern identification of pulsating variable stars and has special significance for the study of asteroseismology.

4 Conclusions and Outlook

Chinese astronomers plan to build 2.5-m Kunlun dark space optical infrared survey telescope (KDUST) using the unique astronomical observation environment of Dome A Kunlun Station. KDUST has certain scientific objectives: the measurement of cosmological parameters and the study of exoplanets. KDUST is far more powerful than AST3, combining wide-field optical infrared survey with extremely high-spatial resolution, which is impossible for any other ground-based observatory. KDUST has a very high-spatial resolution in the optical band and will have the unique capability of a quasi-space telescope. It can survey the sky faster than any existing ground-based telescope and will play important role in future astronomical observations. This unique quasi-spatial capability enables the 2.5-m optical near-infrared telescope KDUST to effectively study the fundamental astrophysical problems of dark matter and dark energy, the structure of galaxies and clusters of galaxies, the structure and dynamics of the Milky Way, star formation, and so on.

In the era of multi-messenger astronomy initiated by LIGO/VIRGO, KDUST will cooperate with the large telescope equipment around the world to build a powerful astronomical observation information network and set its eyes on the distant space, while the construction of China's Antarctic observatory will open up a new observation platform for people to understand our universe.

References

1. Cui, Xiangqun; Yuan, Xiangyan; Gong, Xuefei, Antarctic Schmidt Telescopes (AST3) for Dome A, Ground-based and Airborne Telescopes II. Edited by Stepp, Larry M.; Gilmozzi, Roberto. Proceedings of the SPIE, Volume 7012, article id. 70122D, 8 pp. (2008).
2. X. Yuan, X. Cui, X. Gong, D. Wang, and Z. Yao, Progress of Antarctic Schmidt Telescopes (AST3) for Dome A, Proc. SPIE, 7733, (2010), pp. 77331V-77331V-7.
3. Li, Zhengyang; Yuan, Xiangyan; Cui, Xiangqun; Wang, Daxing; Gong, Xuefei; Du, Fujia; Zhang, Yi; Hu, Yi; Wen, Haikun; Li, Xiaoyan; Xu, Lingzhe; Shang, Zhaohui; Wang, Lifan, Status of the first Antarctic survey telescopes for Dome A, Proc. SPIE 8444, Ground-based and Airborne Telescopes IV, 84441O (September 17, 2012).
4. Yuan, Xiangyan; Cui, Xiangqun; Gu, Bozhong; Yang, Shihai; Du, Fujia; Li, Xiaoyan; Wang, Daxing; Li, Xinnan; Gong, Xuefei; Wen, Haikun; Li, Zhengyang; Lu, Haiping; Xu, Lingzhe; Zhang, Ru; Zhang, Yi; Wang, Lifan; Shang, Zhaohui; Hu, Yi; Ma, Bin; Liu, Qiang; Wei, Peng, The AST3 project: Antarctic Survey Telescopes for Dome A, Proceedings of the SPIE, Volume 9145, id. 91450F 8 pp. (2014).
5. The LIGO Scientific Collaboration; The Virgo Collaboration, GW170817: Observation of Gravitational Waves from a Binary Neutron Star Inspiral, eprint arXiv:1710.05832
6. The LIGO Scientific Collaboration; GW150914: First results from the search for binary black hole coalescence with Advanced LIGO, Physical Review D, Volume 93, Issue 12, id.122003
7. Kasliwal, Mansi M., Transients in the local Universe: systematically bridging the gap between novae and supernovae, Bulletin of the Astronmical Society of India, Vol. 39, No. 3, p. 375–385, (2011)
8. LIGO Scientific Collaboration; Virgo Collaboration; GBM, Fermi et al, Multi-messenger Observations of a Binary Neutron Star Merger, eprint arXiv:1710.05833
9. Hu, Lei; Wu, Xuefeng; Andreoni, I. et al, Optical Observations of LIGO Source GW 170817 by the Antarctic Survey Telescopes at Dome A, Antarctica, eprint arXiv:1710.05462
10. Kilpatrick, Charles D.; Foley, Ryan J.; Kasen, Daniel et al, Electromagnetic Evidence that SSS17a is the Result of a Binary Neutron Star Merger, eprint arXiv:1710.05434
11. Wang, Lingzhi; Ma, Bin; Li, Gang; Hu, Yi; Fu, Jianning; Wang, Lifan; Ashley, Michael C. B.; Cui, Xiangqun; Du, Fujia; Gong, Xuefei; Li, Xiaoyan; Li, Zhengyang; Liu, Qiang; Pennypacker, Carl R.; Shang, Zhaohui; Yuan, Xiangyan; York, Donald G.; Zhou, Jilin, Variable Stars Observed in the Galactic Disk by AST3-1 from Dome A, Antarctica, The Astronomical Journal, Volume 153, Issue 3, article id. 104, 24 pp. (2017).

12. Michael G. Burton, Astronomy in Antarctica, The Astronomy and Astrophysics Review, Volume 18, Issue 4, pp. 417–469
13. H Zou, X Zhou, Z Jiang etal, The sky brightness and transparency in i-band at Dome A, Antarctica, Astronomical Journal, 140(2):2146–2146, 2010.
14. Yuan, X. and Su, D-q, Optical system of the Three Antarctic Survey Telescopes, Monthly Notices of the Royal Astronmical Society, 424, (2012), 23–30.
15. Yuan, Xiangyan; Cui, Xiangqun; Su, Ding-qiang; Zhu, Yongtian; Wang, Lifan; Gu, Bozhong; Gong, Xuefei; Li, Xinnan, Preliminary design of the Kunlun Dark Universe Survey Telescope (KDUST), Astrophysics from Antarctica, Proceedings of the International Astronomical Union, IAU Symposium, Volume 288, pp. 271-274.
16. Zhu, Yongtian; Wang, Lifan; Yuan, Xiangyan; Gu, Bozhong; Li, Xinnan; Yang, Shihai; Gong, Xuefei; Du, Fujia; Qi, Yongjun; Xu, Lingzhe, Kunlun Dark Universe Survey Telescope, Proceedings of the SPIE, Volume 9145, id. 91450E 17 pp. (2014).
17. Li, Xiaoyan; Wang, Daxing; Xu, Lingzhe; Zhao, Jianlin; Du, Fujia; Zhang, Yue, Control system for the first three Antarctic Survey Telescopes (AST3-1), Ground-based and Airborne Telescopes IV. Proceedings of the SPIE, Volume 8444, article id. 84445 M, 8 pp. (2012).
18. Ma, Bin; Shang, Zhaohui; Hu, Yi; Liu, Qiang; Wang, Lifan; Wei, Peng, A new method of CCD dark current correction via extracting the dark Information from scientific images, Proceedings of the SPIE, Volume 9154, id. 91541T 8 pp. (2014).
19. Ma, Bin; Shang, Zhaohui; Wang, Lifan; Hu, Yi; Liu, Qiang; Wei, Peng, The nonlinear photon transfer curve of CCDs and its effects on photometry, Proceedings of the SPIE, Volume 9154, id. 91541U 10 pp. (2014).
20. Shang, Zhaohui; Hu, Keliang; Hu, Yi; Li, Jiliang; Li, Jin; Liu, Qiang; Ma, Bin; Quinn, Jason L.; Sun, Jizhou; Wang, Lifan; Xiao, Jian; Yu, Jia; Yu, Ce; Yang, Mujin; Yuan, Xiangyan; Zeng, Zhen, Operation, control, and data system for Antarctic Survey Telescope (AST3), Observatory Operations: Strategies, Processes, and Systems IV. Proceedings of the SPIE, Volume 8448, article id. 844826, 7 pp. (2012).
21. Ma, Bin; Shang, Zhaohui; Wang, Lifan; Doggs, Kasey; Hu, Yi; Liu, Qiang; Song, Qian, Xue, Suijian, The test of the 10 k x 10 k CCD for Antarctic Survey Telescopes (AST3), Ground-based and Airborne Instrumentation for Astronomy IV. Proceedings of the SPIE, Volume 8446, article id. 84466R, 7 pp. (2012).

Lifan Wang is the director of the Chinese Center for Antarctic Astronomy, and a researcher at Purple Mountain Observatory of Chinese Academy of Sciences, also a Ph.D. tutor, as an innovative talent in the "Thousand Talents Plan"; he is the chief scientist of the high redshift transient source research program, and has published more than 200 articles. He is mainly engaged in Supernovae and Cosmology study, also the chief scientist of Antarctic Survey Telescopes AST3 project and Chinese Antarctic Observatory during the "12th Five-Year Plan." He has been hosting the International Cooperative Research Project "Polarization study of three-dimensional supernova" for a long time, and he is also a leader of International polarization Observation of Supernova. For recent years, his main researches are utilizing the geographical advantages of Antarctic Kunlun Station to develop optical infrared telescope and to explore the nature of dark energy and explore the transient source in the reionization era.

Retrospect and Prospect of Artificial Intelligence Research in China

Jie Tang, Sha Yuan and Yuan Zhou

Abstract With the rapid development and application of artificial intelligence (AI), the computer technology has entered the era of new Information Technology (IT) called Intelligent Technology. AI can accelerate the information construction of science and technology. In the past two years, the AI research has been promoted to the level of the national development strategy in China. This chapter explores the origin and development of AI and the AI development in China. AMiner, a big data analysis and service platform for science and technology, is independently developed by China. It is a successful case in the informatization of science and technology in China. Based on the open dataset of AI in AMiner, we give the classification of the AI research in China. We overview the AI research situation in China based on the experts, chapters, and patents analysis. The AI applications, such as speech recognition, face recognition, automatic driving, and so on, are introduced in the chapter. We also discuss the opportunities and challenges of AI in China. In general, this chapter fills the gaps in the authoritative analysis of the AI research situation in China.

Keywords Artificial intelligence · Research situation · Application · Opportunity and challenge

1 Introduction

The artificial intelligence (AI) field is a comprehensive discipline [1] that draws upon computer science, mathematics, psychology, linguistics, philosophy, neuroscience, artificial psychology, and many others. Artificial intelligence is difficult to define precisely. In this field, a machine is considered to have intelligence [2] if it passes the Turing test. The Turing test, proposed by Alan Turing (1950), is a test method used to determine whether a computer has the power of human thinking or

J. Tang (✉) · S. Yuan
Department of Computer Science and Technology, Tsinghua University, Beijing, China

Y. Zhou
School of Public Policy and Management, Tsinghua University, Beijing, China

© Publishing House of Electronics Industry 2020
China's e-Science Blue Book 2018,
https://doi.org/10.1007/978-981-13-9390-7_6

not. A computer passes the test if a human interrogator makes some written questions, without physical interaction, and the interrogator cannot know if the responses come from a person or from a computer. The computer needs to have the following capabilities: natural language processing, knowledge representation, automated reasoning, machine learning, computer vision, and robotics. The six disciplines compose most of the artificial intelligence domain. Thus, artificial intelligence usually refers to an artificial machine or program that simulates the human thinking. It exhibits similar intelligent behavior to the human mind.

AI techniques develop fast in China. After years of continuous accumulation, the number of international scientific and technological chapters and invention patents in China is the second in the world. The key technologies in some areas of artificial intelligence have achieved breakthroughs. Natural language processing, language recognition, computer vision, and other fields are at the world's leading level. Autopilot, UAV, biometric recognition, and other fields have entered practical application. China's AI enterprises are accelerating their growth and gaining wide recognition and attention in the international capital market. At the same time, there is still a significant gap between the overall development level of AI in China and that in the USA. Thus, in the process of AI development in China, achievements and gaps coexist, and opportunities and challenges coexist.

Starting from the development history of AI, this chapter reviews the three ups and downs of AI development and sorts out the development history of AI in China. Subsequently, based on the AI field classification in China, this chapter presents an in-depth statistical analysis of the current AI research in China, including the number of AI experts in China, the distribution of AI experts in China, and the number of chapters and patents published in the field of AI in China. Based on the analysis of the current research situation of AI in China, the opportunities and challenges of AI in China are discussed in order to provide reference and guidance for the development of AI in China.

2 History of AI

The history of AI should not be confined only to the emergence of AI concepts. Philosophy, mathematics, economics, neuroscience, psychology, computer science, cybernetics, and linguistics all contribute to the emergence and development of AI. In the following paragraphs, we chronologically describe several milestones that constitute the foundations of the current artificial intelligence.

2.1 Before the Advent of AI

The Greek philosopher Aristotle (384 B.C.–322 B.C.) was one of the first erudite to study reasoning processes. He proposed the deductive method, a process of reasoning from one or more statements to reach a certain conclusion. His representative work "Organon," a collection of six works on logic, laid the foundation of formal logic.

Francis Bacon (1561–1626) updated the Aristotle's Organon, and started the empiricism movement, characterized by the dictum "Nothing is in the understanding, which was not first in the senses."

David Hume's (1711–1776) proposed what is now the principle of induction, a form of reasoning whereby general statements are derived from a collection of singular observations.

Wilhelm Leibnitz (1646–1716) put forward mathematical logic to symbolize the formal logic, so that the operation and reasoning of human thinking could be realized.

While the first ideas about logic were started by the philosophers of ancient Greece, George Boole (1815–1864) began its mathematical development. He proposed Boolean algebra to solve problems that could not be handled by the traditional logic.

Kurt Gödel (1906–1978) studied on the completeness and decidability of the formal system. It pointed out some limitations of formalization and mechanization of human thinking. It proved that some things could not be achieved in theory and points out the limitation of AI research.

Alan Turing (1912–1954) proposed the Turing machine, which established a theoretical basis for the emergence of electronic computers. In 1950, he put forward the Turing test, which gave a clear definition of artificial intelligence for the first time.

McCulloch (1898–1969) established the first mathematical model of neural networks, which laid the foundation for the intelligent realization through human brain simulation. His work created microscopic AI.

Mauchly (1907–1980) invented the first electronic digital computer called ENIAC, which laid the material foundation for the AI research.

Shannon (1916–2001) founded the information theory in 1948. The combination of information theory and psychology formed the trend of macroscopic AI research. Based on psychology and information theory, cognitive psychology proposed various mathematical models to describe human mental activities.

2.2 Emergence of AI

In the summer of 1956, the Dartmouth workshop [3] proposed by McCarthy, Marvin Minsky, Nathaniel Rochester, and Shannon was held at Dartmouth College. The proposal of the conference stated to proceed based on the conjecture that every aspect of learning or any other feature of intelligence could in principle be so precisely described that a machine could be made to simulate it. It is a seminal event for artificial intelligence as an independent field. Since then, artificial intelligence is formally founded as an academic discipline. The conference focused on the areas of computer, natural language processing, neural network, theory of computation, abstraction, and creativity. These areas are still considered relevant to the work of the artificial intelligence field.

This conference promoted the first upsurge of AI research in the 1960s. At this time, the research on AI was mainly based on reasoning and search, and some theorems were proved. In 1958, the emergence of perceptron pushed artificial intelligence research to its first peak. However, due to the limitation of manual rule-making and computational ability, artificial intelligence in this period has no way to solve complex practical problems. Then in the 1970s, the research of AI entered a low ebb.

In the 1980s, the introduction of knowledge description and knowledge management expert system promote the second research upsurge of artificial intelligence. However, due to the limitations of knowledge description and knowledge management, as well as Minsky's proof that traditional perceptron cannot solve XOR problems, AI entered the second trough in the mid-1990s. In 1991, IBM's supercomputer Deep Blue won the chess championship. Although it is also a logical reasoning model based on rules, it has powerful computing power based on the development of computer hardware.

In the late 1990s, with the birth of search engines, the Internet has been explosively popularized all over the world. In the twenty-first century, with the development of Web technology, machine learning technology using massive data has risen rapidly. Hinton's deep learning technology, proposed in 2006, has made breakthroughs in image processing and speech recognition, opening the third upsurge of artificial intelligence research. With the development of new technologies such as big data, cloud computing, and deep learning, the research upsurge of artificial intelligence has reached an unprecedented height.

2.3   AI in China

Compared with the development of AI in developed countries, the research of AI in our country started late, and the road of development was tortuous, even once suppressed. Although the development process of AI research in China is very difficult, the research of AI in China has been progressing in a tortuous way and has achieved fruitful results.

In the 1960s, when the first upsurge of AI research appeared in the world, there was almost no AI research in China under the influence of the Soviet Union's criticism of AI and cybernetics. In the 1980s, when the second upsurge of AI research appeared, China caught up with the reform and opening, and the research of AI in China was unlocked and began a difficult initial stage. During this period, knowledge engineering and expert system developed rapidly in developed countries and achieved significant economic benefits. Since 1980, China has sent many foreign students to study the new achievements of modern science and technology in developed countries in Europe and America, including artificial intelligence and pattern recognition. In 1981, the Chinese society of artificial intelligence was established, which played a great role in promoting the research and development of artificial intelligence in China. In the mid and late 1980s, AI research in China began to develop normally and rapidly, and related research projects of AI appeared in the national scientific research plan.

After entering the twenty-first century, many AI research projects have been supported by national science and technology major projects, national high-tech research and development plan (863 plan), national key basic research and development plan (973 plan), national natural science fund key and major projects, and other national research funds. Supported by these national fund plans, universities have fostered AI disciplines, offered various levels of AI courses, and trained many scientific and technological people to engage in the research of AI disciplines.

In the past two years, the research of artificial intelligence has been promoted to the height of national development strategy in China. The State Council's "New Generation of Artificial Intelligence Development Plan" points out that the strategic objectives of artificial intelligence are divided into three steps. The first step is that by 2020, the overall technology and application of AI will keep pace with the advanced level of the world. AI industry will become a new important economic growth point. The application of AI technology will become a new way to improve people's livelihood. It will strongly support the entry into the ranks of innovative countries and achieve the goal of building a well-off society in an all-round way. The second step is to achieve a breakthrough in the basic theory of artificial intelligence by 2025. Some technologies and applications have reached the world leading level. Artificial intelligence has become the main driving force for industrial upgrading and economic transformation in China, and positive progress has been made in the construction of an intelligent society. By 2030, the theory, technology, and application of AI have reached the world's leading level and become the world's main AI innovation center. Intelligent economy and intelligent society have achieved remarkable results, laying an important foundation for ranking in the forefront of innovative countries and economic powers.

3 Research of AI

AI is a new field of interdisciplinary integration, which brings controversy and conflict, revision and improvement of multiple perspectives. In the development of AI research, scholars have different understandings and understandings of the essence of AI. They have adopted different research methods and means to study AI and formed different schools of AI research. At present, the research methods of artificial intelligence are mainly divided into three disciplines: symbolism (also known as logic), behaviorism (also known as cybernetics), and connectionism (also known as bionics).

Symbolism is an intelligent simulation method based on logical reasoning. Its principle is mainly the hypothesis of physical symbolic system and the principle of finite rationality. The discipline holds that the basic unit of human cognition and thinking is symbols, and the cognitive process is an operation on the representation of symbols. Symbolism is devoted to simulate the cognitive process of human beings with the symbolic operation of computers. By studying the functional mechanism of human cognitive system, it describes the cognitive process of human beings with some symbols, and inputs such symbols into the computer that can process symbols, thus simulate the cognitive process of human beings and realize artificial intelligence. The hypothesis of physical symbolic systems is guaranteed by methodology: the use of symbols and the use of symbolic systems as intermediaries for describing the world; the design of search mechanisms, especially heuristic search, to explore the possible reasoning space that these symbolic systems can support; the separation of cognitive architecture; and the assumption that a reasonably designed symbolic system can provide intelligent and complete causal reasons. Regardless of the way, it is implemented. Based on this point of view, AI finally becomes an empirical and constructive discipline, which attempts to understand intelligence by building an intelligent working model.

Behaviorism holds that the basis of intelligent behavior is the reaction mechanism of "perception-action." Based on the theory, method and technology of intelligent control system, the discipline studies anthropomorphic intelligent control behavior. The basic viewpoints of behaviorism can be summarized as follows. Formal expression of knowledge and modeling methods are one of the important obstacles to AI. Intelligence should be prototyped by the response of the environment to the action after the machine acts directly on the environment. Intelligent behavior can only be embodied in the world through interaction with the surrounding environment. AI can be like the intelligence of human, which evolves gradually and develops and strengthens in stages. Early research work focused on simulating human's intelligent behavior and role in the control process and researching cybernetics systems such as self-optimization, self-adaptation, self-tuning, self-stabilization, self-organization and self-learning. By the 1960s and 1970s, some progress had been made in the research of cybernetics systems mentioned above, and intelligent control and intelligent robot systems were born in the 1980s. With the rise of behaviorism, cybernetics and systems engineering have been integrated into the study of artificial intelligence.

Connectivism is an intelligent simulation method based on the connection mechanism and learning algorithm between neural networks. Its principle is mainly the connection mechanism and learning algorithm between neural network and neural network. The main point of view of this discipline is that the brain is the basis of all intelligent activities. Therefore, starting from the brain neurons and their connection mechanism, the study of the structure of the brain and its information processing process and mechanism is expected to reveal the mystery of human intelligence, thus truly realizing the simulation of human intelligence on the machine. The main features of this method are: storing information in a distributed way, processing information in a parallel way, and having the ability of self-organization and self-learning. It is these characteristics that make the neural network provide a new possibility for the realization of artificial intelligence. At present, due to the rapid development of computational storage technology, new technologies such as deep neural network and production network countermeasure push the research of artificial intelligence based on connectionism to a new climax.

Although different disciplines have different basic theories, research methods and technological routes, they are constantly promoting the development of artificial intelligence science in practice.

4 Classification of AI in China

Association for Computing Machinery (ACM) Computing Classification System (CCS), which has been integrated into search capabilities of the ACM Digital Library, is the de facto standard classification system for the computing field. It relies on a semantic vocabulary as the single source of categories and concepts that reflect the state of the art of the computing discipline and is receptive to structural change as it evolves in the future. ACM CCS contains 13 domains, such as theory of computing, information systems, computing methodologies, and so on. According to the ACM CCS and AMiner, we construct the Chinese AI domain tree, in which we also add some other areas of computing and application in interdisciplinary fields including computational economics, brain science, and so on.

As illustrated in the following picture, there are 27 concepts in the first level of the Chinese AI domain tree, including multi-agent system, machine learning, deep learning, and so on. These research domains represent the most active AI research areas in China. In the second level of the Chinese AI domain tree, we remove the categories in ACM CCS that have no Chinese research scholars and add the hot research fields, such as Chinese processing according to AMiner. Finally, there are 27 concepts in the first level of the Chinese AI domain tree and 44 concepts in the second level. The domain tree covers all areas in Chinese AI research (Fig. 1).

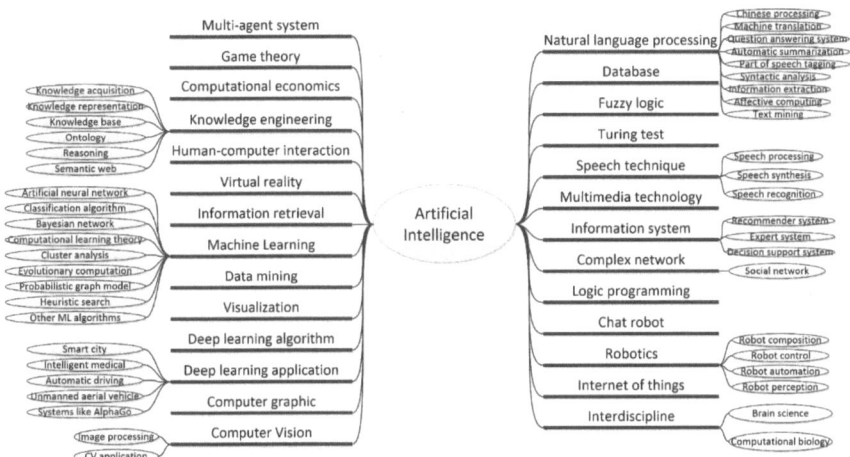

Fig. 1 AI domain tree in China

5   Statistic of AI Academic Data in China

AMiner [4] is the largest data mining and analysis system for academic science and technology in China. AMiner has established a science and technology think tank with more than 200 million academic chapters, patents, and 130 million scholars. According to the Chinese AI domain tree, we statistically analyze the AI academic data in AMiner. Among them, the download address of AMiner Open Data Set is: https://www.aminer.cn/data.

5.1   Global Distribution of AI Experts

There are 18107 AI experts in the world, in which:

1. There are 17231 males and 876 females. The males account for 95.16%.
2. The number of senior AI experts, whose H index is larger than 30, is 4918. This number accounts for 27.16% in the total AI experts.
3. The number of leading AI experts, whose H index is larger than 60, is 742. The number accounts for 4.10%.
4. Among all AI experts in the world, Chinese scientists account for 16.43%.

The proportion of AI experts in each country or region is illustrated in the following table. Among the Chinese AI experts, the number of mainland experts accounts for 91.13%, the Hong Kong and Macao Special Administrative Region experts account for 8.87%.

Among the global AI experts, senior AI experts are distributed in 43 different countries and regions worldwide. The proportion of senior AI experts in each country or region is shown in the following table. Among the senior AI experts in the world, Chinese scientists account for 10.94%. Among the Chinese senior AI experts, the number of mainland experts accounts for 82.72%, the Hong Kong and Macao Special Administrative Region experts account for 17.28%.

Among the global AI experts, leading AI experts are distributed in 25 different countries and regions worldwide. The proportion of leading AI experts in each country or region is shown in the following table. Among the leading AI experts in the world, Chinese scientists account for 6.36%. Among the Chinese leading AI experts, the number of mainland experts accounts for 74.06%, the Hong Kong and Macao Special Administrative Region experts account for 25.94% (Table 1).

5.2   Chinese AI Patents

The Chinese AI patents analysis is mainly based on the relevant data of State Intellectual Property Office of China. We randomly sample 15222 AI-related

Table 1 Global distribution of AI expert

Nationality	Class		
	AI experts (%)	Senior AI experts	Leading AI experts
USA	39.71	54.13%	68%
China	**16.43**	**10.94%**	**6.36%**
UK	6.3	6.21%	5.18%
Italy	3.69	2.41%	1.18%
Canada	3.66	3.97%	3.29%
Japan	3.61	1.67%	–
Germany	2.92	2.27%	1.65%
Australia	2.58	2.01%	1.41%
Holland	2.57	2.35%	1.41%
France	1.93	1.1%	–
Singapore	1.86	1.85%	–
India	1.61	1.02%	–
Greece	1.36	–	–
Israel	1.19	1.86%	2.12%
Spain	1.17	–	–
Switzerland	1.15	1.33%	1.4%
Austria	1.06	–	–
Belgium	1.03	–	–
Others	6.17	6.88%	8%

the bold part is the situation of China

patents since January 1, 2015. 41.47% of these patents are in the field of computer. 13.64% of them are in the field of electronic communication, and 5% of them are in the field of control. All these patents are widely used in all aspects of the social production and life.

We also analyze the type of institutions from which these patents come from. 61.62% of them come from a diverse set of companies. 31.97% of them come from universities and colleges. 5.12% of them come from research institutions. The others come from the government. As illustrated in the following figure, in the top 20 institutions with the largest number of AI patent, there are 11 enterprises and nine universities, so that the proportion of enterprises is relatively higher. In the whole dataset, the AI patents of enterprises are the main force.

In the figure, UESTC is the abbreviation of University of Electronic Science and Technology of China. NUPT is the abbreviation of Nanjing University of Posts and Telecommunications. SCUT is the abbreviation of South China University of Technology. SJTU is the abbreviation of Shanghai Jiao Tong University.

In the geographical distribution of Chinese AI distribution, Beijing accounts for 20.41%, Shenzhen accounts for 8.98%, and Shanghai accounts for 7.30%. These cities are the first three cities with most Chinese AI patents, while the IT industry is also very dense in the three cities (Fig. 2).

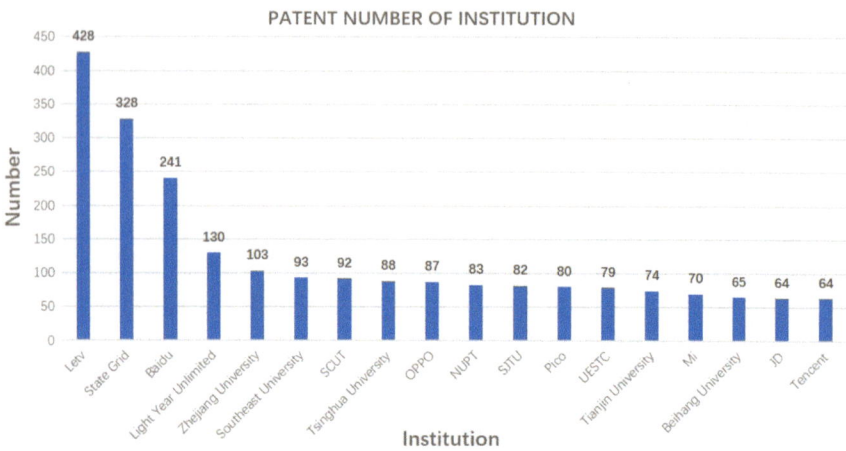

Fig. 2 Number of patents

The research areas of all these Chinese AI patents cover virtually all the spectra of AI research. The most active areas of Chinese AI patents are listed as follows. There are 13746 patents about machine learning and its application, 8929 patents about Internet of Things, 4948 about computer vision, 4835 about speech technology, 3577 about virtual reality, 1422 about data mining, 786 about deep learning, 819 about natural language processing, 1547 about robotics. In addition, there are many patents related to the AI application, such as 9356 about unmanned aerial vehicle, 3207 about face recognition, 710 about social network, 647 about automatic driving, and so on. Due to the maturity of technology, there are only three related to the intelligent computing chips and systems and zero about quantum intelligent computing.

Finally, we take deep learning as an example to analyze the development trends of the Chinese AI patents. There is no patent related to deep learning in 2012. There are 31 in 2013, 80 in 2014, 237 in 2015, and 465 in 2016. As one of the most popular research areas in AI field, the number of DL patents has been blowout and is still on the upward trend since 2012. In the US patent database—USPTO, the total number of DL patents is only 242. In recent years, the number of patents in China has surpassed the USA in many AI fields, such as deep learning.

5.3 Chinese AI Chapters

In this subsection, we analyze the number of the international top conference chapters (CCF-A and CCF-B) published in the field of artificial intelligence in recent two years. We collect 5573 chapters, of which 1554 were Chinese scholars.

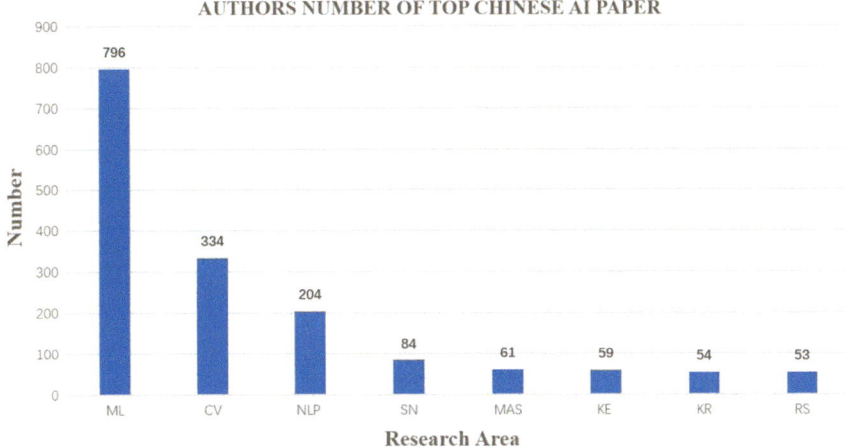

Fig. 3 The number of top AI chapters in China

According to the classification tree of artificial intelligence field, each chapter is classified into related fields, so that this field is the research field of every author in the chapter. Figure 3 shows that there are more than 50 Chinese scholars in the fields of machine learning (796 persons, 51.22% of the total), computer vision (334 persons, 21.49% of the total), natural language processing (204 persons, 13.13% of the total), social network (84 persons, 5.41% of the total), multi-agent system (61 persons, 3.93% of the total), knowledge engineering (59 persons, accounting for 3.80% of the total), knowledge representation (54 persons, accounting for 3.47% of the total), and recommendation system (53 persons, accounting for 3.41% of the total). The top chapters published by Chinese scholars are also subfields of AI which are more active in the world.

Next, we focus on the distribution of students in the authors. With the help of personal data of scholars in AMiner system and scholar information searched on the Internet, 452 students, 1102 professors or researchers were tagged. After having the position information of the scholar, the scholar is classified according to the organization. Figure 4 shows that in most domestic universities, there are more professors than students. For example, there are 64 teachers and 30 students in Tsinghua University, and 40 teachers and 11 students in Peking University. For foreign universities, there are often more students than teachers. For example, Carnegie Mellon University has four teachers and 20 students.

Then, we select 12 representative sub-domains of AI, including computer vision, machine learning, natural language processing, robotics, neural networks, uncertain knowledge and reasoning, multi-agent systems, computational learning theory, evolutionary computing, knowledge engineering, automated decision-making and scheduling, and reasoning. In AMiner system, the number of chapters published by

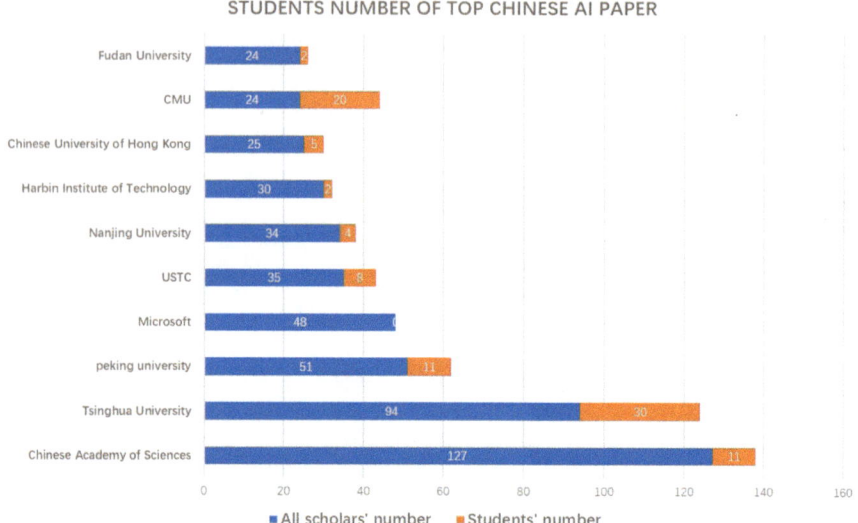

Fig. 4 The number of professors versus students

Chinese and foreign scholars in these fields is counted from 2015 to the present, as shown in Table 2 below. Among them, the published chapters are ranked according to the number of people and the proportion of Chinese scholars.

6   Application of AI in China

In the AI application, Baidu, Alibaba, and Tencent (BAT), the three giants of domestic Internet enterprises, have set up AI research institutes, such as Baidu Institute of Deep Learning, Alibaba, and Tencent Laboratory of AI. Chinese enterprises represented by BAT have achieved remarkable results in application practice in specific areas of artificial intelligence. These achievements are widely used in daily life, such as transportation, banking, e-commerce, security (including fingerprint recognition, speech recognition, face recognition, iris recognition, and gait recognition), robot vision, and navigation.

Based on the artificial intelligence domain tree, this section lists the successful cases of the application of artificial intelligence in China, including virtual reality, deep learning applications (including autopilot, UAV), computer vision applications (including face recognition), speech technology (including language recognition), and so on. In these applications, Chinese speech recognition, face recognition, and other technologies have been in the leading position in the world.

Table 2 Number of chapters

	No.	Foreign	Chinese	Chinese/foreign ratio
Computer vision	13200	9003	4197	31.79545
Machine learning	5701	3999	1702	29.85441
Natural language processing	6471	4622	1849	28.57364
Robotics	7251	5666	1585	21.85905
Neural networks	2927	2346	581	19.84968
Uncertain knowledge and reasoning	599	487	112	18.69783
Multi-agent systems	2312	1974	338	14.61938
Computational learning theory	381	327	54	14.17323
Evolutionary computing	596	546	50	8.389262
Knowledge engineering	459	428	31	6.753813
Automated decision-making and scheduling	360	338	22	6.111111
Reasoning	382	368	14	3.664921

6.1　Virtual Reality

Virtual Reality (VR) technology is a new interdisciplinary subject based on computer graphics, multimedia technology, computer simulation technology, human machine interface technology, and sensor technology. The virtual reality technology can create and experience the computer simulation system of virtual world, so that users can immerse in the interactive simulation environment of three dimensional dynamic visual scene and entity behavior. Virtual reality technology has four main characteristics, including sense of existence, interaction, autonomy, and multi-perception. Although the research of virtual reality technology in our country started late and lagged the developed international countries, it has attracted great attention from national departments and scientific researchers. According to the national conditions of our country, a series of research plans of virtual reality technology have been formulated and carried out.

At present, the application of virtual reality in China has been extended to many fields, such as scientific research, education and training, engineering design, commerce, military, aerospace, medicine, film and television, art and entertainment, and has comprehensively improved the quality of daily life of our people. According to the analysis of the status of Chinese AI patent, Beijing Bird View Technology Co., Ltd. (Pico) is in the leading position in the field of virtual reality technology in China. Pico now has 110 licensed patents covering core technologies in the fields of virtual reality, such as image, acoustics, optics, hardware and structure design, underlying optimization of operating systems, spatial positioning, and motion tracking. In addition, Beijing Xiaomi Technology Co., Ltd., referred to as Xiaomi, as a domestic mobile Internet company focusing on intelligent hardware and electronic product research and development, also has many innovations in the field of virtual reality technology. In 2017, the Xiaomi VR glasses developed by the Xiaomi Exploration Laboratory won one of the three most authoritative design awards in the world, the Japanese Excellent Design Award.

6.2 Deep Learning Application

At present, automatic driving and UAV are well-known in the field of deep learning and application in China. The realization of automatic driving depends on the maturity of deep learning technology. In the traditional "vehicle-road-person" closed-loop control mode, 92% of traffic accidents are caused by human factors, and traffic jams are mostly related to drivers' violation of traffic rules. Successful implementation of automatic driving will fundamentally change the traditional "vehicle-road-human" closed-loop control mode, and form a "vehicle-road" closed-loop, which can enhance highway safety, alleviate traffic congestion, and greatly improve the efficiency and safety of the traffic system. In recent years, the research of automatic driving has developed vigorously. Either high-tech Internet companies or major traditional automobile manufacturers have entered the research field of automatic driving. In April 2017, Baidu launched the Apollo program, open source its autopilot platform, and dropped a blockbuster in the field of autopilot at home and around the world. Baidu Apollo plans to complete a complete set of software, hardware, and service solutions for automatic driving, including vehicle platform, hardware platform, software platform, cloud data service, and so on. The Apollo autopilot open platform provides an open software platform for Baidu partners in the automotive industry and autopilot field to help them build a complete autonomous driving system combined with vehicle and hardware system.

As early as the 1980s, unmanned aerial vehicle (UAV) has been applied in the military field. In recent years, UAV has become a hot spot in military and civil research, which has great military and economic significance. In China, Dajiang Innovation Technology Co., Ltd. (DJI) is the world's leading research and development and manufacturer of UAV control systems and UAV solutions. Through continuous innovation, Dajiang is committed to providing revolutionary intelligent flight control products and solutions with the strongest performance and best experience for UAV industry, industry users, and professional aerial photography applications. Now UAVs have been applied to agriculture, energy, public safety, infrastructure, and construction. For example, through the UAV industry application platform, energy companies can real-time control equipment status in remote areas (including wind power stations, oil and gas equipment testing, power line patrol and nuclear power plant patrol, etc.), timely remove hidden troubles, so as to make maintenance work more efficient and safer. UAVs have been widely used in search and rescue, firefighting, disaster relief and law enforcement to monitor and search emergencies in public places, quickly transmit the situation on the scene and improve response efficiency.

Shangtang Science and Technology focuses on the original technology of deep learning and has established the top self-developed deep learning Supercomputing Center in China. It is the first-class supplier of artificial intelligence algorithms in China. At present, Shangtang Technologies have established cooperation with more than 400 leading enterprises in various industries at home and abroad, including Honda, Qualcomm, Yingda, China Mobile, UnionPay, Wanda, Suning, Haihang, CNN Information Office, Huawei, Xiaomi, OPPO, Vivo, Weibo, and other well-known enterprises and government agencies, covering security, finance, smartphones, mobile Internet, automobiles, and intelligence. Hui retail, robotics, and many other industries provide complete artificial intelligence solutions based on face recognition, image recognition, video analysis, unmanned driving, medical image recognition, and other technologies.

6.3   Computer Vision

Face recognition is the most widely used method in the field of computer vision. Face recognition is a kind of biometric recognition technology, which is based on automatically collected human face feature information without human intervention. The application fields of face recognition involve identification and authentication of public security organs and banking systems, video surveillance of public places, security verification systems, etc.

Face recognition technology has developed from the 1960s to the present. Face recognition algorithm has developed from the initial need to rely on operators to realize the gray-scale model of face to the full-automatic three-dimensional dynamic recognition. Face recognition technology has entered a commercial and practical stage. According to the patent analysis, Beijing Wide-sighted Technology Co., Ltd. has a leading position in the number of patents in the direction of face recognition in China. Kuang Shi technology is the most commercially applied and most mature face recognition company in China. It provides face recognition services to Alipay, China CITIC Bank, and Wuxi Public Security Bureau. Face++ is a cloud service platform that provides free face detection, face recognition, face attribute analysis, and other services.

6.4   Speech

Voice is the means of human communication and information acquisition. Intelligent voice technology has broad prospects for industrialization in all fields of society. The core technologies of speech technology include speech recognition technology, speech synthesis technology, speech evaluation technology, voiceprint language technology, etc. China has reached a leading international level in speech recognition. Among the authorized patents in the field of speech recognition, Xunfei is in the leading position. As the leader of China's intelligent voice industry,

Xunfei has international leading achievements in voice synthesis, speech recognition, spoken language assessment, natural language processing, and other technologies. As the team leader of the Chinese Voice Interaction Technical Standards Working Group, Xunfei took the lead in formulating the Chinese Voice Technical Standards. At present, the speech recognition rate of Xunfei input method has reached 98%, and the handwriting recognition rate is the first in the industry. In 2017, it flew into the top 50 most innovative companies in the world and ranked first in China.

7　Challenges of AI Development in China

The emergence of machine learning has stimulated the third upsurge of artificial intelligence research, and the rise of deep learning has pushed the third upsurge of artificial intelligence research to a climax. With the development and maturity of modern computer information technology such as cloud computing technology and big data technology, the third upsurge of artificial intelligence research is becoming more and more intense in China. There is no doubt that AI research is in the best era in history, full of vitality, bringing huge opportunities to scientific research and industry. However, while we are fortunate to be in such a good time, we also need to be vigilant and avoid blindly optimistic. In this section, we list the challenges of AI research in China.

(1) **Policy standard** The policy standard issued by the government is the social enabling support to ensure the quality and legitimacy of AI application. The scientific and technological revolution brought about by artificial intelligence has changed people's lifestyle; at the same time, it has eliminated many traditional industries' labors force and brought about the instability of social employment structure. While formulating the development strategy of AI, the state must strengthen the formulation of AI standards and norms, promulgate laws and regulations and formulate ethical and moral framework to ensure the healthy development of AI, carry out the identification of civil and criminal liability related to AI application, establish traceability and accountability system, and clarify the rights and obligations of AI legal subjects and their corresponding social responsibility. Especially in the fields of virtual reality, autopilot, UAV, face recognition, and speech recognition which are widely used at present, we should speed up the research and formulation of relevant safety management laws and regulations, promulgate the ethical rules and, evaluate the benefits and risks of AI applications in advance, and study the countermeasures of emergencies.

(2) **Privacy security** The revolutionary of the AI technology has had an important impact on infringing personal privacy, challenging economic security and social stability. At present, the monitoring system for AI has not been established, and privacy security cannot be effectively guaranteed. Under the

function of policy standards and supervision system, we should establish the evaluation measures of the impact of AI on national security and confidentiality, establish a security monitoring and early warning mechanism, and introduce penalties for abuse of data, violation of personal privacy, violation of ethics, and other acts. The security and privacy risks involved in AI exist in data acquisition, storage, and application. With the progress of AI technology, the ability of extracting data and knowledge is also rapidly improving. Standardizing privacy and security protection are one of the main challenges faced by AI technology.

(3) **Technical difficulties** Knowledge representation and emotional expression are the bottlenecks of breaking through the development of AI. At present, AI is still in a period of weak AI. For example, when the host comes home, the robot will turn on the air conditioner or heating in advance according to the temperature of the day, but the difference between the air conditioner and the heating has not yet been recognized. In order to enter the stage of strong artificial intelligence, we must use the research of cognitive science to crack the conscious phenomena such as autonomy and emotion. In traditional psychology, cognition refers to the process of knowing external things or information processing of external things acting on human sensory organs. Cognitive science aims to study the working principle of human brain and mind. It is a new interdisciplinary subject of psychology, computer science, neuroscience, linguistics, anthropology, and so on. In order to break through the challenge of knowledge representation and emotional expression, artificial intelligence must be combined with cognitive science in order to further develop to the stage of strong artificial intelligence.

8 Conclusion

In this chapter, the domain classification tree of artificial intelligence research in China is given. Based on the authoritative scientific data in AMiner, which is one of the largest data mining and service platform for scientific and technological information with Chinese independent intellectual property rights, the current research situation of artificial intelligence in China is summarized from three aspects: experts, chapters, and patents. Based on the patent data obtained by statistics, this chapter lists the practical applications of AI in speech recognition, face recognition, automatic driving, and UAV. Finally, the challenges of AI are discussed from three aspects: policy standards, privacy security, and technical difficulties.

References

1. Santo Fortunato, Carl T Bergstrom, Katy Börner, James A Evans, Dirk Helbing, Staša Milojević, et al. 2018. Science of science. Science 359, 6379 (2018).
2. Rosenblatt F. The perceptron: A probabilistic model for information storage and organization in the brain [J]. Psychological Review, 1958, 65(6):386.
3. McCarthy J, Minsky M L, Rochester N, et al. A Proposal for the Dartmouth Summer Research Project on Artificial Intelligence [J]. Journal of Molecular Biology, 2006, 278(1):279–289.
4. Jie Tang, Jing Zhang, Limin Yao, Juanzi Li, Li Zhang, and Zhong Su. 2008. Arnetminer: extraction and mining of academic social networks. In Proceedings of the 14th ACM SIGKDD international conference on Knowledge discovery and data mining. ACM, 990–998.

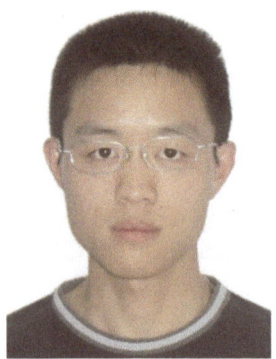

Jie Tang is an associate professor and doctoral supervisor in Tsinghua University. His research interests are social network and data mining, has won the National Science Fund for Distinguished Young Scholars in 2018, the first prize of Beijing Science and Technology Progress in 2017 (ranked first), Newton Senior Scholars Fund in 2015, the first prize of China Artificial Intelligence Society for scientific and technological progress in 2013 (ranked first), etc. He has published more than 70 Top chapters, cited 11727 times by Google Scholar, and had a personal H index of 55.

Progress of e-Science in Major Projects

Implementation and Application of National Science and Technology Information System

Shaohua Hu, Zhuohao Wang, Xuepeng Song, Jing Chen, Haosong Men, Dong Wang, Qing Li and Jing Zhang

Abstract The National Science and Technology Information System (NSTIS) is a complicated information system which aims to support the management and implementation of national science and technology research and development (S&T R&D) programs and to provide all-round services for various groups of users. To fulfill the requirements of the current reform in the field of national science and technology development, with an in-depth analysis of service requirements and key technologies, we propose a scalable and high-available architecture which supports the management and implementation of national S&T R&D programs to meet the demands of the public, departmental joint conferences of national S&T R&D programs, program management institutes, as well as supervision and evaluation institutions. We also propose a distributed and multi-node operation system which is suitable for multi-department and multi-level coordination. Moreover, we establish the core resources and functions, such as basic service, platform service, resource service, operation support, and propose relevant key technologies and corresponding solutions based on a four-layer architecture. Finally, we analyze and summarize the performance and achievements of NSTIS, and provide an outlook on future lines of research.

Keywords Reform of Science and Technology programs · Science and Technology Management Platform · Management of Scientific Research Projects · Application integration · Data fusion

1 Introduction

A series of reform measures have been adopted in the field of science and technology to further promote the innovation-driven development in recent years. According to *Opinions on Strengthening the Management of Scientific Research*

S. Hu (✉) · Z. Wang · X. Song · J. Chen · H. Men · D. Wang · Q. Li · J. Zhang
Information Center of Ministry of Science and Technology of the People's Republic of China, Beijing, People's Republic of China

© Publishing House of Electronics Industry 2020
China's e-Science Blue Book 2018,
https://doi.org/10.1007/978-981-13-9390-7_7

139

Programs Funded by Central Government and Related Funds (GF (2014) 11) [1] and *Program on Deepening the Reform of the Management of Science and Technology Programs (such as special programs and funds) Funded by Central Government* (GF (2014) 64) [2] released by the State Council in December 2014, the National Science and Technology Information System (NSTIS) shall be improved to include all implementation processes of five types of national level science and technology programs. NSTIS takes care of the whole process of S&T R&D program management, including demand collection, guideline release, program application, proposal preparation, budget arrangement, supervision and inspection, and completion acceptance. With years of efforts in informatization, several national science and technology management and service platforms have been built to support various S&T R&D programs and funds. To achieve an unified management of various S&T R&D programs and avoid unnecessary waste of resources caused by multiple platforms, two national online application service platforms for S&T R&D programs, namely National Scientific and Technological Program Application Center and Budget Application and Management Center [3], are built by the Information Center of Ministry of Science and Technology (MOSTIC) in 2006 to support unified online applications and management services for several S&T R&D programs, such as the National High Technology Research and Development Program of China (863 Program), the National Key Basic Research Program of China (973 Program), the National Basic Research Program of China, the National Key Technology R&D Program, the China Spark Program, the China Torch Program and the International S&T Cooperation Program of China, addressing issues like excessive online application channels and unorganized information systems by building a unified public service framework. The NSTIS was put into services in 2014 in the Ministry of Science and Technology (MOST), which substitutes the original national scientific and technological program management system. NSTIS provides a unified platform for program management, operation synergy and data resource services, removes management barriers among different organizations at various levels, and realizes the integration and coordination from the application, approval and organization of projects to the post-stage management and service.

The National Natural Science Foundation of China (NSFC) has built the Internet-based Science Information System (ISIS) to deal with application and management operations of natural science fund programs covering key processes like application, acceptance and review. The Chinese Academy of Sciences (CAS) has developed the Academia Resource Planning (ARP) system to manage resources such as scientific and technological programs and related budgets by removing the obstacles of information flow and operations between the CAS and its affiliated institutes through a unified platform and resource center. However, inter-connectivity and resource coordination have not been realized among those systems due to the differences in the functions, service scopes, technical architectures and management subjects, which makes it hard to directly serve the National Science and Technology Management Platform. Therefore, in response to the requirements of MOST for the implementation of reform, MOSTIC launched the project to construct NSTIS with a variety of technical resources from 2014. The Publice Service Platform of the National Science and Technology Information System was put into services officially in September 2015, which becomes a key service platform for the promotion of S&T system reform, and the strenghen of S&T resource allocation.

2 Demand Analysis

NSTIS is a comprehensive information service system and information technology application platform which operates across multiple governmental levels based on the idea of designing an integrated, highly efficient and intelligent system that covers all processes of S&T R&D programs. In this chapter, we analyze the demands and challenges to construct NSTIS from the aspects of project duration, management complexity, subject diversity and personnel mobility.

2.1 Service Demands

Service demands for NSTIS mainly come from *Program on Deepening the Reform of the Management of Science and Technology Program (such as special programs and funds) Funded by Central Government* (GF (2014) 64) and subsequent policies and regulations. According to the requirements of the reform, NSTIS should be used to realize the whole process management of the national S&T R&D programs by providing the following services: unified application acceptance, information query, organization and implementation, operation consulting services to the scientific research personnel, comprehensive services and the decision-making information servicse to the joint-department conference, professional process management sevices to the management institutions, and unified service evaluation and supervision mechanism. Through the integration of various operating processes and data resources, the NSTIS has become a state-level technology platform for scientific decision-making, collaborative management, optimization of resource allocation and service of science and technology innovation.

2.2 Application Demands

1. Unified User System and Extensible Authorization Management

The user system needs to satisfy the requirements and the restraints from diversified operations at different levels. NSTIS serves massive users from various fields that ranges from governmental organization to research institutes. The users are from diverse categories and the number of users is massive, which indicates great possibility of unexpected growth.

The access control units are diversified to meet different scenarios. NSTIS shall provide differentiated operation functions and information services to various users, including basic accessing grants, function controls and roles, data accessing permits, platform controls and related fine-granularity functions.

2. Unified Portal System and Verification Management

NSTIS involves public service module, comprehensive management module, project management module, expert service module and review service module, which are composed of several independent operation systems and their supporting environment. The modules store comprehensive management information, application data, operation management data and technical management records. Therefore, the user system and the access control system are well designed to support mutual recognition, to achieve simple and quick login and mutual accesses of users of multiple systems and to solve problems such as unified identity authentication and single sign-on.

Meanwhile, the user system design should have a simple maintenance schema and high scalability. For instance, the system shall be easily adaptable to social management reforms such as integration of the operation license, the organization code certificate, and the certificate of taxation registration into one document, which shall response to the demands of enterprise information update and maintenance and adapt to the consistency assurance requirements caused by the maintenance of enterprise information change during project management.

3. Comprehensive Coordination and Project Management Services for Management Units

NSTIS shall collect data from multiple sources to support comprehensive coordination and management and whole-process management of scientific and technological programs. By collecting, sorting out and updating program information, project data, budget data and service data maintained by several operation systems and departments, NSTIS is able to provide data statistics, information support and comparative evaluation, thus striking balance between comprehensive operations and realizing operation collaboration.

4. One-stop Application Services of the S&T R&D Programs for the Public

NSTIS shall provide researchers, research institutions, review experts, professional institutions, evaluation institutions and administrative bodies of science and technology with one-stop services, including project publicity, project application, information feedback, consulting, guideline publishing, application acceptance, and organization and application.

Pressure from intensive access by massive users shall be considered when providing online services for the public. Distributed system shall be taken into consideration to ensure fast, steady, efficient and safe public services. Due to the complicated operation process and data of traditional science and technology information system, the user experience is to be improved. Therefore, by making the services more intelligent and convenient, the system shall provide targeted and accurate services for the users.

3 Overall Architecture

3.1 System

Science and technology information management involves multiple management departments and local management institutions. It deals with five types of national level scientific and technological programs and offers unified and comprehensive information management services related to S&T R&D programs for joint conferences, strategic consulting, management departments, management institutions, research institutions, researchers, experts and the public. The system of science and technology information management is as shown in Fig. 1.

On top of the system are the service layer for users. The operation platform and the service platform of NSTIS provide operation management support services,

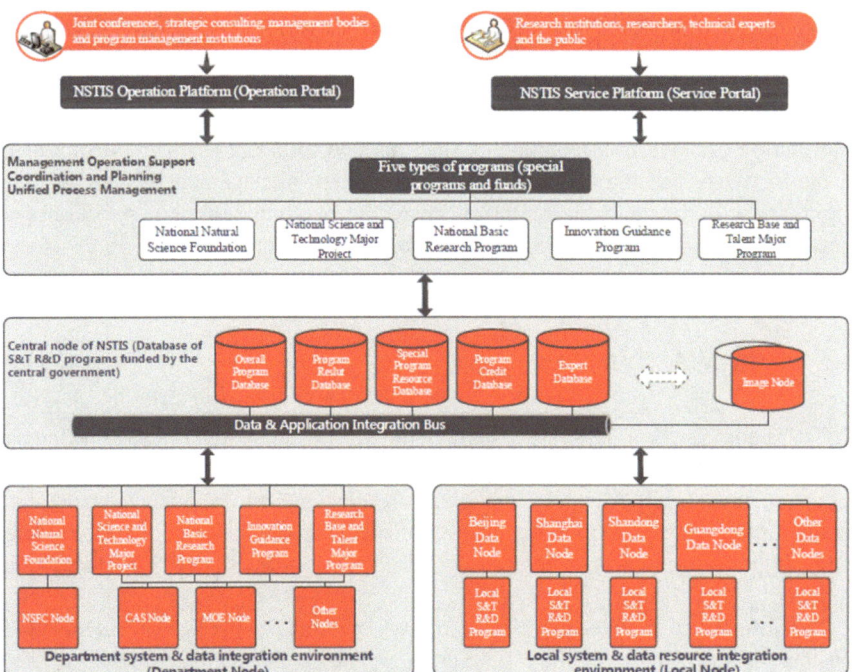

Fig. 1 Operation system of S&T management

coordination and planning services and unified process management services based on data resources and operation processes of five types of science and technology programs.

The core of the system is the central node of NSTIS operating in the network environment of MOST. The central node is used by the public service platform, project management module, technical management module, key application systems, S&T R&D program database funded by the central government, and related data and resource environment. The central node integrates various application systems with application integration buses. With the program database as the data resources, the central node includes program data from multiple organizations. The Database of S&T R&D programs funded by the central government provides information services to S&T administrative units at all levels, research institutes and research staff, involving scientific research projects, scientific funds and the information of project processes. It mainly provides macro-coordination and operation regulation services, science and technology operation management services and public science and technology services.

At the bottom of the system are data resources from the nodes of related departments and local management units of science and technology. These nodes operate in the network of related units and are responsible for the implementation process management module, the data resources of science and technology programs and related application systems. The central node of the system is connected to the nodes of related departments and local units with data integration bus, realizing inter-connectivity of data resources.

The corresponding department of each node is responsible for the compilation, registration, review and release of program data and achievements, as well as the registration and update of sharing management information on corresponding information system. The departments shall also be responsible for the deployment of the hardware and the software of the information management system, and the deployment, update and the maintenance of the external service application systems. All nodes are interconnected with a unified standard interface.

3.2 System Architecture

NSTIS is composed of four layers, namely the infrastructure layer, the service layer, the application layer and the access channel layer. Users complete operations at the access channel layer. All system modules are designed in line with related safety requirements and protection service by following operation management requirements in post-stage maintenance process. The architecture of NSTIS is as shown in Fig. 2.

The infrastructure layer provides hardware support for the system architecture. On the basis of servers, storage devices, network, it also includes cloud environment management module, resource management module and task management module.

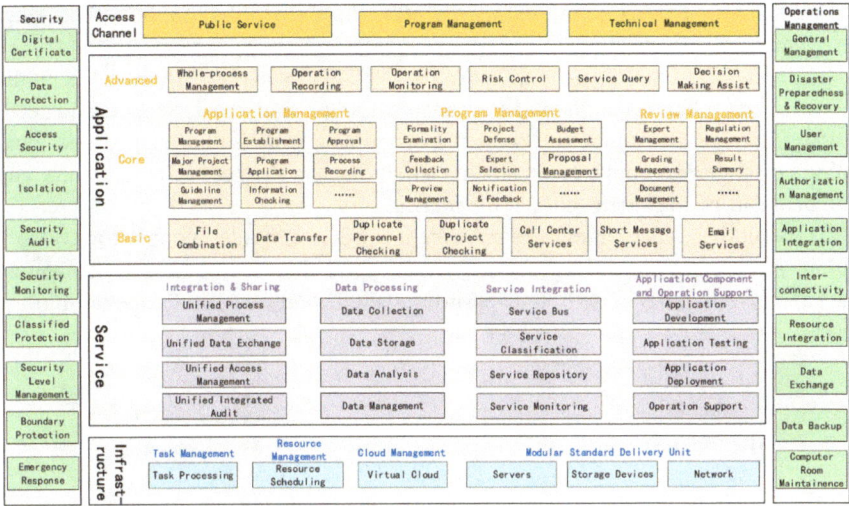

Fig. 2 Architecture of NSTIS

The service platform includes service integration module, integration sharing module, data processing module, application component module and operation support module. With the service bus, the integration module integrates various services in the architecture and connects related service repositories to realize service synergies and provide platform and interface for the operation and monitoring of all services. As the basic module for applications, services, data and users integration, the integration sharing module provides unified interfaces and standards for various operation process management, application data exchange, user rights management and application integration audit. The data processing module is the data center of the architecture which provides data structure, interface specifications, transformats and communication modes for data collection, interaction, storage, analysis and management among application systems. The application components and operation support module offer necessary service interfaces and management platforms for the development, testing, deployment, operation and maintenance of application systems.

The application layer gathers various application systems, which are divided into three sub-layers, namely advanced application layer, core application layer and basic application layer, based on the invocation relations. The advanced application layer carries the key functions of the architecture, including whole-process management, operation remains, operation monitoring, internal risk control, functional operation retrieval and assisting decision-making support. The core application layer maintains the core operation application systems and includes the program application and management module for researchers, the program management module for management staff and the evaluation management module for experts. The basic application layer supports the core application layer with basic application services, including file composition, data push, staff duplicate checking, program duplicate checking, call services, SMS, and email services.

The access channel layer is the portal of NSTIS for users, including the public service portals for researchers, research institutions and local science and technology management units, the project management portals for science and technology management units, professional institutions and budget evaluation units, and the technical management portals for related units.

The security and operation management module provide support for collaboration, safety and steadiness, service sharing and data inter-connectivity among different layers in the system, thus ensuring the efficient and steady operation of all layers.

3.3 Core Functions

1. User Authorization Management System

NSTIS provides support and information services for operations at different levels. NSTIS serves massive users from various fields. Each type of users plays different role and is allowed to access differentiated systems and functions. By establishing standard user authorization system, NSTIS can manage the modules accessed by users, limit the data can be accessed, control user operations over data and record user behaviors, thus ensuring the safety of data and the orderly use of the system on demand.

Due to strong adaptability, fine application granularity and flexible building strategy, the role-based access control (RBAC) technology is an ideal choice as the standard authorization system model in NSTIS. RBAC is divided into RBAC0, RBAC1, RBAC2 and RBAC3, among which RBAC0 is the basis and the latter three are the expansions of RBAC0. Since the users of NSTIS are from a wide range of units and departments at different levels and NSTIS requires the functions of application, review, recommendation, approval and management which requires complicated access control, RBAC3 is chosen as the access management model of NSTIS.

2. Unified Single Sign-on and Verification Services

NSTIS involves macro-decision-making information, special program information, integrated coordination and management operation, special plan management operation, project management operation and process information of science and technology credit, achievements and performance assessment, and is composed of multiple independent operation systems along with their supporting environment. Existing user verification systems fail to meet the development demands due to the system management challenges brought by the massive passwords which also causes troubles for users. Unified single sign-on and verification services is an effective solution to this problem.

Unified verification center makes unified user management and identity verification among multiple application systems possible. Seamless access and interactions across several systems can be realized as long as the user complete verification. The user information is managed in one verification center. Verification notes are transferred among application systems, which not only ensures information safety but realizes the isolation of verification information and data.

OAuth2.0 defines a safe, open and simple standard to authorize user resources. The third-party systems can access the authorization information of users without user accounts and passwords. OAuth2.0 makes cross-domain access among information systems possible. To deal with the pressure from concurrent access, NSTIS adopts the Nginx-based distributed unified identity verification mechanism, decouples the user verification information from the verification nodes with OAuth2.0 verification server and reduces response time and handles the great number of concurrent users with multi-node task scheduling, thus ensuring the smooth operation of the verification system.

3. Comprehensive Operation Management

Supported by unified user management and authorization access system, unified application integration and service interface and unified data storage and interaction platform, NSTIS supports the planning, coordination, special program management, project organization and process management under science and technology systems, and provides technical services and operation guarantee such as acceptance management, data browsing, form inspection, information feedback, evaluation management, online evaluation, defense evaluation, expert selection, proposal confirmation, process management and acceptance management. With rigorous application system (module) integration, access control and operation tracking, NSTIS supports various operation applications and comprehensive management.

Focusing on supporting data management and operation services of scientific and technological programs, NSTIS, on the basis of safe data storage and reasonable data classification, provides efficient data support for system operations. NSTIS extracts unified data exchange and computing services from public operation services, generating a system construction model that covers data, services and operations.

Based on multi-layer software system and service-oriented architecture (SOA), NSTIS widely adopts service components, modular technologies and highly flexible configurable technologies, thus realizing information services sharing among sub-systems based on the loose coupling principle by establishing SOA with application integration technologies. By integrating services rapidly, NSTIS develops combined applications so as to meet the flexible and expandable demands on comprehensive operation management, making fast response to demands possible.

4. Unified Public Services

The public service platform of NSTIS serves research institutions, researchers, the public and all management bodies in relation to scientific and technological program application. It provides application acceptance services, operation environments and technical services with functions like unified service portal, user information verification, information disclosure, guideline publication, online project application, online operation settling, data resource integration and interactive information synergy, thus supporting operation synergies, data exchange and steady operation of public service application systems (modules).

To provide unified one-stop application services, operation integration has become one of the major goals of system integration. The public service platform integrates application systems based on their logic and manages to provide a unified view and access portal for users to interact with the information system so that they can sign on the system and interact with other users, contents, applications and processes in a safe environment. With concentrated page display, the system integrates user interfaces to support the switch and skip of the whole operation process.

A safety protection system that covers the Internet layer, the Web layer, the application layer and the data layer is established. Based on the user authorization management system of NSTIS, a user identification and verification system and the safety protection for data transmission channel are built in the Internet layer. A multi-level firewall against common attacks is built in the Web layer. An access control system of user authorization, service authorization and access auditing that deals with operation processes in relation to program application is built in the application layer. The data layer provides terminal access authorization as well as encryption and back-up (high availability) of programs, operations and processes.

3.4   Key Technologies

1. Application Integration and Service Inter-connection

To address the challenges from diverse application system architectures, data resource structures and services operations, in this chapter, we propose the approach of application integration and service inter-connection to realize the operation integration of S&T R&D programs. The service integration layer is built based on data integration to support data exchange, data retrieval, statistical report and data analysis, thus realizing operation integration.

Program application, proposal management, process management and acceptance management are supported by several systems. To deal with such complicated operation model, a unified service model that encapsulates user services, project services, budget services, expert services, technology literature services, planned data services, scientific instrument and technological resource services should be firstly designed. With unified portal system, service bus and service integration specifications, the program-related operations and data are processed and integrated after the portal system and application system services are connected with the service bus. The underlying service bus components are also employed to register, classify, index and maintain services, thus realizing routing, invocation and exchange of services, increasing the value of existing systems, reducing repetitive construction, improving construction efficiency and lowering the construction difficulty.

The general management data are stored in extensive existing systems and involve all core operations in relation to scientific and technological program management. Technically, the integration of heterogeneous data constitutes an important part of operation integration. Common data integration approaches include Fedrated Database System (FDBS), data exchange system and data repository. FDBS is composed of semi-autonomous database systems that are involved in the federation. All databases within the federation have added each other's access interfaces (such as the gateway), mapping the database model with other databases. The data exchange system as the most basic component of the unified application platform provides unified data sharing and exchange services. It adopts existing message middleware, application integration middleware and J2EE application server to provide basic service features, including unified data and service expression methods, data description tool, Web Services-based application system connection tools, standard data processing features and standard data transmission components. Data repository employs ETL tools to regularly extract, clean and transform data from several heterogeneous databases in accordance with the unified and concentrated view requirements, and then stores the data into repositories. With the data integration approaches above taken into consideration, the unified data model and specifications are established. Program data and budget data scattered in different information systems are integrated with the specific operations taken into consideration.

2. Data System and Management

The data resources of NSTIS include program data, operation management data and operation sharing data. The data resource system structure made up of major data types and their relations is as shown in Fig. 3.

The structural, semi-structural and non-structural databases are built, with the database building methods and strategies soundly formulated based on the data structures of the data resources of NSTIS. In light of the structure features and storage scale of three types of resources, different technical approaches and strategies are formulated to pool and integrate different data resources, with the storage structure and database building approaches for each type of data defined. Unified meta-data description standards are designed. The data are processed in line of specific standards and their structure features, thus defining the data standard catalog for unified meta-data description. Based on the traditional relational databases, a big data storage and management mechanism are formulated to support the fast storage, read/write, collection and back-up of massive semi-structural and non-structural data with hierarchical storage, partitioned management and cache optimization mechanism.

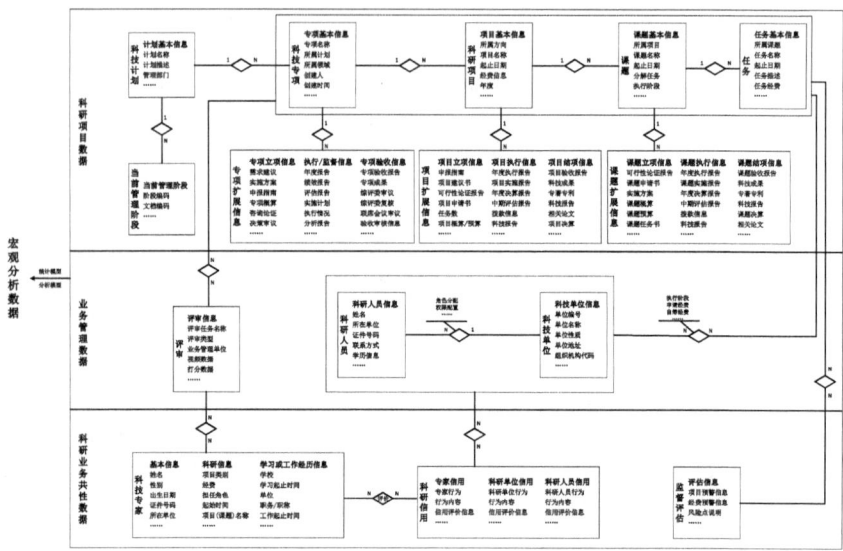

Fig. 3 Data resource system structure

Science and technology management element correlation model is established with distributed multi-dimensional data repositories to further integrate science and technological resources, thus correlating structural, semi-structural and non-structural data, tapping the value and facilitating thorough application. The data repository model is designed with three models and two correlation methods to integrate data resources for invocation. The basic element model (star-shaped structure) integrates basic data such as programs, special programs, projects, subjects, employers, staff and budget; the hierarchical model (tree-shaped structure), with the core data as pre-application, application, tasks, annual report, technical report and acceptance report, covers all information in relation to national science and technology management; and the multi-core correlation model (network model) realizes the closed-loop correlation and integration of data resources such as project data, channel review, form review, defense review, expert grading, results, documents, operation remains, user data and role access. The integration and sharing of structural, non-structural and semi-structural data is an efficient way to explore the repository structure system of science and technology management data.

3. Operation and Maintenance on Cloud

The operation of NSTIS relies on host, storage, networks, middleware, database and load balancer. Traditional application deployment requires manual installation, configuration and deployment, which consumes much time and cannot support resource sharing, with higher chances of errors.

An event-driven scheduling method of application operation support environment is designed based on existing cloud computing infrastructure management technologies. The host, storage, networks, middleware, database and load balancer imaging in the application operation environment is standardized; the key steps of the deployment process are then sorted out to be classic events; experience of implementation staff is generalized as automatic scripts; and the events and scripts are bound as needed when deploying applications. The automatic scripts would be triggered when the events take place. For instance, if an application needs to use the middleware and database and the middleware needs to access the database, then the execution script shall be triggered to write the IP and port of the database into the middleware and start it. With this approach, all related components in the operation environment shall be scheduled by events and scripts when deploying applications, thus realizing independent creation of the whole application environment and improving the deployment efficiency.

An application elastic scaling approach based on load pressure is designed by monitoring the cloud environment where the operation systems operate in real time. Load-related metrics (CPU use rate, memory use rate, request concurrency count and average response time) are monitored and collected when the operation application runs. The collected metrics would then be calculated with elastic scaling algorithm. The calculation results would be compared with the threshold values set in the elastic scaling strategies to determine whether to trigger elastic scaling or not, then to increase or decrease application nodes with IaaS and to trigger the script with events, integrating the increased or decreased application nodes into the whole application environment, which is transparent for the users.

4. Intelligent User Information Processing Technologies

Key technologies including big data and machine learning are employed to capture the operations, behaviors and preferences of users and achieve accurate user profiling, describing the properties of users from multiple dimensions and extracting all user information so as to support accurate, efficient and timely monitoring of scientific and technological events and to provide data support of management and decision-making.

The user profiling system is composed of the labeling system and the corresponding analysis model. Labels define the perspective in which we observe, describe and learn about users. User profiles are the collection of labels. Labels as a whole are inter-connected. The user profile in NSTIS depicts users from aspects of application, review and management, which including the basic information, application information, review information and proposal information of the users.

User profiling involves information correlation and extraction, profile modeling and theme analysis, and profile expression. User behavior abstraction, rule inference and machine learning technologies are employed to analyze and mine user labels, thus mapping out and improving the relation network among users so as to support monitoring of scientific and technological events and to provide data support of management and decision-making.

4 Achievements

First, NSTIS is successfully supporting the operation, management and implementation of current national S&T R&D programs. The system covers a series of
key procedure in line with the reform requirements, such as application acceptance,
application approval, proposal management, process management and acceptance
management, which is a key support for the launch and operation of the new
science and technology program system of 13th Five-Year Plan.

Second, the Public Service Platform of the National Science and Technology
System [4] was build and become a one-stop portal to serve all researchers within
china on September 30, 2015. The NSTIS Public Service Platform (shown in Fig. 4)
was put into service which realizes the synchronous design, development and
application of the system construction process and management reform. Currently,

Fig. 4 NSTIS public service system

the platform can support the whole-process publicity, unified online application, program process services, operation retrieval and service interactions of national key research programs and national key scientific and technological special programs.

Third, NSTIS supports the whole process management of special S&T R&D programs. NSTIS serves the joint conferences, management bodies and professional management institutions. It supports the application acceptance, form review, feedback, expert selection, program review, proposal arrangement and task signing of national key research programs and national special science and technology programs. NSTIS also supports trace remain throughout operation management, feedback and management recalling, which realizes whole-process and one-stop management.

Fourth, NSTIS supports the whole-process tracing of management operations. With unified tracking features among the whole management process, NSTIS records the operations in program recommendation, form review, online review, video review and expert selection. The video review covers defense scenes, defense presentations and review scenes, thus supporting management reflection, arbitration, petition and public opinion handling. A total of nearly 200 video review has been supported, with the number of programs and experts involved being about 30,000 and 26,000.

5   Conclusions and Prospects

In the future, NSTIS will be further refined to keep in line with the general requests from the government and public.

First, we will build a better data system to offer more efficient and reliable data services. We will enhance data integration, management and analysis in light of the construction and application of NSTIS.

Second, we will further optimize the architecture of NSTIS to make best use of new technology and build a more robust and efficient and safer platform.

Third, we will step up our efforts to further improve service efficiency and effectiveness of NSTIS to better meet the demands from the users of all levels.

References

1. http://www.gov.cn/zhengce/content/2014-03/12/content_8711.htm
2. http://www.gov.cn/zhengce/content/2015-01/12/content_9383.htm
3. http://program.most.gov.cn
4. http://service.most.gov.cn

Shaohua Hu is a researcher and executive deputy director of the Information Center of MOST. His research interests include information system architecture, data integration, data mining and analysis, knowledge management and knowledge service technologies. As a chief architect and team leader, he has built a series of major information systems including NSTIS.

China Science and Technology Cloud Situation and Prospects

Jun Li, Jingjing Li and Shanshan Shi

Abstract China Science and Technology Cloud (CSTC) is an important part of the "13th Five-Year Plan" informatization program of Chinese Academy of Sciences. Guidance with "Pioneer Initiative" and "Smart Chinese Academy of Sciences", "China Science and Technology Cloud" construction project will build an informatization resource management and service cloud platform, with science and technology professionals as the center, and unified resource scheduling and user self-service as the distinctive characteristics. This chapter gives a brief introduction of "China Science and Technology Cloud", including the development, technical architecture, achievements, and future work.

Keywords Cloud services · Scientific research network · Supercomputing · Scientific data

1 Introduction

According to the national innovation-driven development strategy, Chinese Academy of Sciences has proposed the "Fours Leads" as the goal of the development strategy, and informatization has become an important part of the strategic planning. "13th Five-Year Plan" has proposed that "China Science and Technology Cloud" will build a cloud platform of information resource management and service, which is science and technology professionals centric, features unified resource scheduling and user self-service.

In summary, based on more stronger network construction, "China Science and Technology Cloud" provides better "hard" informatization services and infrastructure, including supercomputing environment, vast cloud storage environment,

J. Li (✉) · J. Li · S. Shi
Computer Network Information Center, Chinese Academy of Sciences, Beijing, China

S. Shi
University of Chinese Academy of Sciences, Beijing, China

© Publishing House of Electronics Industry 2020
China's e-Science Blue Book 2018,
https://doi.org/10.1007/978-981-13-9390-7_8

and big data processing. Furthermore, it provides "soft" informatization services, including information resources, software resources, etc.

Through 20 years of unremitting construction, and following the development wave of world informatization technology, "China Science and Technology Cloud" environment has already initially formed. It provides public sharing, accessible and seamless availability for scientific research staff of CAS. Through deep integration and unified management of resources, it has formed various of informatization infrastructure resources and services, including informatization infrastructure (IaaS), platform (PaaS), software (SaaS), etc. inside and outside of Chinese Academy of Sciences. Furthermore, it provides a unified resource management platform, network environment, supercomputing environment, and cloud computing, cloud storage, and big data processing environment, informatization resource service and software resource service. This chapter introduces the current situation, construction achievements, and prospects of "China Science and Technology Cloud".

2   Development Ideas and Technical Architecture

The "China Science and Technology Cloud" Function Architecture is described as follows. The "hard" capabilities of "China Science and Technology Cloud" are composed of the network resource pool, supercomputer resource pool, cloud computing resource pool, cloud storage resource pool, and big data processing environment. Network resource pool (high-speed network and testbed) provides network environment support for "China Science and Technology Cloud". The supercomputing resource pool provides high-performance computing service and application capabilities. Cloud computing resource pool and cloud storage resource provide cloud computing, Data-intensive processing, cloud storage, and cloud disaster recovery capabilities. The "hard" capabilities provide infrastructure-level resource support for software resource pool, information resource pool, and other projects, and provide resources service for customers in science and technology communities through resource management and funding service platform. As the "soft" capabilities, software resource pool and information resource pool provide information resources service and software resources service for the "13th Five-Year Plan" of the CAS informatization project construction and science and technology users. The overall "China Science and Technology Cloud" architecture is as shown in Fig. 1 [1].

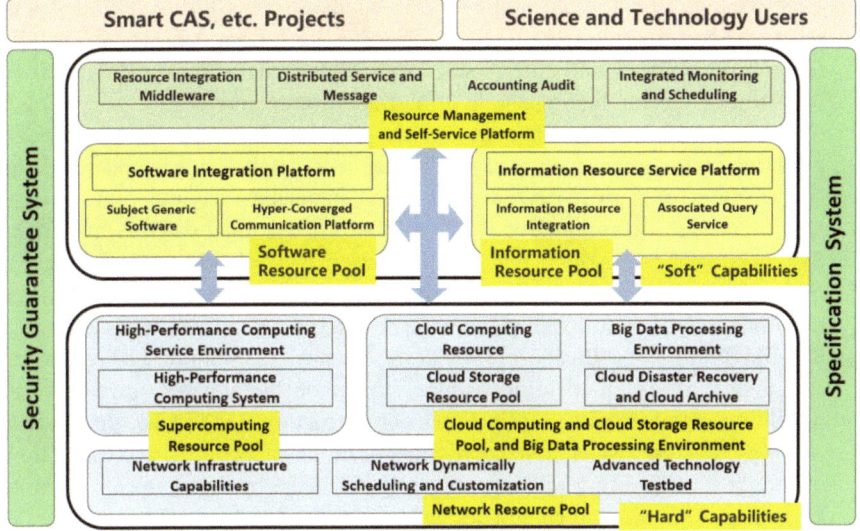

Fig. 1 The overall "China Science and Technology Cloud" architecture

3 "China Science and Technology Cloud" Resource and Service Construction

3.1 Unified Resource Management Platform

To improve resource utilization, reduce operation & maintenance cost, and improve data sharing, the CSTC has built a resource management and unified scheduling platform. The management and scheduling objects include computing resources, network resources, storage resources, data and document resources, etc. Further, through the self-service platform construction, it provides scientific research users in the form of cloud services.

At present, in order to ensure normal and effective in the unified resource management platform use, "China Science and Technology Cloud" has divided four major businesses based on function, as follows.

(1) Operation & Management and Monitoring Business: Provide stable, safe and reliable cloud infrastructure services, implement infrastructure management and monitoring of various of systems of national and the Chinese Academy of Sciences informatization, undertake the task of special network construction of informatization of CAS, and improve the automation and intelligence level of the of cloud infrastructure in overall operation and monitoring.

(2) Cloud Technology Development and Application Business: Provide research & development of cloud infrastructure and application services. Undertake the construction task of special high-performance computing, cloud computing and

cloud storage of informatization of CAS, undertake the research & development and construction task of cloud infrastructure for national, local scientific research institutions and enterprises, and facilitate the improvement of cloud infrastructure technology capabilities and application services.

(3) Software Platform Development and Application Business: Build a cloud platform of unified resource scheduling and user self-service, undertake the construction task of special resource management and self-service, hyper-converged communication, etc. of the informational of CAS, undertake the research & development and construction task of relevant software for national, local scientific research institutions and enterprises, and realize the deep integration of software platform services and technology innovation.

(4) General Technology Business: Provide research and system design of general technology of the cloud infrastructure, strengthen the application research & development of future informational infrastructure, undertake the overall planning, design task and standard specifications of the CSTC construction, undertake the planning and overall design task for national, local scientific research institutions and enterprises, enhance the technology leadership and the Science & Technology Innovation service capability of the Chinese Science and Technology Cloud infrastructure.

3.2   Network Environment

China Science & Technology Network (CSTNET) is the main part and important foundation of network construction. It plays a supporter role—"internally carrying scientific research information transmission and application services, and externally extending information resources sharing and cooperation". It shows the new progress from 4 dimensions, the overall basic network service capability construction, strategic construction of Hong Kong open exchange node, inter-network interconnection construction and new services facing the wireless technology trend. Through the CNGI constructions, including "Core Network Project", "Customer Premises Network Project" and the "Construction and Application Demonstration of CNGI Information Infrastructure Project", China Science & Technology Network has a core network with 100G channels, a long-distance backbone network with 155 M–2.5Gbit/s transmission capability, and a metropolitan area network with 1G–10Gbit/s channels. It provides the comprehensive IPv4/IPv6 dual stack access service, and the international network exports have reached 52Gbit/s, of which Chinese-American network and Chinese-Europe network are 10Gbit/s.

Meanwhile, via fibers circuit, digital circuits, etc., CSTNET provides customers of scientific research institutes, enterprises public institutions with IPv4/IPv6 Internet access, operation & maintenance and high-speed data transmission services, including high-speed, secure and convenient access to the Internet. Direct-connection channel with the backbone network of domestic mainstream operators. Excellent overall network quality. Sufficient independent export bandwidth. Support customer Internet application, including voice, video, data transmission, web meeting, office automation, etc.

Furthermore, the CSTC has taken a series of measures to further strengthen the network environment construction. Improve the network exchanging capability, network reliability and stability through upgrading network infrastructure capabilities. Improve the network operation control and flexible service delivery capability through the cloud network management and service system construction. Based on the advanced technology testbed, China Science & Technology Cloud launches the research on service quality assurance and enhancement, vast data transmission solutions and forward-looking new technology applications.

Meanwhile, we deeply integrate and gather the informatization infrastructures of inside and outside CAS using cloud computing, build and operate a cloud infrastructure supervision platform, and moderately developed the remote supervision on basic resources of scientific research institutes. Build and operate & maintain a cloud service platform, realize on-demand scheduling, real-time monitoring and comprehensive auditing of distributed heterogeneous resources, and provide cloud resource services for individual users. We launch the research& development and provide operation guarantee of featured services based on cloud resources, such as the E-mail system, video conference, CSTNET Passport, team document libraries, conference service platform, etc. we put more efforts on the training of institutes on aspects of operation & maintenance and application of the informatization infrastructure.

With CSTNET as the fundamental and Cloud Technology as the important supporter, the CSTC has come into the stage of steady development.

3.3 Supercomputing Environment

Supercomputing Environment is an important part of the CSTC construction. CAS continued to construct based on the "three-layer architecture" supercomputing environment from the "12th Five-Year Plan" project, and it has achieved steady progress on the aspects of supercomputing capability and application effect, and the computing capability has improved to the "petaflop", the number of sub-centers has increased to nine, more completely covers applied disciplines, and the breadth and depth of applying have been further improved. The main supercomputer of the Supercomputing Center of the CAS has been upgraded to the sixth generation, the "Era" supercomputing system, with a peak of 2.36 petaflops. Table 1 shows the overview of the supercomputing center and sub-centers.

Meanwhile, CAS has constructed super-computing resource pools for better service for the CSTC, as follows. Build high-performance computing systems with computing capability of more than 10P to support the application requirements of key science and technology infrastructures effectively. With different discipline areas as the core, we have researched and optimized key technologies of application

Table 1 The overview of the supercomputing center and sub-centers

Supercomputing Center	Institute	Featured Application Direction	Main Computing Resources
Main Center	CNIC	Provide large-scale, high-quality supercomputing services in the whole CAS and nationwide	2.3 petaflops, "Era" supercomputing
Lanzhou sub-center	CAREERI	Geoscience, ecology research	15 trillion calculations, 200 TB
Qingdao sub-center	IOCAS	Marine science, bioenergy and process research	26 trillion calculations, 300 TB
Hefei sub-center	CASHIPS, USTC	Physical science computing	106.3 trillion calculations, 227 TB, 132.6 trillion calculations, 64 TB,
Kunming sub-center	KIB	Biology, natural medicine chemistry calculation field	10.9 trillion calculations, 100 TB,
Shenzhen sub-center	SIAT	Industrial computing, cloud computing	10.9 trillion calculations, 150 TB
Dalian sub-center	DICP	Computational chemistry science	11.6 trillion calculations, 30 TB
Wuhan sub-center	IHB	Biological information science	10 trillion calculations, 100 TB
Shenyang sub-center	IMR	Computational material science	10 trillion calculations, 30 TB
Guangzhou sub-center	GIBH	Stem cells and regenerative medicine, innovative drug design, etc.	50 trillion calculations, 77 TB

services, and build a national high-performance computing basic service environment. Take the application and computing tasks as traction to build a heterogeneous platform for migration, deployment, optimization and testing of applications in multidisciplinary areas, integrate high-quality computing resources of national supercomputing centers and local supercomputing centers, and extend the available computing resources for multidisciplinary computing applications. With the goal of serving users, actively research and promote various forms of high-performance computing application services. Extend functions and optimize the performance of development interfaces to support constructions and development of the application community in more discipline areas. Promote the capability and method of technical support and training to promote the development of high-performance computing in more discipline areas. Combine cloud computing and Internet thinking, actively use new technologies to develop scientific research ideas, use new methods to break through scientific research bottlenecks, improve the high-performance computing services of the CAS, and paly the high-performance computing's a driving role in innovation.

In addition, based on the supercomputing environment, our ongoing work includes enhancing the research and development of computing science application software in subject areas with better foundations such as computational chemistry and CFD. We have carried on research and development of computing applications in new subject areas such as precision medicine, brain science, etc.

3.4 Cloud Computing, Cloud Storage and Big Data Processing Environment

As another infrastructure of the CSTC, cloud computing and cloud storage provide a processing and application environment for scientific big data.

"China Science and Technology Cloud" already has a hybrid cloud environment with private cloud and public cloud properties, and storage environment, and it supports a series of e-Science services such as smart CAS, science big data, etc. It meets self-service of public cloud research users, such as timely access to resources, elastic expansion, and pay-on-demand, as well as the needs of proprietary cloud users, such as proprietary service model, high security and high business continuity, etc. "China Science and Technology Cloud" hybrid cloud service platform has computing capability of public services with 12,000 cores, provides unified operation & maintenance, on-demand resources and services for scientific research users. Scientific research users can create their own virtual data center (Virtual Data Center, VDC), which helps scientific research users save costs of construction and operation & maintenance, and improve support and services level provided to scientific research activities.

With the continuous launch of cloud computing and cloud storage systems, CAS has almost completed scientific data resource integration and vast storage environment construction, and integrated storage resources and scientific data into a data-oriented data infrastructure using "data cloud" service model. The total storage of capacity "data cloud" has reached 3 PB, provided 456 TB sharing scientific data, total increase storage capacity has reached 50 PB monthly, and data infrastructure services are increasingly becoming the basic public services for CAS scientific research informatization.

In order to further enhance cloud computing resources of the CSTC construction, "13th Five-Year Plan" project has planned the "Cloud Computing Resource Pool" construction for including cloud computing infrastructure environment construction and data-intensive processing environment construction. Cloud infrastructure environment has constructed a hybrid cloud service platform based on open Openstack technology architecture, and it provides on-demand acquisition, flexible expansion, high-level, such as high security, etc. and high-business continuity resources usage for researchers of CAS through rational infrastructure resource integration. Data-intensive processing environment construction includes creating a foundational operation platform of distributed cloud computing/storage cluster based on a standard X86-based server, generic network devices, etc., deploying big data processing platforms such as Hadoop, spark, storm etc., and it provides effective support services for various applications of scientific big data.

3.5 Informatization Resource Service

CAS informatization has completed the constructions of management cloud, education cloud, and domain-specific cloud. Others significant achievements have been made on aspects of "Instrument and Equipment Share Management System of Chinese Academy of Sciences (the Shared Network)", the CAS website cluster, science communication, news propaganda and security guarantee, etc. capabilities.

"Scientific Data Resource Integration and Sharing Project" completed during the 12th Five-Year Plan period has integrated and constructed 13 subject area libraries, 7 research institute databases, and 20 professional databases, covering multiple subject areas. These data play an important role in scientific research projects and even economic society. These data resources will be integrated into the information resource pool in accordance with the standard specification of the information resource pool.

At present, we are carrying out a comprehensive construction of information resource pool based on "China Science and Technology Cloud" infrastructure for further enhancement of information resource shared and open level of CAS. With scientific data achievements during the "12th Five-Year Plan" period, information resource pool construction integrates existing resources, and it will be supported by "China Science and Technology Cloud" and cover various of information resources, such as scientific data, literature patents, science popularization education, project reports, instrument and equipment, and individual documents, etc. With resource management and self-service platform as the portal, research informatization and big data development—oriented, we integrate the incremental information data which is constantly generated from scientific research activities of CAS and various of stock information resources which are accumulated over a long period of time, and create information resource catalogues, core resources and main service models of CAS. The main works include:

(1) Build an information resource service platform, realize cross-database retrieve and resource recommendation functions, and support Web and mainstream mobile terminal access.
(2) Implement docking with more than 15 resource databases in different fields such as scientific data, literature patents, science popularization education, etc.
(3) Select 1 to 2 secondary subject areas and implement data literature correlation services.

3.6 Software Resource Service

"China Science and Technology Cloud" software resource pool will provide shared software resource cloud services in public and subject areas, greatly enhance integrated communication service capability, and promote the formation of software applications and shared ecosystem. Researchers can directly use software resource pool for data analysis, simulation and other scientific researches, avoid repeat deployment and build of software operating environment, and seamlessly use the "China Science and Technology Cloud" hardware facilities. CAS will build a "China Science and Technology Cloud" software resource pool with the following objectives:

(1) Build a software integration platform, provide shared software resource cloud services in public and subject areas, and achieve software integration, sharing, evaluation, and service. Develop scientific software release and service guides, support software of scientific research fields online integration and application, support software release and software information management to provide support services in form of on-demand services and service push, provide centralized and unified software maintenance, upgrade and other services, deploy scientific software and provide software lifecycle management services.

(2) Support the construction of generic software of discipline areas, form a generic software library to serve disciplines, and improve the software application level for discipline areas. To countering the common problems and key technologies of informatization application, the software resource pool will support the construction of a series of generic software supporting various of disciplines, including service support software, framework software, basic algorithm library, visualization framework software, etc. and form informatization application support system.

(3) Build a hyper-converged communication platform, achieve centralized integration of E-mail, documents, conferences, and messages, and support online communication and collaboration anywhere. The completed platform will build a centralized and unified multi-user terminal, achieves centralization, unification, and integration of E-mail, files, conference messages, etc. Based on the open platform, seamlessly access to next-generation ARP, website cluster, and application systems, as well as the output of the converged communication capability will be realized.

Figure 2 shows the overall architecture of the software resource pool. Based on the network, computing, storage, cloud computing, and other resources of the CSTC, the bottom layer of the software resource pool provides the operating environment, computing resources, and storage environment. The upper layer achieves the docking based the interface specifications provided by the resource management and the self-service platform, supports research and development of common software of the discipline areas, support scientific software users to use software resources online, and communication and collaboration among scientific research users are realized through the hyper-converged communication platform. The main functions of software platform of public and discipline areas include software classification management, software user evaluation management, software integration specification interface, software audit management, software user service, knowledge base system, and software operation environment. The main functions of professional software of discipline areas include service support software, framework software, basic algorithm library and visualization framework software. The main functions of hyper-converged communication platform include

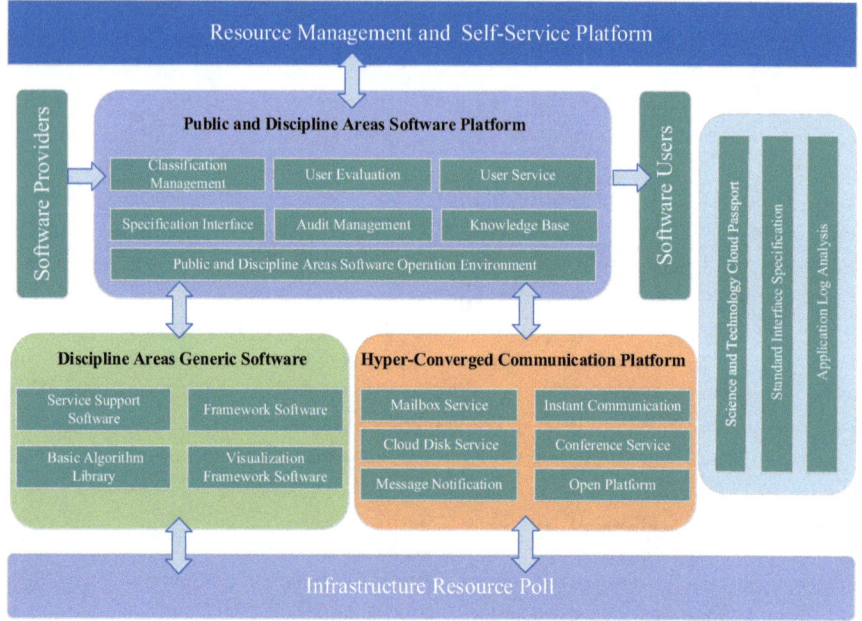

Fig. 2 The overall architecture of the software resource pool

mailbox service, cloud disk service, instant communication, conference service, message notification and open platform, etc. As the basic service of software resource pool, the CSTC Passport provides unified authentication services for applications of software resource pool platform, and lay the foundation for docking of other resource pools. The standard interface specification will provide API and SDK standard specifications for software resource pool capability output and docking of other resource pools. Application log analysis is used as a basic support application for software resource pools.

4 China Science and Technology Cloud Development Status

The "13th Five-Year Plan" China Science and Technology Cloud construction project has already been fully launched. After the initial planning and rapid iterative development, it has achieved great results in six aspects, including resource management and self-service platform construction, network resource pool construction, super-computing resource pool construction, cloud computing, cloud storage resource pool and big data processing environment construction, information resource pool construction, and software resource pool construction.

4.1 Current Service Capabilities and Application Effects

CSTNET has a rich experience and history in network operation. It provides services for 216 scientific research institutions of the CAS, 137 scientific research institutions outside of CAS including National Bureau of Statistical Computing Center, China Meteorological Administration Information Center, China Earthquake Networks Center, etc. and new and high technology enterprises. It provides services for more than 1 million scientific research users nationwide. In 2017, CSTNET's international connectivity, domestic connectivity, and core network service availability percent have all reached 99.99%, and the backbone network service availability percent has reached 99.99%. It effectively supported the implementation and promotion of national-level major projects and institute-level major projects undertaken by institutes from CAS and other institutes, including the "The Synchrotron Radiation Source Station Remote Monitoring and Data Transfer Application in IHEP", "The Very Long Baseline Interferometry (VLBI) Application" and "The Space Science Pilot Project, Space Science Satellite Ground Communication Network" "ITER International High-Speed Data Network Transmission".

As of November 20, 2017, the CSTC supercomputing resource pool had opened 599 user accounts. Users cover 21 provinces, municipalities, and Hong Kong Special Administrative Region. User disciplines cover chemistry, life sciences, information technology, materials, mechanics, physics, geophysics, aerospace, astronomy, mathematics, energy, and industry. As of November 30, 2017, the "Era" system has finished more than 6 million submitted jobs and has served 180 million CPU hours. From December 2016 to November 2017, the facility utilization at full load by user-specific queue had researched more than 70%. In 2017, the data storage service environment availability percent have all reached 99.99%, and it had supported 292 application systems. In 2017 and in the earthquake disaster relief work between the two Iraqi borders, the CSTC effectively cooperated with ChinaGEOSS Data Sharing Network National to assist the Iranian and Iraqi government disaster reduction agencies to promote the earthquake relief work, and effectively ensure that emergency data can be timely pushed to disaster reduction agencies of the countries. We had received high praise and a letter of thanks from THE ChinaGEOSS Data Sharing Network project.

As of November 20, 2017, the total number of users of the CSTC mail system had exceeded 310,000, and the number of system organizations had exceeded 300, including more than 230 institutes of CAS, 58 laboratories and research groups, and 23 companies of CAS. The number of users of CAS has reached more than 300,000, and the percentage of active users on workdays had exceeded 31%. In 2017, the system availability had reached 99.99%. In 2017, the video conference system had served more than 140 institutes of CAS, with about 240 video conference terminals. In 2017, more than 100 video conferences were held, including more than 50,000 participants, and almost 2,000 nodes. The desktop conference system had initiated about 2,000 conferences and more than 8,500 participants.

4.2 Resource Management and Self-Service Platform Construction

On the aspect of resource management and self-service platform construction, the main works we have accomplished include resource management and intelligent scheduling platform prototype design. Cloud platform survey, testing, analysis, and self-service platform prototype design. User identity unified management system prototype design. Resources monitor platform prototype design. Identify resources and services that require auditing and accounting, and design a database.

4.3 Network Resource Pool Construction

Realize the domestically upgrading of the core equipment. Complete the deployment of flow cleaning equipment. Realize three basic functions of the network basic performance monitoring system, including network monitoring, link quality detection, and equipment log monitoring. Build a small-scale network test environment in laboratories, and carry out some pre-development and small-scale test verification for some technologies, such as SDN, 5G MEC, ICN, etc.

4.4 Supercomputing Resource Pool Construction

Research distributed service architecture, and support resource integration with capability above 200P. Research application compilation tools, and research and develop continuous construction of software of container-based environment. Integrate application-centric computing resources. Build the test platform for published domestic software and hardware. Test performance of a number of applications. Provide test environment for high-performance computing services. Build the software set of one application field, develop rapid application deployment tools, and carry out deployment on multiple computing nodes. Collect software features. Multiple methods of job submission and management. Research and develop the spatiotemporal geographic information of supercomputing resources, science computing data visualization tool libraries, and realize basic visualization module, and optimized visualization module for the scalar field. Optimize performance of the existing interfaces. Optimize and upgrade the general computing portal. Open cloud platform and develop online documentation, online testing Function. Research and develop the hourly pay function. Develop platform technical solutions, and build the prototype verification system.

4.5 Cloud Computing and Cloud Storage Resource Pool, Big Data Processing Environment Construction

We have completed the sorting and division of the infrastructure equipment. Initially complete the cloud platform testing of multiple vendors. Complete the cloud computing implementation plan. The selection of cloud computing platforms has been determined; The new generation of cloud computing platform has been initially online, and tested and promoted partially. The CAS website cluster system has been migrated from the computer room of the headquarter of CAS to the Huairou IDC environment, and deployed in the proprietary cloud environment. We have completed the procurement and deployment of the new generation cloud computing environment security system, and it provides the security foundation for the proprietary cloud environment construction. Completion of the whole solution design of the data processing environment. Initially completion of the cloud storage technology solutions programming. The selection of cloud storage platforms has been determined. The new generation of cloud storage platform has initially been online, and tested and promoted partially. We have completed the procurement and deployment of the new generation cloud storage environment security system, and it provides the security foundation for the proprietary cloud environment construction. The disaster recovery backup technology solution has been programmed.

4.6 Information Resource Pool Construction

The information resource service platform has been initially built, and all functional modules of the platform have been almost completed. Classify and arrange the information resource types included in the information resource pool, provide preliminary communication to relevant resource providers, and determine the relevant pooling specifications. Associated query engines of data, literature, and patent have been initially designed.

4.7 Software Resource Pool Construction

The platform solution design has been completed, and the software resources have been classified and gathered. Continuously integrate, test, release and deploy the container-based environment software. The resource access control based on user permissions has been initially implemented. More than 10 common research applications have been deployed, and it can provide the online access software service. The preparation and the requirements analysis of the selection work have been completed.

The initial design has been completed and some functions have been implemented. We have completed the writing and refinement of the requirements analysis and the implementation solution, and we have already started to build the website, APP, and client. The E-mail system version and partial hardware have been upgraded. We have completed the desktop conference cloud implementation solution, and build the new generation desktop conference cloud (it can simultaneously support 20 conferences of pilot environments). Cloud disk selection and construction plans, test system construction have been completed.

5 Conclusion and Prospect

Informationization is an inevitable trend to the world development. The CSTC has already experienced 20 years from its birth to construction. Informatization construction of CAS has certain service capabilities, promotes the integration of informationization and scientific research activities, and it has become an irreplaceable supporter of the research and innovation activities of CAS. "China Science and Technology Cloud" environment has been initially formed and it provides public sharing, accessible and seamless availability for scientific researches of CAS. The CSTC has achieved significant achievements on aspects of network infrastructure, supercomputing, cloud computing, and cloud storage services, and it is progressing steadily. However, with the rapid emergence of information technology and the explosive demand brought by the scientific research paradigm transformation, the CSTC needs to be further improved on aspects of infrastructure resources core capabilities, comprehensive service capabilities, and deep integration with technological innovation activities.

Meanwhile, the scientific research informatization development asks all innovation subjects to take participate in the global environment of scientific research competition and collaboration, and informatization infrastructure construction has become an indispensable condition for scientific research and innovation. With the rapid development of information technology, informationization has become a crucial means to support national scientific and technological innovation. However, there is still an increasing gap between the level of scientific research and information infrastructure of the CAS and the level of development of the industry.

To satisfy the demands of informatization development, based on the "13th Five-year Plan" of CAS, "China Science and Technology Cloud" has been fully launched. Guidance with the requirements of "Pioneer Initiative" and "Smart CAS", technology resources of CAS have been deeply integrated. The CSTC will further strengthen the "hard" capability construction of public Informatization such as network, supercomputing environment and vast cloud storage. It will also accelerate the "soft" capability construction, including information resource pool construction and software resource pool construction. We will build a resource management and service cloud platform, with science and technology workers as the center, and unified resource scheduling, and user self-service as the distinctive characteristics. According to the "13th Five-year Plan" of CAS, the CSTC will improve the service capability of various of informatization resources, and effectively support the national-level and institute-level scientific research projects to be carried out.

Reference

1. Computer Network Information Center, Chinese Academy of Sciences. Implementation of 13th Five-Year Plan of China science and technology cloud. 2017.

Jun Li is a professor-level researcher and doctoral supervisor in Computer Network Information Center, Chinese Academy of Sciences. His research focus on computer networks. As the project leader, He had undertaken the National Computing and Networking Facility of China (NCFC Project) loaned by the World Bank. He had led the research team to develop the first domestic router (Oct. 1992), and build China's first international Internet line (Apr. 1994). He had undertaken the construction, operation and management of China Science and Technology Network (CSTNET). He has hosted and participated multiple 863 and National Natural Science Foundation projects. He is currently the chief engineer of CNIC.

Acoustic System of Jiaolong Manned Submersible and Its Future Development

Min Zhu

Abstract Deep sea manned submersible is a key technology for deep sea exploring and exploitation. Chinese "Jiaolong" is the deepest one among manned submersible on duty around the word. The functions of Jiaolong's acoustic system include underwater communication, positioning, obstacle avoidance, target searching, high resolution bathymetric scanning, etc. Its advanced high speed underwater acoustic communication system can transmit images, data, text, voice and Morse codes, and its advanced high resolution bathymetric side scan sonar can acquire high resolution 3D map of the sea bottom. The overall performance of Jiaolong acoustic system is superior to any other manned submersible. In future, manned submersible acoustic system will develop to full ocean depth, higher underwater communication capability, higher navigation and position accuracy, more measurement functions and modules. And for China, acoustic technologies and instruments will be localized.

Keywords Manned submersible · Acoustic system · Underwater acoustic communication

1 Introduction of Jiaolong Manned Submersible

There are huge resources beneath deep water in vast oceans. These resources are tactical reservation for sustainable development of human society. Main countries around world are all working hard on this field. China must master those deep sea key technologies in its own hand. It's the only way for China to possess the capability to go down into deep sea to research, explore and exploit those resources so as to guarantee Chinese benefit in deep sea.

M. Zhu (✉)
Institute of Acoustic, Chinese Academy of Sciences, Beijing, China
e-mail: zhumin@mail.ioa.ac.cn

M. Zhu
Beijing Engineering Research Center of Underwater Acoustic Instrument, Beijing, China

© Publishing House of Electronics Industry 2020
China's e-Science Blue Book 2018,
https://doi.org/10.1007/978-981-13-9390-7_9

Fig. 1 Jiaolong manned submersible

Jiaolong (Fig. 1) manned submersible's development is supported by Chinese national high technology development project (863 project). It started from the project "7000 m manned submersible" in tenth 5-year national plan. Several following projects support its sea trial and upgrade. From 2002 to 2012, it took about 10 years to achieve the final result.

The commitment enterprise of the development projects is China Ocean Mineral Resources R&D Association (COMRA) who is the user of Jiaolong.

The submersible is developed by China Ship Scientific Research Center (CSSRC), Shenyang Institute of Automation (SIA), Institute of Acoustic Chinese Academy of Sciences (IACAS) and other many organizations.

Now National Deep Sea Center (NDSC) is in charge of Jiaolong's maintains and operation.

Jiaolong is designed to work at depth of 7000 ms. During the sea trial at June–July 2012 [1], it dived to the maximum depth of 7062 m, and created a new world record of diving depth within similar type of manned submersible.

Jiaolong can load 3 persons including 1 pilot and 2 scientists, and take them down to 7000 m deep sea bed to carry out observation and operation. It's a valuable platform for deep sea scientific research, resources exploration and exploitation, engineering operations, etc. Since 2013, Jiaolong has been used in South China Sea, Pacific Ocean, and southwest Indian Ocean, in cold spring area, manganese nodule area, cobalt-rich crust area, and hydrothermal sulfide ore area, to carry out mineral resource exploration, biology and chemical research, etc. A large number of samples, photos, videos and data were acquired and some new species were found. Till

(a) ALVIN of USA   (b) MIR1&MIR2 of Russia

(c) Nautile of France   (d) Shinkai 6500 of Japan

Fig. 2 Similar type deep sea manned submersible around world

the end of 2017, the dive number of Jiaolong had been more than 150. These achievements indicate Chinese manned submersible technology and deep sea resource exploration capability has reached internal advance level.

There four other countries around world built deep sea manned submersibles with similar depth ranking and capability. They include ALVIN of USA (Fig. 2a), MIR 1/2 of Russia (Fig. 2b, retired), Nautile of France (Fig. 2c) and Shinkai 6500 of Japan (Fig. 2d). Comparing with them, Jiaolong are in leading position in diving depth, battery capacity, control system, underwater acoustic communication and operation capability. Its other technologies are all internal advance level. Its overall technology level is the best among them.

Jiaolong has won several awards. "Sea trial of Jiaolong manned submersible" won the First Prize of 2013 Ocean Engineering Science and Technology Award. "7000 m manned submersible Jiaolong" won Grand prize of 2014 Science and Technology Award form Chinese Society of Naval Architects and Marine Engineers (CSNAME). "Development & Application of Jiaolong Submersible" won the First Prize of 2017 State Science and Technology Progress Award.

The success of Jiaolong push China into the top rank in the field of deep sea manned submersible. Jiaolong has become one great pillar to support Chinese maritime power strategy and Chinese revive. Now manned deep sea diving is a well-known symbol of leading technology and as popular as manned space flight.

2   The Functions and Composition of Jiaolong's Acoustic System

Because of strong absorption effect, radio and light decrease rapidly in sea water so they cannot propagate to long distance. The best of them are long wave radio and blue-green laser with only no more than 500 m range. So long rang communication, positioning, detection and measurement are all rely on acoustic technology. Acoustic system is a key system of manned submersible.

Jiaolong carries 3 people down to 7000-m depth, safety is its top priority. It has many different mission types including exploring manganese nodule, cobalt-rich crust and hydrothermal vent, biology and chemical research, etc. It needs to work at many complex environments such as sea mountains, ridges, trenches, hydrothermal vents. All these things pose very high requirement on acoustic system's function, specification and reliability.

Institute of Acoustic, Chinese Academy of Sciences developed acoustic instruments and integrated acoustic system of Jiaolong manned submersible. According to mission requirements, the advance acoustic system of Jiaolong insists of underwater acoustic system, high resolution bathymetry side scan sonar, positioning system, imaging sonar, Doppler velocity log, obstacle avoidance sonar. Illustrated in Fig. 3 is the configuration of 6 acoustic instruments. They play the roles as Jiaolong's eyes, mouth, ear, and realized functions of communication, positioning, velocity measurement, obstacle avoidance, object searching, bathymetry observation. Relies on the support of acoustic system, Jiaolong can cruise in mysterious deep ocean safely and accomplish various missions.

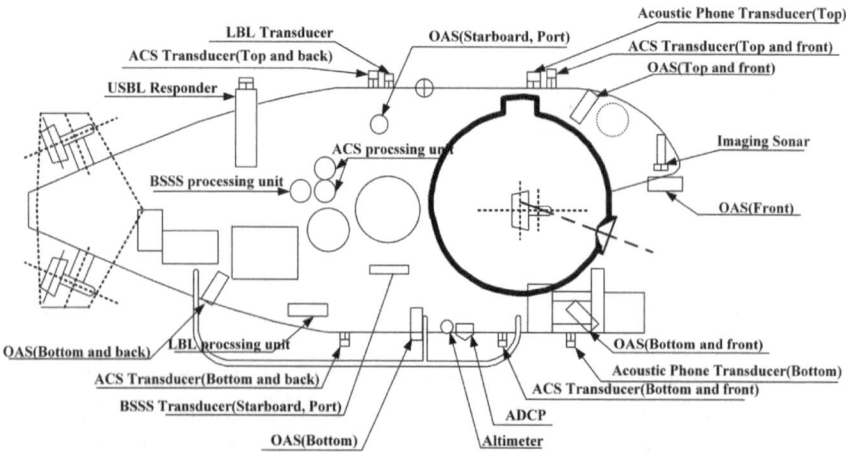

Fig. 3 Configuration of acoustic instruments in Jiaolong

2.1 Underwater Acoustic Communication System

For the safety and convenience of operation, there is no cable linking deep sea manned submersible and its mother ship. Instead, the link relies on underwater acoustic communication system. Such link is vital to the safety of the submersible.

The acoustic communication system of Jiaolong includes a high speed digital communication system [2] and an underwater telephone as a backup.

2.2 High Resolution Bathymetry Side Scan Sonar

High-resolution bathymetric side scan sonar system (HRBSSS) is mounted on both side of manned submersible. It can measure micro-topography of seafloor and targets on seabed and in water volume. It also can produce real time, high-resolution three-dimensional map of the seafloor. It can work under complex terrain conditions and measure height of targets, so it's suitable for the investigation in cobalt-rich crust area and the measurement of geometrical dimensions of hydrothermal vent "chimney" in hydrothermal vent field. No other manned submersibles in the world installs HRBSSS.

2.3 Obstacle Avoidance Sonar

Seven obstacle avoidance (OAS) sonars are mounted in different direction to measure the existence and distance to obstacles at fore-up, fore, fore-down, down, down-back, left and right direction. With OAS information, pilot can avoid dangerous collision. OAS can also provide height data for control system.

2.4 Forward Looking Imaging Sonar

A mechanical scanning imaging sonar is mounted at the forehead of the submersible to scan foreside seabed and floating objects. It provides a much further distance observation than light can reach. Pilots can use it to search anything they interest in and to avoid obstacles that may endanger the submersible.

2.5 Acoustic Doppler Velocity Log

Acoustic Doppler velocity log is down-looking mounted at the bottom of the submersible. It is used to measure the 3D velocity of the submersible which is essential for the control and navigation of the submersible. It is also used to measure the current velocity profile of the water volume beneath the submersible.

2.6 Acoustic Position System

Acoustic position system of Jiaolong includes ultra-short baseline position system (USBL) and long baseline position system (LBL). USBL system includes a sonar system at mother ship and a responder at submersible. LBL system includes a transceiver at submersible and several beacons deployed at sea bed with much better accuracy than USBL in the cost of time and inconvenience.

Acoustic position system can measure the position of the underwater submersible. With USBL surface group can monitor the submersible's coordination and depth in real time. USBL data need to be sent from mother ship to submersible via acoustic communication system for its underwater navigation.

2.7 Comparison of Acoustic System of Deep-Sea Submersibles

Listed in Table 1 is the comparison of acoustic system of the 5 similar type deep-sea submersibles around world [3–8]. Jiaolong submersible's acoustic system is the best. Its advantages are high speed digital acoustic communication system and high resolution bathymetry side scan sonar. Both are innovative achievements of Institute of Acoustic with several China and US patented key technologies.

Table 1 Comparison of acoustic system of the 5 deep-sea submersibles

Acoustic instruments	China Jiaolong	USA ALVIN	Russia MIR I/II	Japan Shinkai 6500	France Nautile
Digital acoustic communication	Full functional digital communication system	Data transmission only	–	Image transmission only	Image transmission only
Underwater telephone	✓	✓	✓	✓	✓
High resolution bathymetry side scan sonar	✓	–	–	–	–
Acoustic Doppler velocity log	✓	✓	✓	✓	✓
Forward looking image sonar	✓	✓	✓	✓	✓
Obstacle avoidance sonar	✓	✓	✓	✓	✓
Position system	✓	✓	✓	✓	✓

3 High Speed Digital Acoustic Communication System of Jiaolong

Jiaolong's high speed underwater acoustic communication (UWAC) system adopts advanced underwater acoustic communication and signal processing technologies. It has perfect capability of data, image, text, voice and command transmission. During whole diving, detailed information of submersible are sent to command center at surface ship. The commanders can get all sensor data in real time, including depth, velocity, cabin temperature, humidity, air pressure and oxygen concentration, etc. With these information, commanders can monitor Jiaolong's status in real time (Fig. 4), evaluate the progress and give instructions to Jiaolong, so as to improve safety and efficiency. The crew member in Jiaolong can chat with surface in voice and text, send back photos they shoot during operation, discuss technic issues. Comparing with other deep sea manned submersibles around the world, Jiaolong's UWAC system has more functions and better performance. This

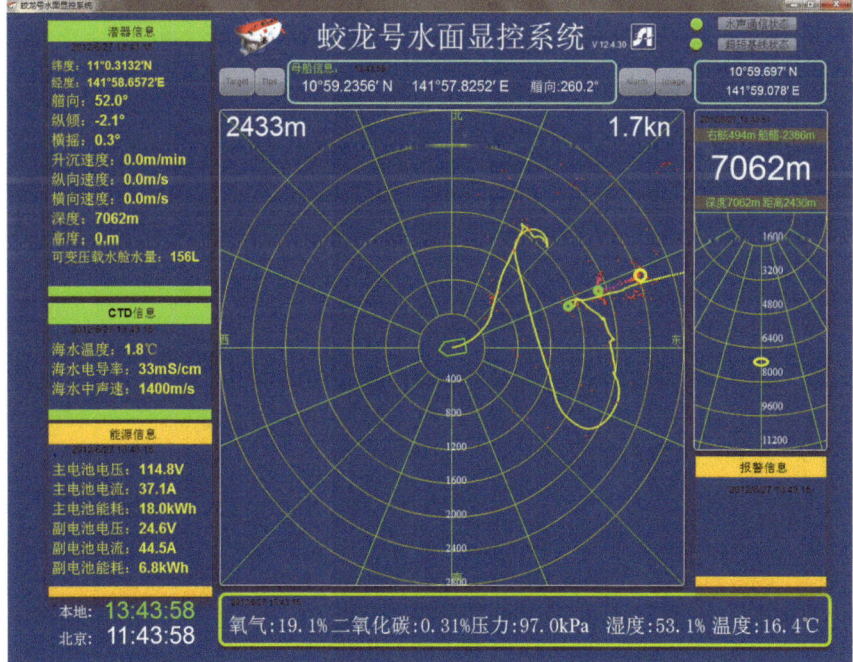

Fig. 4 Real-time monitoring Jiaolong's status at mother ship

advanced high speed UWAC system is appraised as one of the 3 leading technologies of Jiaolong manned submersible.

During argument stage of Jiaolong project, the development team analyzed the communication systems of deep sea manned submersible around the world. These systems were all using analog underwater telephone as their main communication method. American Alvin installed a low speed acoustic modem but seldom used. No report was found on Russian "MIR 1/2" acoustic modem. Both Japanese "Shinkai 6500" and French Nautile installed acoustic modem developed by JAMSTEC. That was dedicate designed for image transmission from submersible to surface ship with 16–24 kHz frequency, DPSK modulation, 16 kbps maximum data rate and 6.5 km maximum range. It use narrow beam angle of 35 degree and vertical path. Such configuration has relative simple channel and high signal to noise ratio (SNR). So it can work in relative high frequency with wider band width to realize high speed easier.

There were 3 options for the development team. The first is using analog underwater telephone as main communication method and use digital modem as a supplemental, just like Alvin. The second is using analog underwater telephone as main communication method and use dedicate designed digital modem to transmit picture to surface, just like "Shinkai 6500" and Nautile. The third is an overtaking. We need to design an innovative communication system based on Jiaolong's mission requirements with a high speed digital communication system as its core and an underwater telephone as a supplemental. The first two options use proven technology. Chosen one of them has low rick but would not do much help in improving our technology level in this field. Chosen the last option has much higher risk but could push the development in high speed digital communication technology.

After carefully consideration on Jiaolong's mission requirements, the needs for technology development and the technology basis they had at that time, based on the confidence that they could solve the key technologies in single carrier high speed digital communication in recent years, the team decided to design an innovative full function underwater communication system capable of high speed image transmission, medium speed data and text transmission, low speed but highly reliable command transmission and analog voice transmission. MPSK (multiple phase shift keying) single carrier coherent communication technology, MFSK (multiple frequency shift keying) non-coherent communication technology, FH (frequency hopping) spread frequency communication technology and SSB (single side-band) modulation technology were integrated into the system. It needed courage to use MPSK as its main communication method because it was still under developing at that time and seldom used in applications. This design is proven to be prospective. Even today, 15 years later, this communication system is still the most advanced one.

In order to realize reliable MPSK communication in complex underwater channel with delay-Doppler double spreading, the team developed an algorithm concatenating self-optimized multi-channel decision feedback adaptive equalizer and Turbo decoder. The key technology is Fast Self-Optimized Least Mean Square (FOLMS) algorithm. It has good performance proven in sea trials in transmission photos and low computational complexity.

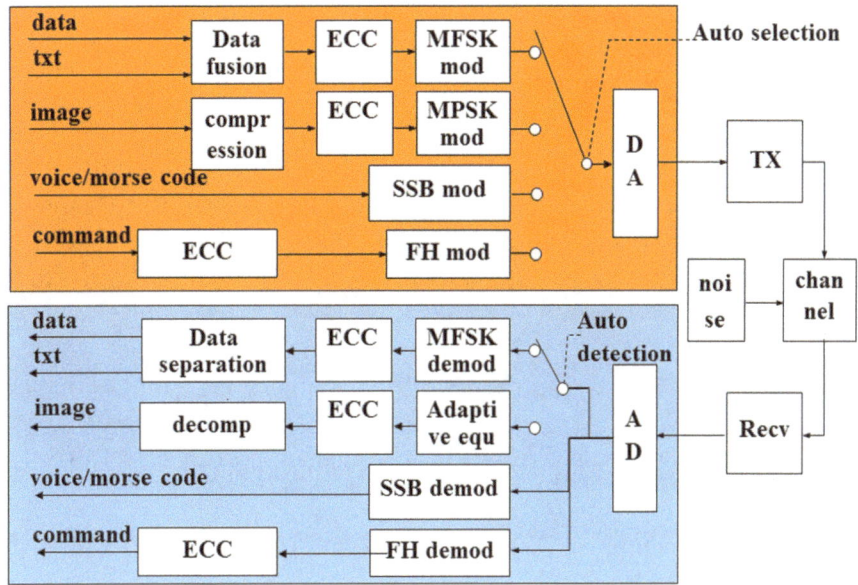

Fig. 5 System diagram of underwater communication system

The communication system needs to handle many different types of information including images, data, text, commands and voice with the four communication technologies. The processing is quite complex. It needs to reduce the operation complexity to make the operator easier. The team designed a user friendly data-driven & auto-detection scheme as illustrated in Fig. 5. In sender sides, each types of information active its own sending processing module. In receiver side, information type is automatically detected, and then corresponding receiving processing module decodes and dispatches information. So the operators could concert on dialogue without to concern the communication technology details.

June 24, 2012, a dialogue was established between pilots at Jiaolong manned submersible and astronauts at Tiangong-1 space station (Fig. 6). Such dialogue is the first one in world history. Reliable underwater communication system played a key role in this dialogue.

Jiaolong has successfully dived more than 150 times in its sea trials and successive experimental applications. Underwater communication system always played a key role in these dives by guarantee smooth link between surface ship and submersible, realizing real time monitoring the status of submersible and sending back many photos shot at deep sea (Fig. 7).

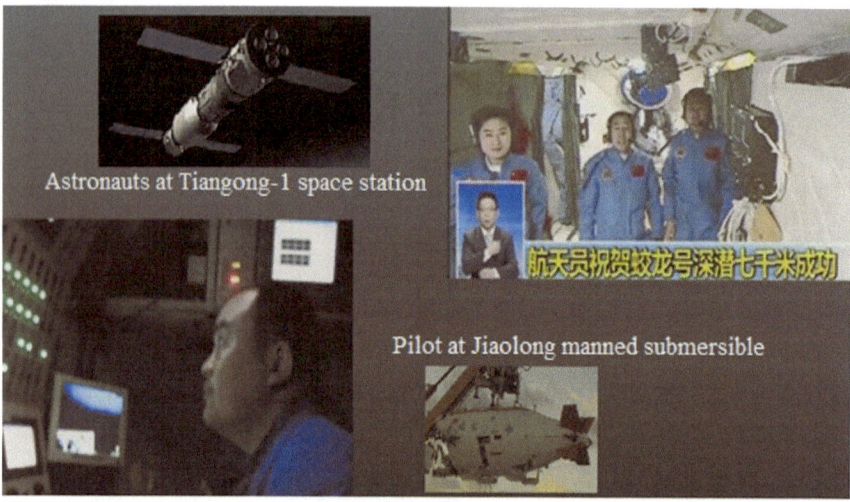

Fig. 6 Jiaolong dialogue with TIANGONG-1 space station with its acoustic communication system

Fig. 7 Photos shot by Jiaolong at sea bed sent back to surface via acoustic link

4 HRBSSS for Jiaolong Manned Submersible

HRBSSS for Jiaolong manned submersible is another advanced acoustic equipment in the world. It can simultaneously obtain high-resolution topographic map and geomorphologic map of the seafloor. There is no manned submersible abroad which install HRBSSS and have the ability to detect seabed topography and geomorphology.

The working principle of HRBSSS is shown in Fig. 8. Sonar signals are gradually "illuminated" to the seafloor from near to far and the echoes scattered by seafloor are received by the receiving arrays respectively. The direction-of-arrival (DOA) of echoes can be estimated by DOA estimation technique, then the horizontal distance and depth of seabed scatters can be obtained. Its difficulties lie in the fact that even though there may be multiple signals including interference signals

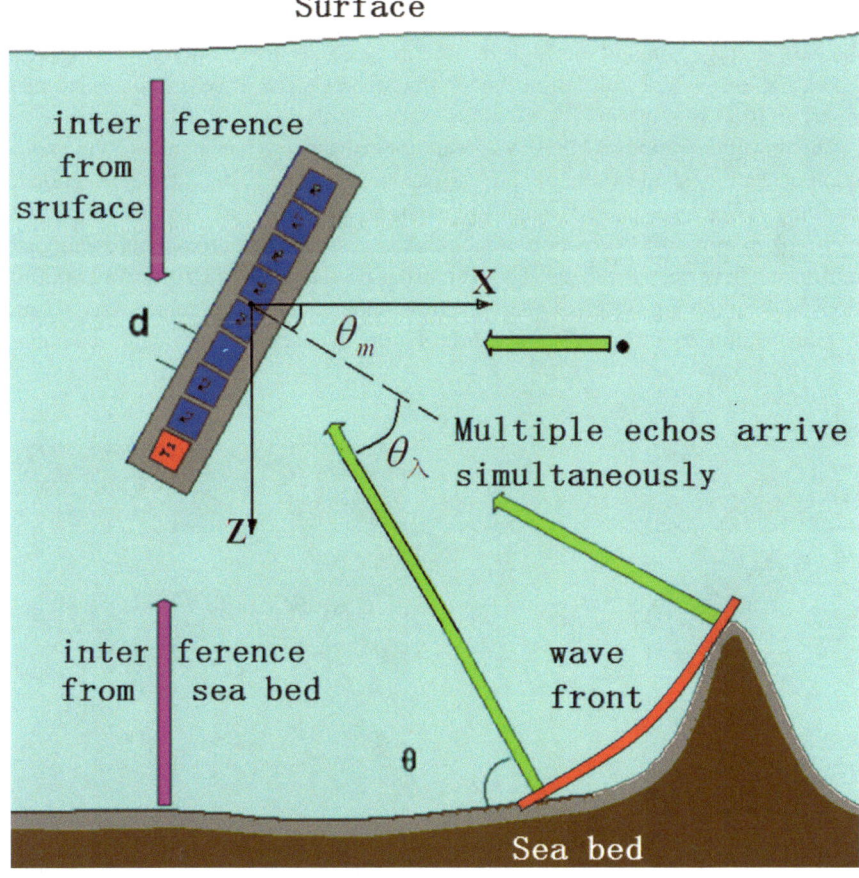

Fig. 8 Schematic diagram of HRBSSS working principle

from seabed and sea surface arriving at sonar arrays simultaneously, the position of seafloor scatters have to be calculated accurately.

Based on the study of underwater acoustic physics, the Institute of Acoustics, Chinese Academy of Sciences explained the main error sources in the original bathymetric side scan algorithm by using the thin-layer model of the seabed, then proposed the solution. The first generation of HRBSSS technological scheme was formed, which was applied to CR02 6000 m AUV and won the second prize of National Science and Technology Award in 2003.

Based on this technology, Jiaolong acoustic technology team further developed a set of technical schemes with the multi-subarrays seabed automatic detection-DOA estimation technology signal processing method as the core, also have got a series of research achievement on the estimation algorithm of targets number, broadband signal technology, position and posture correction of submersible, and the technology of results image mosaic. Then the second generation of HRBSSS technological scheme was formed, which was applied in many projects such as Jiaolong manned submersible system, 6000 m deep towing system, the security system of Qingdao Olympic Sailing Race. HRBSSS for Jiaolong manned submersible has obtained a large amount of high-resolution seafloor detection results. Figure 9 shows Jiaolong's high-resolution topographic map (left) and geomorphologic map (right) at 7000 m seafloor.

The third generation of HRBSSS technological scheme has increased the exploration of seafloor shallow stratum profile, can simultaneously realize the detection of sounding, side scan and shallow stratum. It can be called integrated micro-geomorphology detection system. All software, electronic hardware and transducers have been localized. The system has been applied in Exploratory 6000 m acoustic deep towing system, Qianlong autonomous underwater vehicle (AUV) and 4500 m autonomous survey system. Figure 10 shows the high-resolution

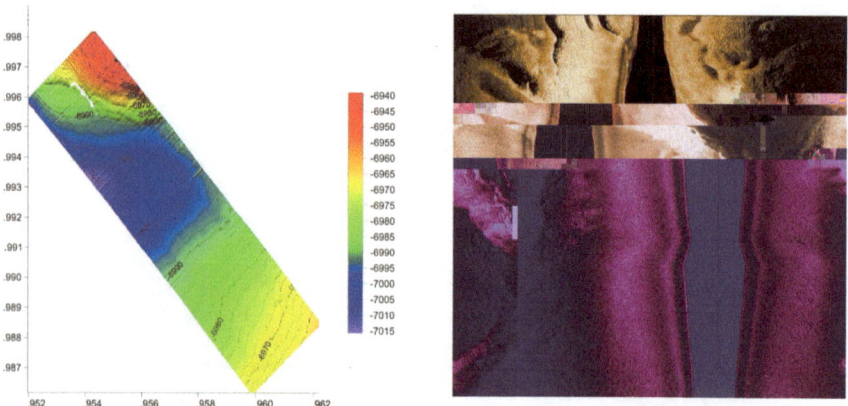

Fig. 9 Jiaolong's high-resolution topographic map (left) and geomorphologic map (right) at 7000 m seafloor

Fig. 10 High resolution bathymetry of Qiandao lake bottom

topographic map of the Qiandao Lake obtained by the third generation HRBSSS technology. Its resolution and mosaic quality have been significantly improved.

Institute of Acoustics, Chinese Academy of Sciences is improving the product line of HRBSSS on the basis of the third generation of HRBSSS hardware, and is developing dual-frequency and low-frequency wide-coverage transducers in order to meet the detection requirements under different circumstances.

5 Prospection of Acoustic System of Deep Sea Manned Submersible

From the development of manned submersibles abroad in recent ten years, including Deep Sea Challenger which dived to 11,000 m depth in 2012, we can see that the composition of their deep sea manned submersible's acoustic system remain unchanged. It includes communication system, positioning system, OAS, altimeter, forward look imaging sonar and velocity log as it was before. In some special applications, multi-beam echo sounder is used. Underwater telephone is still their main communication method. The breakthrough in communication and measurement of Jiaolong's acoustic system is one highlight in international manned submersible technology development.

Listed below are the future developments of manned submersible acoustic system.

5.1 Diving Deeper

Both manned and unmanned submersibles are going to full ocean depth, 11,000 m. China is developing full ocean depth manned and unmanned submersibles. To support these projects, acoustic system must realize higher depth rating and longer range of communication and positioning.

5.2 Bette Underwater Communication Capability

It need to keep improving underwater communication technology to realize longer range, higher data rate and lower error rate.

5.3 Higher Navigation and Positioning Accuracy

It needs to keep improving USBL, LBL technology and multi-sensor integration navigation algorithm to realize higher underwater navigation and positioning accuracy.

5.4 Multi-platform Cooperation

Today deep sea manned submersibles are working alone. Cooperation with other underwater platform is very limited. In future, cooperating with multiple other underwater platforms is becoming standard mode. China has already built Jiaolong 7000 m manned submersible, "Shenhai Yongshi" 4500 m manned submersible, is building full depth rating manned and unmanned submersible and many other underwater platforms such as landers, moorings. To realize multiple platform cooperation in the same working area, they must be link into a wireless network via acoustic communication. Effective networking protocol and standards need to be established.

5.5 Modularized Survey System with More Capabilities

Today, the survey capability of manned submersible is quite limited. Abroad manned submersible seldom install survey sonar besides those needed for safety. Jiaolong has a BSSSS for survey but only used in some of the dives.

In future, more types of survey sonar are required. They need to be modularized so the submersible could be re-configured to meet different mission requirement, such as searching wrecks in large area, high resolution mapping, sub-bottom profiling.

5.6 Totally Made in China

Many instruments of Jiaolong's acoustic system including forward looking imaging sonar, velocity log, positioning system were imported. Transducer of communication system, BSSS and OAS were imported too at the beginning.

China experts are working hard in improving their technical and engineering level of acoustic instruments. So the above mentioned transducers were replaced by Chinese made parts years later. The home-made rate of "Shenhai Yongshi" acoustic system is much higher. All of its acoustic instruments except forward looking imaging sonar are made in China. The home-made rate of the full depth rating manned submersible acoustic system is 100%.

References

1. Cui Weicheng, Liu Feng, Hu Zhen, Zhu Min, Guo Wei, Liu Chenggang. 7000 m sea trials test of the deep manned submersible "JIAOLONG[J]. Journal of Ship Mechanics, 2012, 10:1131–1143.
2. ZhuWeiqing, Zhu Min, Wu Yanbo, Yang Bo, Xu Lijun, Fu Xiang, Pan Feng. Signal Processing in Underwater Acoustic Communication System for Manned Deep Submersible Jiaolong [J]. Chinese Journal of Acoustics, 2012,06:565–573.
3. Michiya Suzuki, etc. "Digital Acoustic Image Transmission System for Deep-Sea Research Submersible", Oceans'92 Conference Proceedings, Oct., 1992, Vol. 2, pp. 567–570
4. William Kohnen, "Manned Research Submersibles: State of Technology 2004/2005, Marine Technology Society Journal[J], 2015, 39(3):121–126
5. Kudo K. Overseas trends in the development of human occupied deep submersibles and a proposal for Japan's way to take. Science and Technology Trends, 2008, 26: 104–123
6. http://www.whoi.edu/main/hov-alvin
7. http://www.jamstec.go.jp/e/about/equipment/ships/shinkai6500.html
8. https://en.wikipedia.org/wiki/Mir_(submersible)

Min Zhu is a professor of Institute of IACAS (Institute of Acoustics, Chinese Academy of Sciences), a doctor director, the director of Ocean Acoustic Technology Laboratory of IACAS, and the leader of Underwater Acoustic Communication Innovation Team in key areas of Chinese Ministry of Science and Technology. His research fields include underwater acoustic communication, acoustic Doppler velocity measurement, acoustic instruments development and acoustic system integration for underwater platform. He is one of the deputy chief designers of deep sea manned submersible "JIAOLONG" and the principal investigator of an Underwater Acoustic communication nodes and networking technology research program. He has accomplished several research and development programs as team leader or a main team member. He won Second Prize of 2003 Annual National invention award, First Prize of 2018 Scientific Technological Progress Award, Special government allowance and eleventh "Ten Outstanding Young Scientists" of Chinese Academy of Sciences. He is supported by national Ten-thousand Talents Program.

Construction and Application of BeiDou Time System

Shaowu Dong, Wenjun Wu and Shougang Zhang

Abstract China's BeiDou satellite navigation system can already provide Position, Navigation and Time(PNT)services in the Asia-Pacific area at present. It is expected that a global system including 35 satellites will be built in 2020 to achieve global coverage. With the construction and steadily improvement of the Beidou system, its application in various fields is also being carried out. Because of its high-precise time reference and synchronization, the global navigation satellite system (GNSS) can realize the positioning, navigation, and timing functions precisely. Every GNSS has independent time reference system, called GNSST, and it is traced to the national official time UTC(k) whose divergence with respect to the Coordinated Universal Time (UTC) has been kept stable within several ns in most time. Based on the synchronization and unification of GNSST and international standard time UTC, its time service function is realized. The BeiDou system reference time, BDT, is traced to UTC through UTC (NTSC) which is maintained by National Time Service Center (NTSC), Chinese Academy of Science. In order to promote the development and application of BeiDou, NTSC implemented the BeiDou international common-view time comparison and got the preliminary result. It is proved that BeiDou can realize the long baseline precise time transfer. In this chapter, BeiDou system, BeiDou time international cooperation, the relationship between BeiDou time and UTC, development of time service of China, and the application of information system in time service work are introduced.

Keywords BeiDou · Information system · Time transfer · UTC · Navigation and positioning · Time service · Traceability

S. Dong (✉) · W. Wu · S. Zhang
National Time Service Center, Chinese Academy of Sciences, Xi'an, Shaanxi, China

S. Dong · W. Wu · S. Zhang
Key Laboratory of Time and Frequency Primary Standards, Chinese Academy of Sciences, Xi'an, Shaanxi, China

S. Dong · S. Zhang
School of Astronomy and Space Science, University of Chinese Academy of Sciences, Beijing, China

© Publishing House of Electronics Industry 2020
China's e-Science Blue Book 2018,
https://doi.org/10.1007/978-981-13-9390-7_10

1   Introduction

Time is the important strategic parameter and resource of the country. Precision time is the basic physical parameter of scientific research, experiment, and engineering technology, and it provides the necessary time coordinate for all dynamic systems and time series process measurement and quantitative research [1]. The symbol of modern timekeeping (generation and maintenance of standard time) is the establishment of atomic timescale based on atomic second and the widespread use of atomic clocks in timekeeping work. With the improvement of the performance of atomic clocks and the emergence of long-distance and high-precision time comparison techniques, the calculation methods of the International Atomic Time (TAI) are continuously improved [2]. With the development of natural science such as astronomy and physics, they have increased requirements on the stability of timescale. In the half past century, the timescale with practical applications has been from the Universal Time (UT) generated by the Earth's rotation with a stability of 10^{-8} (1 day) to the current Atomic Time (AT) with a stability of 10^{-15} (1 day) or even higher, which contains not only a qualitative change in the transformation of the UT to the TAI, but the atomic frequency itself increases by an order of magnitude almost every 7 years, which is the most important development. Providing a highly stable time system for scientific applications is a common task for all time laboratories in the world. At present, the most accurate atomic clock is from the initial accuracy of 10^{-11} of microwave clock to the strontium optical clock of 10^{-19} [3], and the long-distance comparison method has been developed from the LORAN-C system of microseconds level to the current two-way satellite time and frequency transfer (TWSTFT) and GNSS precise point positioning (PPP) of sub-nanoseconds, the comparing precision increased by 10,000 times [4].

The existing global satellite navigation and positioning systems such as GPS, GLONASS, Galileo, and BeiDou all adopt the time measurement strategy. GNSS is a typical example of using precise timing for positioning. Each GNSS satellite is equipped with an on-board atomic clock. Usually, the GNSS system time (GNSST) is obtained by integrated processing of the ground and the satellite atomic clocks [5]. Modern global navigation satellite system based on time measurement not only has the ability of navigation and positioning, but also the precise timing function, namely PNT. Each GNSS system has an independent internal time reference system called system time, named GNSST. Such as the GPS system time is GPST, GLONASS time is GLONASST, Galileo time is GST, and the BeiDou system time is BDT. GNSST is the basis for the coordinated operation of the entire system and the precise synchronization of the navigation signals transmitted by the satellites. At present, most GNSS system time is a uniform atomic timescale for continuous operation, no leap second, only the Russian GLONASS system uses Coordinated Universal Time (UTC) as the reference, and coordinate with UTC by inserting leap second operation. The GNSST must firstly be traced to its national official time UTC(k), where k is the code name of the national timekeeping laboratory in the international atomic time system, which is achieved by routine comparison with international standard time UTC, to realize synchronization and unification of the global time. The International Telecommunications Union (ITU) regulations require that the difference between UTC(k) and UTC must be within $\pm$ 100 ns. At the same time, the establishment times of each GNSS system are different, and the start time and epoch of the GNSST are different, which causes a slight time differences between the GNSSTs, so that the time of each navigation system is different. Therefore, the compatibility and interoperability of GNSST is also one of the research hotspots in GNSS currently.

2 Timekeeping and Atomic Timescale

2.1 From Universal Time to Atomic Time

Time is the physical quantity that achieves the highest measurement accuracy among all physical quantities. The measurement of other physical quantities such as length and voltage can be converted to time and frequency to achieve more precise measurement.

From ancient times, the measurement and definition of time are based on the observation results of astrometry. The time obtained by astrometry is called the Astronomical or Universal Time (UT), and it is also called the earth clock, that is, the time measurement method based on the rotation period of the earth [6]. Until now, time and frequency are still belonging to the discipline of astrometry and celestial mechanics. Astronomical time based on astrometry measurements has played a huge role in human social activities and scientific technological progress. However, due to the instability of the Earth's rotation period, the measurement accuracy of the UT is not high enough to meet the needs of modern science and technology. Therefore, the atomic frequency standard based on quantum physics appeared after the 1950s. In October 1967, the 13th General Conference on Weight and Measures (CGPM) officially defined the atomic second determined by the atomic clock as the international standard second, to replace the definition of astronomical second. The atomic second is defined as: the duration of 9,192,631,770 periods of the radiation corresponding to the transition between the two hyperfine levels of the ground state of the cesium 133 atom. At present, two sets of time measurement system, the atomic time and the astronomical time, are used globally, and the international legal standard time, Coordinated Universal Time UTC, is the product of the coordination of the two timescales.

UTC represents two timescales: the combination of international atomic time TAI and world time UT1, which takes the second of atomic time and is close as possible to UT1 at the moment. UTC is defined as:

$$UTC(t) - TAI(t) = N \text{ seconds } (N \text{ is an integer}),$$
$$UTC(t) - UT1(t) < \pm x \text{ seconds } (x < 1)$$

where t represents the moment, and N is the number of "leap second." The last leap second adjustment was on December 31, 2016. The difference between UTC and TAI is 37 s, that is, TAI-UTC = 37 s, UTC is 37 s behind TAI, it reflects the trend and degree of the long-term slowdown of the earth's rotation.

2.2   Calculation of Coordinated Universal Time

The Coordinated Universal Time (UTC) is a "paper" time. In order to enable users to obtain real-time physical time signals, time centers use atomic clocks to establish their own timekeeping systems refer to the establishment of UTC. By establishing a local time system and UTC traceability relationship, they can obtain a stable local Coordinated Universal Time which is the local realization of UTC, and it is generally recorded as UTC(k). The BIPM integrates the data of the atomic clocks of all the time laboratories worldwide, and uses the weighted average algorithm to calculate the international atomic time scale, so that its frequency stability, accuracy, and reliability are better than any of those generated by a single clock. The production process (Fig. 1) of UTC are as follows:

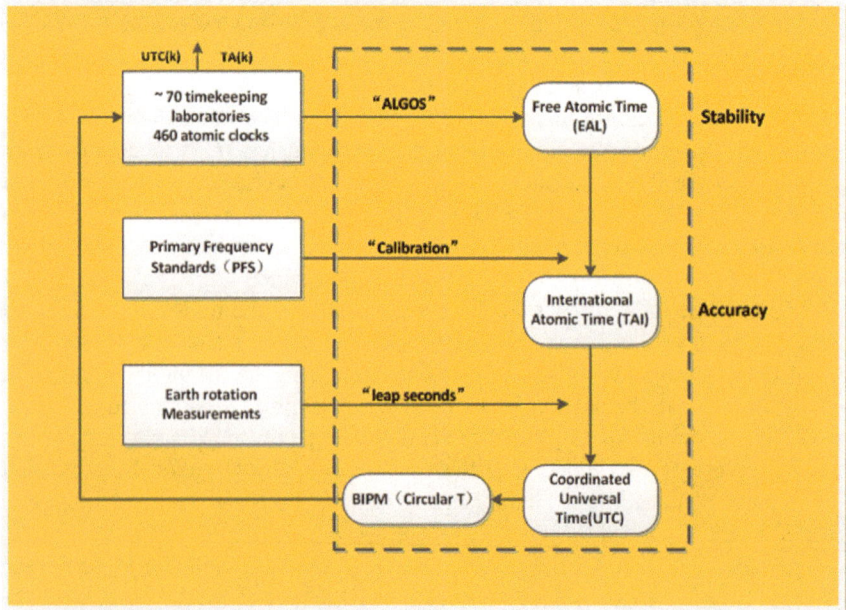

Fig. 1 Schematic diagram of the international time system

(1) Calculation of free atomic timescale EAL. Utilizing the weighted average of about 460 clocks (as of September 2017) participating in TAI cooperation, the weight of each clock mainly depends on its long-term stability of, thus ensuring the long-term stability of the EAL;

(2) Frequency calibration. BIPM obtains the weighted average of the frequencies of primary frequency standards of several time laboratories (after general relativity and blackbody radiation correction) through international time comparison, comparing it with EAL and make frequency calibration for the frequency of EAL to obtain TAI;

(3) Leap second adjustment. While calculating the TAI, BIPM determines the leap second according to the difference between UT1 and UTC provided by the International Earth Rotation Service (IERS), thus obtained the Coordinated Universal Time (UTC).

Local Coordinated Universal Time UTC(k) is usually output from an actual clock, either with frequency correction or without frequency correction. It is the basis for the dissemination of national official time, the generation and maintenance of UTC(k) is the top priority of national time laboratories.

2.3 Progress of China's Official Time System UTC(k)

The UTC (CSAO) system was built for Chinese terrestrial long- and short-wave radio time transmission stations in the late 1970s. CSAO stands for the Shaanxi Observatory of the Chinese Academy of Sciences (CAS) and it is the predecessor of National Time Service Center (NTSC) of CAS. According to the 1980 Bureau International de L'Heure Annual Report (that is 1980 International Annual Report of Time), there were 45 timekeeping laboratories participating in the generation of TAI and UTC in the world. At that time, the UTC (CSAO) was consisted of three Cs clocks and two hydrogen masers. The atomic clock data have officially participated in calculation of TAI, and the international time link uses Loran-C Northwest Pacific Chain (9970-Y). In 1987, the BIH Annual report has officially published the data of Chinese Joint Atomic Time Commission (JATC). At that time, the JATC system was consisted of Shaanxi Astronomical Observatory (CSAO), Shanghai Astronomical Observatory (SO), Beijing Astronomical Observatory (BAO), Wuhan Institute of Physics (WTO), and Beijing Institute of Radio Measurement and Research (BIRM), totally 22 atomic clocks were jointed together to generate UTC (JATC). It is the first example to realize the sharing of atomic clock resources and the joint timekeeping of laboratories located in different places in China. In the early 1990s, CSAO took the lead in using the most advanced long-distance time comparison technology, and CSAO's GPS common-view comparison data were officially used in TAI.

Over the past 40 years, UTC (NTSC) has been running continuously, steadily and reliably, its performance has been continuously improving, and its role has been continuously expanding. At present, UTC (NTSC) is one of the most important timekeeping systems in the world, the weight contribution in the calculation of international atomic time has been ranked the top five in the world, making important contributions to international time work, as shown in Figs. 2 and 3. At the same time, UTC (NTSC) has made an outstanding contribution to the time service work of China, which satisfies the needs of the national economic construction and the scientific research. Entering the new era, UTC (NTSC) also played an important role in the construction and development of Chinese BeiDou satellite navigation system in the test verification, performance evaluation, and operation assistant of BeiDou system.

2.4 Application of Information System in Timekeeping

Informatization and big data are important features of modern time service work. Timekeeping is the process of generating and maintaining time reference signals and information. It involves the process of operation and maintenance and management of atomic clock assembly, process management of measurement and comparison, atomic time data processing, and time information transmission to time transmitting stations and time signal delivery terminals, each of which is highly informative, networked, and intelligent.

In order to realize high accuracy and long-term stability for the atomic timescale obtained, it is usually necessary to analyze the data measured for months, years, or even more. The definition of the atomic second is determined by combining the observation data of the earth's rotation for hundreds of years. As the responsible agency for time service in China, NTSC established information systems and facilities such as local atomic time professional LAN, time and frequency science database and time scientific data service platform, in order to provide better service

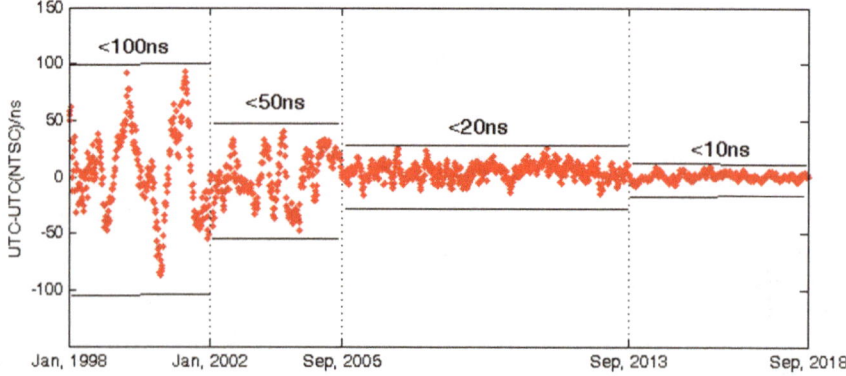

Fig. 2 Maintenance of China's official time in the last 20 years

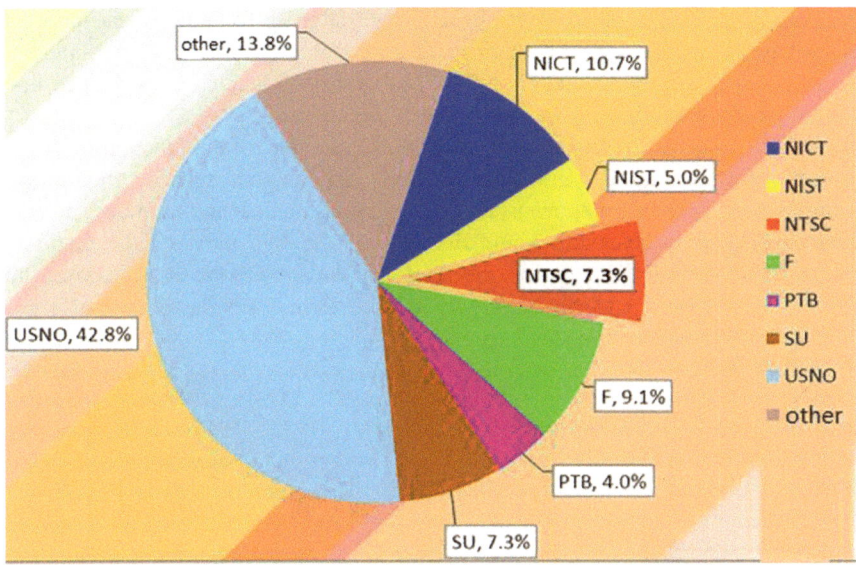

Fig. 3 Weight contribution of timekeeping laboratories to the TAI calculation in 2016

for time service, Fig. 4 is a specialized atomic time scale LAN structure diagram of the NTSC.

In view of the role of scientific data in scientific research and social progress, scientific institutions and organizations both at home and abroad have paid more and more attention to the construction of scientific database, the sharing of scientific data, and the application of scientific data. As the most authoritative natural science research institution in China, the Chinese Academy of Sciences also attaches great importance to the research and establishment of scientific databases. The Chinese Academy of Sciences proposed the construction of scientific database and information system in 1983. The time and frequency science database aims to organize the important time and frequency data and information resources accumulated for

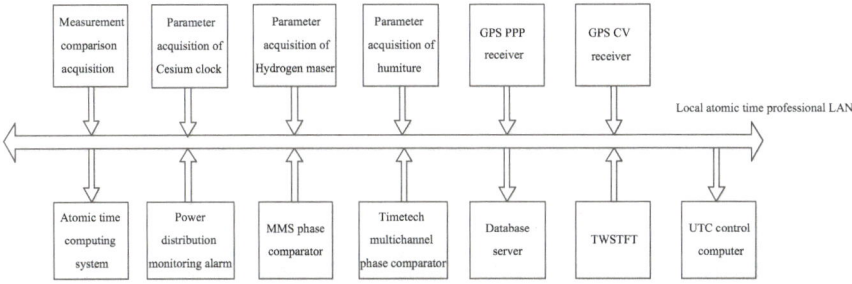

Fig. 4 Atomic time local area network of NTSC

years, and combine with computer technology, database technology and network technology to provide support, sharing and service for scientific research, national economic construction, and so on. NTSC has built and operated a specialized time and frequency science database, as shown in Figs. 5 and 6.

The establishment of the database mainly takes account of the functions of login, data storage, data access, graphical display, and network security monitoring. Figure 5 is the internal components and workflow of the time and frequency science database. Figure 6 gives the frame diagram of the five parts of the database, including the presentation layer, the user authentication, the data service, the internal management, and the underlying module. Figure 7 is the main interface of the network system for time science data service.

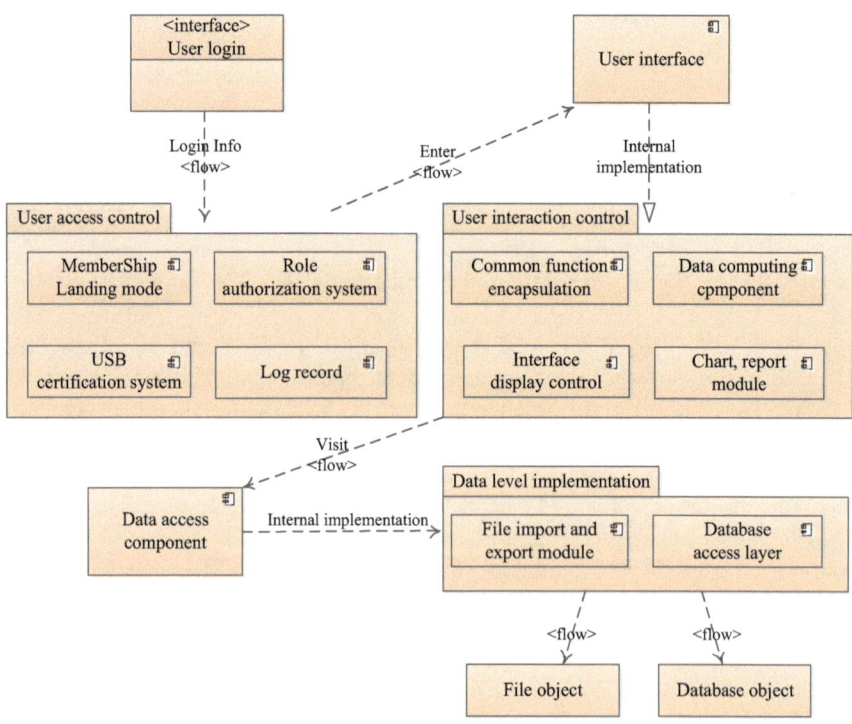

Fig. 5 Internal components and workflow of time and frequency science database

Expressison layer	Ajax page without refresh mode		RIA rich client technology	
User authentication	Role based user rights management	DES password encryption	UsbKey equipment high strength safety	
Data service	Atomic clock state data	Hydrogen maser data	Cesium clock data	
	Comparison data	Integrated atomic time comparison data	Local atomic time comparison data	
		TWSTFT comparison data	GPS data	
	BIPM interactive data	Atomic time calculation result	Atomic clock rate	
		Atomic clock weight	BIPM monthly interaction	
	Bulletin data	Bulletin of earth rotation	Time and frequency bulletin	
Internal management	Device management	Device information	Device state	
	User management	User management	Management of scientific research institution	
		Role management	Module management	
		Access log	Operation log	
	System configuration	Data backup management	Opcration state management	
		System parameter configuration	Atomic clock dictionary	
		National area dictionary	Data dictionary	
Underlying module	File import and export module	Data access module	Logging module	

Fig. 6 Structure of time and frequency science database

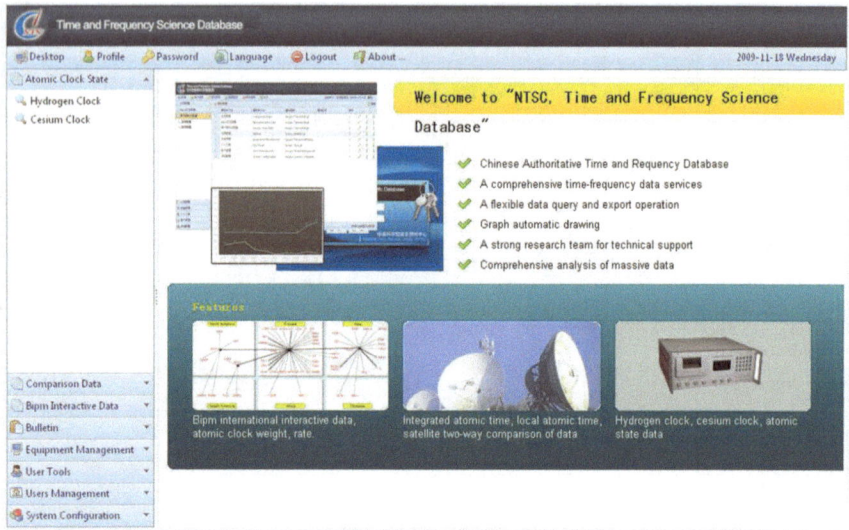

Fig. 7 Time science data service platform

3 Dissemination and Application of Standard Time and Frequency Signal

3.1 Time Service Techniques and Their Application in Various Fields

Standard time and frequency signals can be transmitted in space by multiple techniques and can be directly received and used by users. Modern time service technology mainly includes long and short radio wave, time code, communication satellite, GNSS, network, telephone, and television.

Modern terrestrial-based time transmission technologies mainly include short-wave timing technology, long-wave timing technology based on Loran-C system, and low-frequency time code technology. Short-wave timing is the earlier time transmitting technology to transmit standard time and standard frequency signal by radio signal, its timing precision is in milliseconds level. Because of the wide coverage of the short-wave timing signal, the simple transmission equipment, the low price of the receiver and the convenient use, it is still used in many fields where not very high timing precision is required. Long-wave timing based on Loran-C system has the characteristics of stable propagation and accurate prediction. Long-wave timing has two modes: ground wave and sky wave, and the timing precision of ground wave is better than 1 microsecond. It is the only terrestrial radio timing technology that can reach the magnitude of microseconds at present. Low-frequency time code timing technology is recommended by the International Telecommunication Union (ITU). It works in low-frequency band and can provide standard time and frequency signals in analog and digital simultaneously. The technology makes full use of microelectronics technology, which enables users to make equipment very simple and inexpensive, and has been widely applied in many fields.

Telephone time service provides standard time signals to users via consulting mode. The user dials the telephone timing server through a modem, after receiving the request of the user computer, and the server sends the standard time information (time code) to the user and completes the time service. At present, the timing precision of the telephone timing service is in millisecond level. Network timing is an important way to provide accurate computer network time services, enabling computer clocks to be synchronized to standard time. Network timing is based on Network Time Protocol (NTP), which provides timing precision of 1–50 ms which depends on the synchronization source and network path. Television timing evolves from analog to modern digital television timing (DTV). The digital television timing technology realizes the timing by using the DTV signal. The digital television timing precision is better than 60 ns.

Satellite time service is the most widely used and the most accurate timing technique. It is generally processed in conjunction with the positioning. The receiver receives multiple satellite spread signals at the same time and obtains multiple pseudo-range measurements to obtain the real-time satellite ephemeris. The user position information and local clock difference information are obtained through the navigation and positioning equation. The receiver traces the receiver time to the UTC by the difference between the GNSS system time and the UTC, to complete the real-time timing function. Satellite timing has the advantages of wide coverage and high precision than terrestrial radio technique. This is because a satellite can cover a large area of the ground while a satellite is on the high space orbit. At the same time, the path of satellite signal is relatively simple compared with terrestrial radio signal. On the other hand, long-distance and high-precision time comparison can be achieved by observing the same satellite on both sites of the ground.

Precision time has been widely used in many fields of modern science and technology, such as satellite launch and electronic reconnaissance, which require billions of molecules per second of precision, and power transmission requires millions of molecules per second precision time. Figure 8 shows the application of China's standard time signal in important national infrastructure. Different fields have different requirements for the accuracy and stability of time and frequency, as shown in Table 1.

3.2   Application of Information System in Time Service

Time service is the process of transmitting legal standard time–frequency signals and information to users. According to different application needs, modern time service includes a variety of timing techniques such as satellite-based, terrestrial,

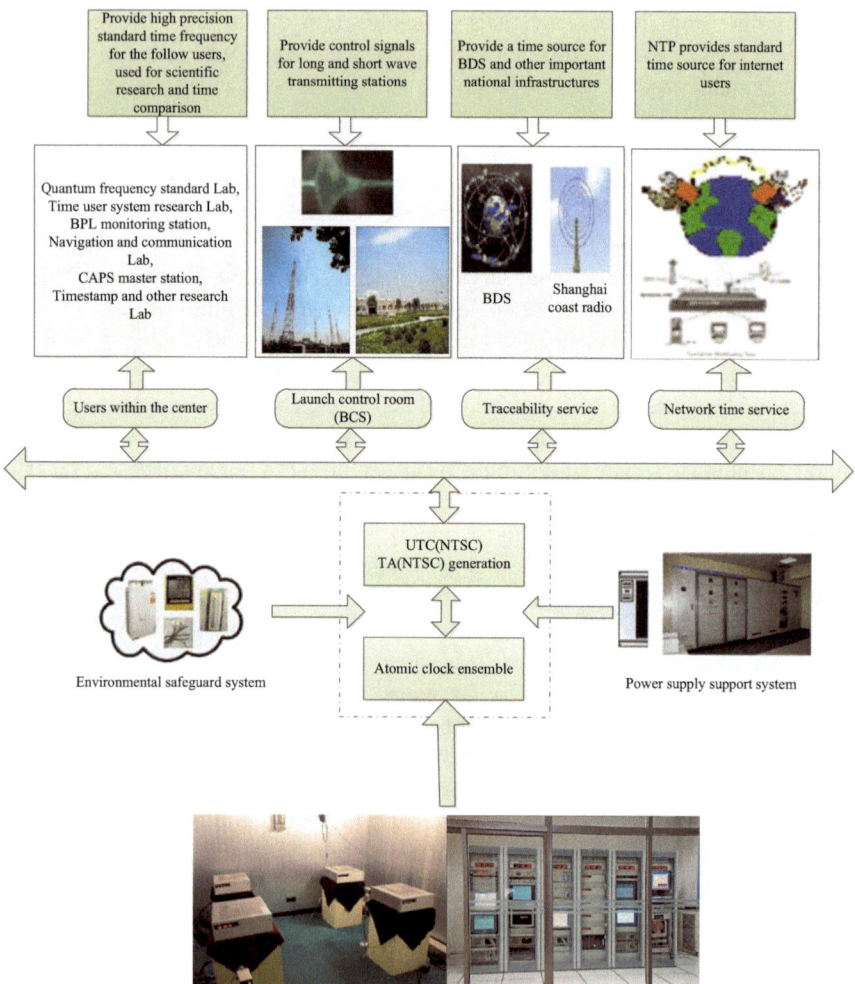

Fig. 8 Applications of time signal

Table 1 Requirements for time and frequency of different applications

	Time accuracy	Frequency stability
Satellite navigation	± ns	$\pm 2 \times 10^{-13}$(1 day)
Electronic reconnaissance satellite	± 10 ns	$\pm 5 \times 10^{-13}$
Cruise missile	± 50 ns	$\pm 5 \times 10^{-13}$
Satellite orbit determination	± 50 ns	$\pm 1 \times 10^{-12}$
High-speed digital communication network	± 0.5 us	$\pm 5 \times 10^{-12}$
Power transmission grids	± 1 us	$\pm 1 \times 10^{-11}$
TV frequency calibration		$\pm 5 \times 10^{-12}$

telephone and network timing. Informatization plays an important role in time service. Satellite timing is to use the communication and navigation satellites to achieve high-precision timing services through space radio waves. Its time information is generated in the ground control station, and the time information generated by the main control station uploads satellite through the satellite data chain and then sends the message to the user with the precise time information through the satellite signals. In terrestrial radio (long- and short-wave time system, Loran system), due to the strong magnetic field of the ground, the time reference system and the control center are usually built in different places. In order to ensure the safe and reliable transmission of time data, virtual private network (VPN) and other specialized networks are built between the control center and the time service stations to transmit time signals and information. Telephone and network time service is a process of realizing time service by using Internet and common communication system as transmission medium. Figure 9 is the informatization of China's time service system.

Along with the arrival of the information age in the twenty-first century, the requirement for precise time is getting increasingly higher. The Internet of Things, intelligent transportation, accurate logistics and distribution, high-speed digital communications, aerospace, and major scientific experiments all depend on precise time and time synchronization. Figure 10 is the information and time system of integration of land-based and space-based. The standard time and frequency signal and information are produced in the base layer. The support layer includes the performance monitoring and environmental monitoring of timing system. The system layer includes the satellite-based, ground-based radio time system and various enhancement systems. The integration system itself is a huge information system. There are various network and communication systems between each part to ensure the real-time, reliable communication and the reliable operation of the whole system.

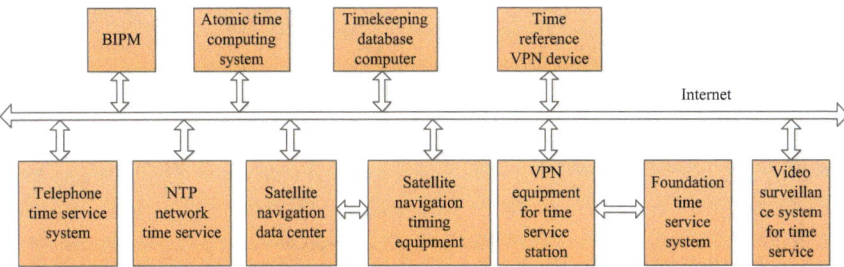

Fig. 9 Informatization of China's time service system

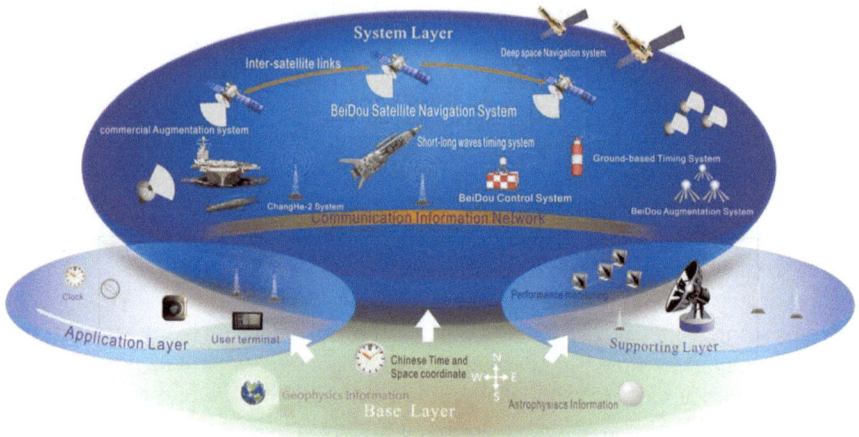

Fig. 10 Time service and informatization system of integrated terrestrial and space

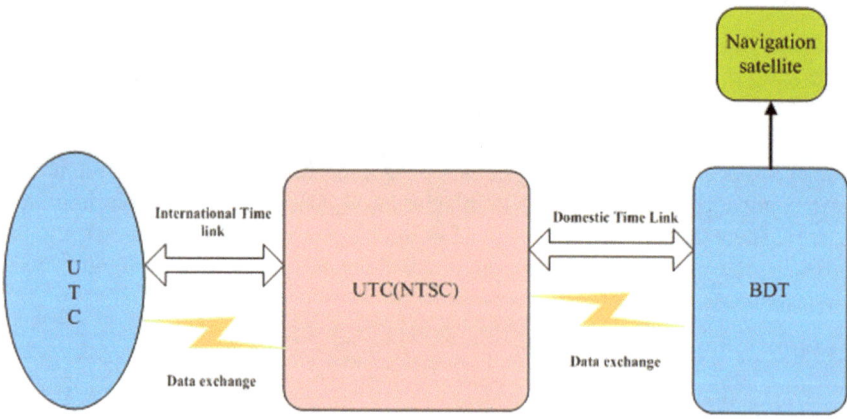

Fig. 11 Traceability of BDT

4   BeiDou Time System

The construction of the second-generation of Beidou satellite navigation system is divided into three stages. From the first stage of 2000–2012, it is called the Beidou verification stage, which provides satellite radio positioning service and short message service for China and its surrounding users. Since 2012, the Asia-Pacific regional navigation has been gradually realized. The satellite constellation is composed of five GEO satellites (58.75 E, 80 E, 110.5 E, 140 E, and 160 E), five IGSO, and four MEO satellites with 55°. In the third stage, the Beidou system satellite will be increased to 35, its positioning accuracy will be better than 10 m, and the timing accuracy will be better than 20 ns [7].

The BeiDou system time, called BDT, was established and maintained by the BeiDou ground control center, which traced to the national official time UTC (NTSC) maintained by NTSC [8], as shown in Fig. 11. The BDT adopts the continuous international atomic time and it is broadcasted through the BeiDou navigation message. Its initial epoch is UTC 00h00m00 s on January 1, 2006 (Sunday). Since BDT is consistent with the UTC at the time in 2006, till June 30, 2018, the difference with TAI and UTC is:

$$[\text{UTC - BDT}] = -4\text{s} + C2, \quad [\text{TAI - BDT}] = 33\,\text{s} + C2$$

where C2 is the difference between BDT and UTC on the second (C2 has not been published in BIPM Circular-T).

The BeiDou system time (BDT) is realized by the jointed "paper time," and is generated and maintained through the master clock of the control station. The atomic clock used for BDT timekeeping is mainly hydrogen atomic clocks. The BDT algorithm is optimized to form a good performance paper timescale, and the frequency deviation, frequency drift, and frequency stability of the free running of the atomic clock are fully considered. The method of weighting in atomic time mainly adopts the Allan variance and the maximum weight restriction that deducts the slope of the clock. In order to keep consistent with UTC, the BDT adjusts the deviation at appropriate time by frequency control, and the adjustment will not exceed 5E^{-15}.

5   International Cooperation on GNSS Time

With the development and application of multi-satellite navigation systems, GNSS interoperability has become one of the important topics. When multiple satellite navigation systems work simultaneously, they will interfere with each other and affect navigation and timing performance. Interoperability will enable multiple navigation systems to be shared, which can achieve better service than using a single system alone without significantly increasing the cost of use. The GNSS time interoperability can better utilize the precise timing function of the GNSS system and realize the international coordination. Based on UTC (NTSC) system, the GNSS time cooperation between China and Russia, China and Europe, and China and the USA can provide important technical support for the compatibility of BeiDou system with other GNSS systems. In the international interoperability of GNSS time system, the time interoperability between Chinese BeiDou and Russian GLONASS is the top priority of the international cooperation for the BeiDou. The interoperability cooperation between GPS and European Galileo has been carried out for many years.

Time service systems and satellite navigation systems are all important infrastructures of a county. In September 11, 2014, President Xi Jinping met with Russian
President Putin, and he emphasized "the cooperation of satellite navigation system
has been started." The importance of the strategic cooperation of satellite navigation
is fully showed by the attention of the leaders of the two countries. In January 27,
2015, both China and Russia held the first meeting of the China–Russia strategic
cooperation on satellite navigation in Beijing, then signed the cooperation agreement, and made clearly the practical cooperation between China and Russia as the
breakthrough point to carry out the two countries' satellite navigation. Both China
and Russia are important members of the TAI. The Russian national official time
UTC (SU) is the reference for GLONASST and UTC (NTSC) the reference of the
BDT. Promoting national official time cooperation between China and Russia is of
significance for the operation of GNSS system. China–Russia satellite navigation
interoperability is an important part of cooperation between the two countries. On
behalf of BeiDou, the NTSC has actively carried out international cooperation in
GNSS time system, especially in China and Russia. According to meetings of the
China–Russian Satellite Navigation International Cooperation Committee and the
China–Russian Compatibility and interoperability working group, we have carried
out: (1) establishment of time comparison links between UTC (NTSC) and UTC
(SU); (2) regular exchange of clock data; and (3) joint calculation of the time
difference between GLONASST and BDT. The cooperation between UTC (NTSC)
and UTC (SU) will provide the necessary technical support for the interoperability
of BeiDou and GLONASS.

6   The Euro-Asian International Time Comparison Based on the BeiDou

With the construction and development of the BeiDou system, NTSC has made the
time comparison with Royal Observatory of Belgium (ORB), Real Institute y
Observatory de la Armada (ROA), Research Institutes of Sweden (RISE or SP) and
Physikalisch-Technische Bundesanstalt (PTB) via BeiDou common view, the
precision is 2–4 ns under the condition that the Beidou system constellation is still
not perfect (at present, only about four BeiDou satellites can be observed in
Europe). The precision of BeiDou common view is a little lower than or at about the
same level compared to that of GPS. With the development of BeiDou system, it
will provide the global open service. The better performance of BeiDou time
comparison could be expected and we also look forward that it could contribute to
the generation of UTC/TAI in next few years. Table 2 is the result of the common
view between NTSC and the four European timekeeping laboratories in 2017.

Table 2 Preliminary results of China-EU common-view time comparison based on BeiDou

Link	Comparison method	Standard deviation (ns)
China–Germany	BeiDou common view	2.25
China–Sweden	BeiDou common view	2.90
China–Belgium	BeiDou common view	2.03
China–Spain	BeiDou common view	4.06

7 Summary

In June 2016, the State Council issued the white paper on China BeiDou satellite navigation system, summarizing the three steps of the BeiDou satellite navigation system: At the end of 2000, the first-genaration BeiDou system was built to provide services to China. At the end of 2012, the second-genaration BeiDou verification system was built to provide services to the Asia-Pacific region. Before and after 2020, the BeiDou global system was built to provide services worldwide. At present, the state has decided to speed up the development of the BeiDou global system and will initially have the ability to provide global services to meet the development needs of the "The Silk Road Economic Belt and the twenty-first-Century Maritime Silk Road" by 2018. In order to achieve the goal of global service, 35 BeiDou satellites are needed in orbit. In the next 2–3 years, China will launch more than 10 BeiDou navigation satellites. At that time, the BeiDou system will provide all-weather, high-precision positioning, navigation, and time service for global users and become a truly global system.

With the completion of the BeiDou system, it will play an increasingly important role in the field of time service. In addition to meeting the needs of China's national economic development, it will also play an important role in international time cooperation and global time unification. NTSC will continue to play a leading role in the construction, development, and development of the BeiDou time system, and this will make it more necessary to promote the application of informatization in time service filed.

References

1. Dong Shaowu, Qu Lili, Yuan Haibo, et al. Outstanding time keeping activities. Journal of time and frequency, 2016, 39(3):129–137.
2. Gianna Panfilo. The coordinated universal time. IEEE instrumentation & measurement magazine, 2016, 19(3):28–33.
3. S.L. Campbell, R.B. Hutson, G.E. Marti, et al. A Fermi-degenerate three-dimensional optical lattice clock. Science, 2017, 358(6359):90–94.
4. Wu Wenjun, Zhang Hong and Guang Wei, et al. The two-way satellite time and frequency transfer by AM22. Journal of time and frequency, 2017, 40(3):155–160.

5. Shaowu Dong, Haitao Wu, Xiaohui Li, et al. The Compass and its time reference system. Metrologia, 2008, 45(6):S47–S50.
6. Dong Yiwen. Time and GNSS. Modern Navigation, 2015, 2:150–152.
7. Yang Y X, Li J L, Wang A B, et al. 2014. Preliminary assessment of the navigation and positioning performance of BeiDou regional navigation satellite system. Science China: Earth Sciences, 2014,44(1):72–81.
8. BeiDou Navigation Satellite System Signal in Space Interface Control Document. Beijing: China Satellite Navigation Office, 2017.

Shaowu Dong received his Ph.D. degree from University of Chinese Academy of Sciences in 2007, and the major is Astrometry and Astromechanics. Now, he is Professor and Ph.D. Supervisor in National Time Service Center of CAS. His main research interests include timekeeping techniques, GNSS time system, etc.

Efficient Acquisition of Geographic Big Data: Domestic Three-Line Stereo Aerial Photography System

Lina Zheng, Guoqin Yuan, Yongming Yang and Haipeng Kuang

Abstract As one of the main approaches to acquire geographic data, aerial remote sensing is of great significance to the development of the geographic information industry in the era of big data. By taking the domestic three-line stereo aerial photography system as the main research object, this chapter analyses the information application demands for the development of aerial photography system by geographic big data acquisition, introduces the development status of aerial remote sensing equipment, describes the working principles and technical indexes of three-line stereo aerial photography system, and discusses the information application prospect of the aerial photography system in aspects such as digital urban construction, earthquake relief works, resource aerial remote control, environmental protection, and homeland security.

Keywords Three-line stereo aerial photography system · Geographic big data · Information application prospect · Independent intellectual property rights

1 Introduction

From 2011 to 2014, big data emerging technologies developed from the burgeoning period to the overheating period and the low period. Countries all over the world paid great attention to the development of big data. In August 2015, China issued the "Outline for Promoting Big Data Development" (hereinafter referred to as "Outline") and officially implemented the big data development strategy. The "Outline" defines the goals and main tasks in the development and application of big data in China in the next 5–10 years and proposes the top ten projects. The issuance of the "Outline" marks the arrival of the era of big data in the development of China's geographic information industry.

L. Zheng (✉) · G. Yuan · Y. Yang · H. Kuang
Changchun Institute of Optics, Fine Mechanics and Physics, Chinese Academy of Sciences, Beijing, China

China's e-Science Blue Book 2018,
https://doi.org/10.1007/978-981-13-9390-7_11

In geographic big data, data acquisition is the key component [1]. One of the main methods to obtain geographic data is aerospace ground observation. The acquisition of geographic big data is greatly supported by ground observation equipment. Ground observation equipment can be categorized as space cameras, aerial cameras, and ultra-low-altitude unmanned aerial vehicles (UAV). Spaceborne cameras can acquire global ground images in orbit. However, these cameras have limited size, volume, focal length, weight, and orbital height, the ground resolution is relatively low, revisit time is long, and real time and manoeuvrability are limited. They are unable to meet the needs of emergency observation of hotspots. Ultra-low-altitude UAVs are low cost and flexible which are suitable for small areas and areas which are difficult to reach by flight. But the area measured in unit time is limited; mapping is less efficient. Ultra-low-altitude UAVs can only be used as a supplement to efficient photogrammetry technology. Aerial cameras can be loaded on general aviation platforms such as Yun-5 and Yun-12, with a flying altitude of 500–5000 m. The high-resolution ground image data can be obtained quickly and flexibly. Aerial cameras have high efficiency, flexibility, high precision, and wide application range, which are more suitable for the acquisition of geographic information in the era of big data. The research and application of aerial mapping cameras are of great significance for the efficient acquisition of geographic big data.

Since 2000, different types of aerial mapping cameras have been developed and applied in the mapping industry in China and oversea, opening up a new era of digital mapping cameras instead of film aerial cameras. Representatives of foreign aerial cameras are DMC, UltraCam, ADS40/80, etc. [2]. The development of aviation mapping camera in China is delayed. The main technical specifications still have a big gap comparing to these of foreign advanced aerial cameras. High-precision digital aerial cameras rely mainly on imports. Therefore, based on the status of China's civil aviation photography, the national high-resolution special (civilian part) issued a guideline on the development of a large field of view three-line stereo aerial camera. The camera uses a civil aviation aircraft as a working platform. It is a multi-functional and digital aerial remote sensor which can simultaneously obtain panchromatic stereoscopic and colour image. It has high precision, high resolution, and wide coverage. Flying height is 2000 m supporting 1:1000 scale. It implements the ability of emergency aviation ground observe mission and meets the application requirements of natural disaster monitoring, urban refinement management, resource survey, and geographic mapping for high-resolution 3D image data, etc.

2 Development of Aviation Remote Sensing Equipment

According to the imaging method, digital aerial mapping cameras are two types: frame type and push-broom type. Frame-type camera has simple data processing flow. It can provide more redundant observations for solving, but its base to height ratio is relatively small, elevation accuracy is low, and the operation requires a high

Fig. 1 Appearance of the
DMC camera

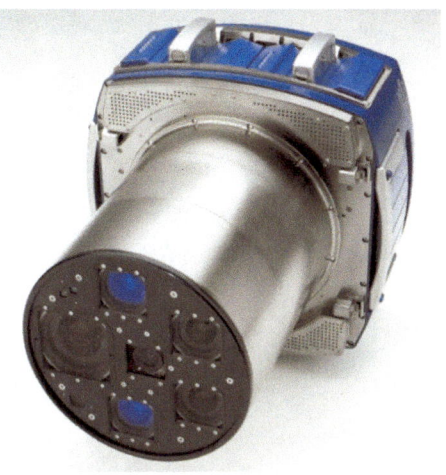

overlap rate, resulting in a large volume of the image data file and long processing time. Push-broom-type camera can ensure strict central projection [3], continuous pixel display, no need for image stitching, no shutter mechanism, high camera reliability, high base to height ratio, and higher elevation precision.

The representative products of the aerial frame-type mapping camera are DMC and UltraCam series cameras. The appearance of the DMC camera is shown in Fig. 1. The main technical specifications are shown in Table 1. The outline of the UltraCam camera is shown in Fig. 2. The main technical specifications are shown in Table 2.

The most advanced foreign push-broom mapping camera is the ADS40/80 camera. The outline of the camera is shown in Fig. 3. New large field of view three-line stereo aerial camera has been developed in China, as shown in Fig. 4. The main technical specifications of a three-line stereo aerial camera and ADS40/80 are compared in Table 3.

Table 1 Main technical specifications of DMC camera

Camera	DMC-II 140	DMC-II 230	DMC-II 250
Pixel number	12 096 × 11 200	15 552 × 14 144	16 768 × 14 016
Pixel size (μm)	7.2	5.6	5.6
Focal length (mm)	90	92	112
Field of view	50.7° × 43.3°	50.7° × 46.6°	45.5° × 38.6°
GSD@500 m	3.9 cm	3.0 cm	2.5 cm
Imaging channel	Panchromatic, RGB, NIR	Panchromatic, RGB, NIR	Panchromatic, RGB, NIR

Fig. 2 Outline of the Ultracam camera

Table 2 UltraCam camera main technical specifications

Camera	UltraCam-XP	UltraCam-Xp WA	UltraCam-Eagle
Pixel number	17,310 × 11,310 (Panchromatic) 5770 × 3770 (Multispectral)	17,310 × 11,310 (Panchromatic) 5770 × 3770 (Multispectral)	20,010 × 13,080 (Panchromatic) 6670 × 4360 (Multispectral)
Pixel size (μm)	6	6	5.2
Focal length (mm)	100 (Panchromatic) 33 (Multispectral)	70 (Panchromatic) 23 (Multispectral)	80 (Panchromatic), 210 (Group 2 Panchromatic) 27 (Multispectral), 70 (Group 2 Multispectral)
Field of view	55° × 37°	73° × 52°	66° × 46° (Group 2:28° × 20°)
Imaging channel	Panchromatic, RGB, NIR	Panchromatic, RGB, NIR	Panchromatic, RGB, NIR

(a) ADS40 camera

(b) ADS80 camera

Fig. 3 ADS40/80 camera

Fig. 4 Domestic three-line
stereo aerial camera

From Table 3, it is seen that the large field of view three-line stereo camera fills the domestic blank, and its main technical index is several times superior than that of the foreign advanced push-broom camera. It represents the advancement of

Table 3 Comparison of main technical specifications of a three-line stereo aerial camera and ADS40/80

Specifications	ADS40/80	Large field of view three-line camera
Focal length (mm)	62.77	130
Pixel number	12,000	32,768
Pixel size (μm)	6.5 × 6.5	5 × 5
2 000 m distance pixel resolution (m)	0.207	0.077
Coverage width at the same resolution (0.1 m)	1206.9	3249.3

China's development of aviation push-broom camera, and it is of great significance for the efficient data acquisition in the era of geographic big data.

3 Composition and Working Principle of Domestic Three-Line Stereo Aerial Camera System

The large field of view three-line stereo aerial camera system is mainly composed of camera body, stable platform, control cabinet, and image data processing software. In order to achieve camera lightweight and fit the installed space, the camera adopts long focal length and large field of view single lens transmission optics. The system realizes stereo mapping by arranging multiple detectors with different angles in the focal plane of the camera.

The camera body includes an optical lens, a focal plane component, an IMU, a temperature control component, a power supply, a main control unit, a photoelectric conversion unit, etc. The camera optical lens adopts an optical system of a tele-centric image, where incident light rays in all directions are approximately normal to the focal plane, to minimize image distortion and spectral distortion. There are three panchromatic and one RGB detector integrated on the focal plane. The four detectors are arranged parallel to each other and perpendicular to the direction of flight. The imaging angle of three panchromatic detectors is different, where the downward vertical view is a downward view detector, the forward tilt image is a forward view detector, and the backward tilt image is a rear-view detector. The camera is equipped with a high-precision IMU to record the external orientation elements of each line of the image when taking picture. The temperature control component can effectively compensate the influence of environmental changes such as temperature and pressure on the performance of the camera and ensure the wide environmental adaptability of the camera. The aircraft power is filtered, converted, and supplied to the camera body. The main control unit controls the camera to take a photo according to the instruction. The photoelectric conversion unit converts the image data and outputs the image data to the storage unit.

The control cabinet includes the control cabinet power supply, the computer, POSPCS, and image storage unit, etc. The power of the control cabinet is filtered by the power supply of the aircraft and supplied to the components of the control cabinet. The control software of the man–machine interface is installed on the control computer. The operator can input the commands and working parameters and display the working state of the camera in real time. The POSPCS is the control system of the POS system which is used to manage the IMU and GPS output data and records the external orientation elements at the exposure moment of the camera. The image storage unit stores the image data output by the camera for post processing.

In order to reduce the impact of aircraft flight attitude changes on imaging quality, the camera uses a three-axis stabilization platform to compensate for the impact of flight attitude.

The working principle of the camera is shown in Fig. 5. When the camera is imaging, the line detectors on the focal plane continuously sample the ground and at the same time obtain a fully overlapping front-view panchromatic PAN band, a down-view panchromatic PAN band, and a rear-view panchromatic PAN. The band, as well as R, G, B single-band images, can directly generate multiple stereo and colour images. The integrated high-precision GPS and IMU provide observations of the external orientation elements for each scan line, reducing the dependence of the ground control points. The camera can perform large-scale mapping of ground targets without ground control points.

4 Efficient Acquisition Technology and Implementation Process of Geographic Big Data

The process of acquiring geographic information in the large-field three-line stereo aerial camera system is shown in Fig. 6. The system can generate 3D DSM, DOM, and other image products through flight planning, aerial photography, data download, aerial triangulation, image correction, editing, and mapping.

The process of acquiring geographic information data can be divided into three stages: first, the ground preparation work before the take-off of the aircraft, including the design of the mapping area, the setting of camera parameters, the setting of the flight mode, etc.; second, take-off and driving into the airway for aerial photography, data recording and storage at the same time; third, completing the aerial photography task and information processing of the acquired image.

Image information processing is the core of geographic data acquisition. The main processing processes include image pre-processing, aerial triangulation, DSM calculation, and DOM production, etc.

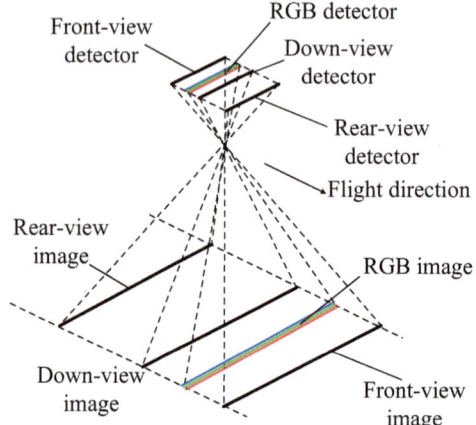

(a) Imaging process of the PAN and RGB detectors

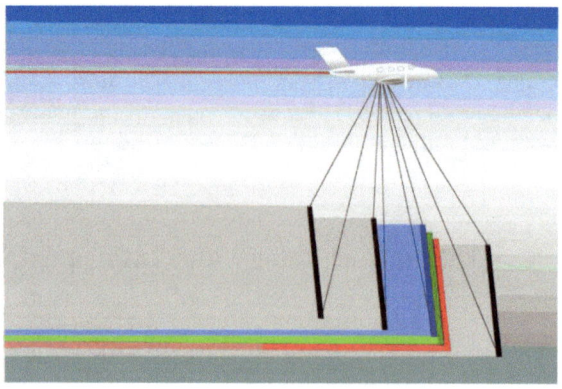

(b) Imaging process of the camera

Fig. 5 Working principle of the camera

Fig. 6 Process of acquiring geographic information

Image pre-processing is mainly to perform radiation correction, atmospheric correction, radiation enhancement, etc., to ensure the consistency of colour tone between image points of the same name.

Aerial triangulation is a key step in image processing. The purpose is to obtain the coordinates and elevation of the encryption points and the exterior orientation elements of the image based on a small number of ground control points. The aerial triangulation generates an air strip based on the POS data of the image, establishes the topological relationship between the images, automatically extracts the tie points in the image overlap range, and performs an operation on a large number of tie points by an iterative method to eliminate the rough points. Using the image point data of the field measurement as the starting data, the three-dimensional coordinates of the pass points and the precise exterior orientation elements of each image are finally obtained by bundle adjustment method.

The DSM calculation is based on the images acquired by the detectors of different viewing angles, and the binocular intersection is performed to obtain the elevation information of the ground point.

The true orthoimage map is based on the elevation information of the ground point and the camera attitude data and geometrically corrects the image acquired by the detector perpendicular to the ground to produce a usable map product.

According to the above production process, the information data products generated by the large field of view three-line stereo aerial photography system mainly include:

(1) Planc mapping remote sensing image products, including geometric precision correction, orthorectification correction, digital orthophoto map, etc.
(2) Stereo mapping remote sensing products, including DSM, DEM, etc.
(3) Rapid image production to meet the needs of users such as emergency rescue and emergency environmental monitoring
(4) Thematic mapping products, including geographical condition monitoring products and typical ground classification products.

5 The Status Quo and Prospects of Information Application

Three-line stereo aerial camera system has advanced features such as high resolution, high work efficiency, and large base to height ratio, high precision, and automation. It can meet the requirement of digital city construction, earthquake relief, resource navigation, environmental protection, and homeland security, etc. The demand for the aspect has broad prospects for information application and will usher in a new space for the geographic information market.

The product classification and users provided by the three-line stereo aerial camera system are as shown in Table 4, which can meet the information requirements of different users.

Table 4 Product classification and users provided by three-line stereo aerial photography system

Product	Grade characteristics	Users
Level 0 product	Only processed by imaging, provided to users who have their own radiation correction and geometric correction capabilities	Mainly distributed to the department of scientific research on camera imaging systems for the detection, calibration, and other processing of camera systems
Level 1 product	Radiation-corrected products based on level 0 products	It is mainly distributed to units and departments engaged in mapping production and scientific research, such as the state bureau of mapping, etc., for stereo mapping to obtain the required geographic, topographical features, spatial location of the target, geometric features, and related attribute data
Level 2A product	Based on the Level 1 product, the geometric rough correction has a certain precision, which is suitable for users who have no geographic information production capacity and has low requirements for positioning accuracy	It is mainly distributed to scientific research units and departments engaged in mapping production, forestry, land, disaster reduction, etc., for stereo mapping to accurately obtain the required geographic, topographical features, spatial location, geometric features and related attribute data of the target, and to map large areas with large scales (1:2 000, 1:1 000) digital topographic maps, digital orthophoto maps, digital elevation models, digital ground models, engineering drawings, city maps, traffic maps, and other special mapping products; land, forestry, disaster reduction, geology, minerals, transportation, water conservancy, and other departments for related applications
Level 2B product	Based on the 2A grade product, geometric precision correction and air strip splicing, with scale accuracy of 1:1 000, users with high positioning accuracy requirements	It is mainly distributed to scientific research units and departments engaged in mapping production, forestry, land, disaster reduction, etc., for stereo mapping to accurately obtain the required geographic, topographical features, spatial location, geometric features and related attribute data of the target, and to map large areas with large scales (1:2 000, 1:1 000) digital topographic maps, digital orthophoto maps, digital elevation models, digital ground models, and special mapping products such as engineering drawings, city maps, and traffic maps; land, forestry, disaster mitigation, geology, minerals, transportation, water conservancy. and other departments for related applications

5.1 Disaster Reduction

China is one of the countries with the most serious natural disasters in the world. The types of natural disasters are numerous, the distribution range is wide, the frequency of occurrence is high, and the characteristics of concurrent, mass and concentrated outbreaks are present. The disaster losses continue to increase [4], which seriously affect the economy development, social progress, improvement of people's livelihood, and national security. The severe disaster situation has put forward new requirements for disaster reduction and disaster relief capabilities in the new era. It is necessary to use high-resolution satellites, aerial remote sensing, and other high-tech means to carry out three-dimensional disaster monitoring industry. To meet the needs of national disaster reduction and emergency response operations, the necessity and urgency of remote sensing technology in the field of disaster reduction and disaster relief in China are also becoming more and more prominent. Most of the satellites launched in China are mainly medium and low spatial resolution data. They are not able to meet the monitoring needs of various disasters, especially high-precision identification, interpretation and precision of buildings, lifeline projects, etc. Foreign high spatial resolution remote sensing data and services are expensive and lack supply guarantee, and the visit time is long, such as the revisit time of GeoEye satellite is 3 days. The image of the disaster area cannot be captured within the most valuable initial 24 h after the disaster occurs, and the emergency response capability is insufficient. Compared with satellite remote sensing, the aeronautical remote sensing platform is flexible, emergency response capability is strong, and the large field of view three-line stereo camera has a base to height ratio of 0.89 and a higher ground resolution. High-resolution panchromatic stereo images and RGB single-band images are available. It can provide effective data for disaster relief and decision making and meet the needs of disaster remote sensing monitoring.

5.2 Resource Navigation

Although China has extensive land and rich resources, it is still a country with a shortage of resources in terms of per capita. Therefore, rational use of resources and detection of the distribution, quantity and quality of resources have become urgent tasks [5]. The use of remote sensing technology for this work is an effective method. Although the resolution of satellite data continues to increase, it is limited by the height of the track, the size of the volume, etc., and the resolution is still relatively low, such as TM data (or ETM data) in the US, SPOT data in France, and CCD data of the China–Brazil Earth Resources Satellite. Computer multi-function processing or fusion processing can only recognize 10-m-level objects and can only meet the production of 1:100,000 or 1:50,000 feature maps. At the same time, the revisiting time is too long to meet the remote sensing survey of geological disasters, the need for high-resolution observations with monitoring and key mineral exploration areas. Compared with satellite remote sensing, the aeronautical remote sensing platform is flexible and has strong emergency response capability. The large field of view three-line stereo camera has the characteristics of large base-high ratio, large field of view, and wide coverage. The main technical specifications are 2.7 times more than the most advanced ADS40/80 in foreign countries. The high-resolution images obtained by the camera can be used for geological mineral surveys, land resources surveys and evaluations, environmental and disaster investigations and evaluations. The camera provides a fast and efficient working mode for land and resources exploration, and the camera's emergency work mode provides real-time work requirements for emergency work tasks.

5.3 Digital City

Digital city refers to computer technology, multimedia technology, and large-scale storage technology, with the broadband network as the link, using 3S technology (remote sensing RS, global positioning system GPS, geographic information system GIS), photogrammetry, simulation virtual technology, etc. The city performs multi-resolution, multi-scale, multi-temporal, and multi-dimensional three-dimensional descriptions. Digital city is a huge system engineering [6], which is an inevitable trend of urban development and social informationization, and a new economic growth point for urban development. Digital mapping reflects digital features in many aspects such as mapping method, mapping process, data integration, data storage and presentation, and spatial analysis. Therefore, digital regional technology has important significance in the research of urban planning, construction, management and service, and real-time monitoring of environment and ecology.

With the deep development of "Digital Earth", the real-time, accuracy and high efficiency of information have become the key to digital technology. How to realize real-time image acquisition and processing to realize its real-time and automatic nature is the main problem the development of the digital city is facing. The large field of view three-line stereo camera has the ability of high resolution, wide coverage, high working efficiency, and images acquired by the camera can provide spatial information for digital city construction.

5.4 Environmental Protection

With the increasingly prominent global environmental problems, frequent environmental disasters and accidents, remote sensing technology has been widely used in environmental monitoring and management, and its role in environmental protection has been highly valued by the international community. Some countries as the US, Japan, and Europe have been vigorously developing environmental remote sensing monitoring technologies in recent years. At present, China's environmental monitoring tasks are very arduous, especially for remote sensing technology-based environmental remote sensing monitoring. Remote sensing technology is widely used in China's environmental monitoring management and practical application scope and field. It is one of the important technical means of environmental protection management and plays an important role in China's environmental sustainable development [7].

The large field of view three-line stereo camera can meet the urgent needs of China's effective remote sensing load for environmental monitoring. The camera has two working modes: normal mode and emergency mode, which can meet the needs of routine environmental monitoring and emergency environmental monitoring. Normal mode could be used for routine remote sensing monitoring of environmental areas, and emergency mode is suitable for differet emergency objects, such as emergency observation of pollutant leakage, oil spills, disasters, mudslides, etc. At the same time, the camera has data timeliness, can produce different grades of products according to user application requirements, can greatly improve the rapid response capability of environmental pollution and ecological change monitoring, and realize quantitative monitoring in major environmental pollution problems such as land water pollution, air pollution, and urban pollution in China. It can meet the needs of environmental protection business dynamic monitoring and emergency response monitoring.

5.5 Public Security

The application research of high-resolution remote sensing data in the public security system has just begun. The application research results are few, the application level is low, the wideness and depth of application need to be further expanded and deepened, and the police are extracted and excavated quickly and accurately from high-resolution remote sensing images. Information and spatial dynamic information can play a positive role in improving the business technology level and emergency response capability of public security work in China. At present, China's high-resolution satellites cannot provide real-time remote sensing data and cannot meet the real-time requirements of data. Therefore, it is of great significance to further expand the breadth and depth of high-resolution remote sensing data services. The large field of view three-line stereo camera has the advantages of high resolution, flexibility, and accurate information. In China's public security work, it can quickly and effectively acquire spatial dynamic information, which can be used for various security accidents, mass incidents, and malignantness. Public security incidents and case evaluations can be used to support decision-making support; geographic information reference can be provided for the scope of impact and loss assessment of major security incidents; historical data comparison services can be provided for intelligence investigations of heavy traffic accidents and sea emergency search and rescue areas, etc.

6   Conclusion

As an efficient means of obtaining geographic big data, the main technical specifications of the domestic large field of view three-line stereo aerial camera are
several times superior than these of the most advanced push-broom aerial camera in
foreign countries. It represents that the development level of China's aviation
remote sensing camera ranks among the top in the world. It has important significance for the efficient acquisition of geographic big data and has broad application
prospects in information disaster reduction, resource navigation, mapping, digital
city construction, and so on.

References

1. Zhou Shunping, Xufeng. Thoughts for Developing Geographic Information Industry under Big
   Data. Geomatics World. 21(1):45–50 (2014). (in Chinese)
2. Zeng Xingyu. Research on development of foreign aerial mapping camera. Jiangxi Cehui.
   3:30–32 (2015) (in Chinese).
3. Li Haixing, Hui Shouwen, Ding yalin. Development and key techniques of optical mapping
   equipment in foreign airborne. Journal of Electronic Easurement and Instrumentation. 28
   (5):469–477 (2014). (in Chinese)
4. YO Kim, SB Seo. Flood risk assessment using regional regression analysis. Natural Hazards.
   63(2):1203–1217 (2012).
5. Giovanni Forzie, LucaTanter, Gabriele Moser. Mapping natural and urban environments using
   airborne multi-sensor ADS40–MIVIS–LiDAR synergies. International Journal of Applied
   Earth Observation and Geoinformation. 23(8):313–323 (2013).
6. LIU Xu.Application of GIS technology in ecological environment protection analysis in
   regional agricultural planning. Guangdong Agricultural Sciences. 9:160–162 (2015). (in
   Chinese)
7. Duan Chunhua. The Application Analyze of GIS Service Security Structure Based on ArcGIS
   Server Token-Taking Digital Yangzhou GIS Public Services for Example. Modern Surveying
   and Mapping. 38(04):41–44(2015). (in Chinese)

Lina Zheng is an associate researcher in Changchun Institute of
Optics Fine Mechanics and Physics and also a supervisor of
postgraduate in the University of the Chinese Academy of
Sciences. She received her B.S. degree in electronics from Jilin
University in 2003, and her M.S. degree in Mechatronics from
the University of the Chinese Academy of Sciences in 2008. She
received her Ph.D. degree in optical engineering from the
University of the Chinese Academy of Sciences in 2013. Her
current research interests include optical imaging and mapping
techniques for the aerial remote camera.

Informatization Promotes Accurate Management and Open Sharing of National Natural Science Fund

Dong Li, Jian Ma, Chang Yao, Wen Chen and Zhaotian Zhang

Abstract The information tidal wave is changing social life and human scientific research activities and scientific research management paradigm. E-Science has become a vital approach of scientific research and management. As an important part of national science research funding system, after many years of informatization construction, National Natural Science Foundation of China (NSFC) has built a whole process management system for research funds and an information opening and sharing environment, which effectively support various NSFC transactions, meanwhile provide advanced information techniques for NSFC to play a guiding and coordinating role in science research.

Keywords E-Science · Accurate management · Opening and sharing · Science fund

"Innovation is the primary driving force behind development; it is the strategic underpinning for building a modernized economy. We should aim for the frontiers of science and technology, strengthen basic research and make major breakthroughs in pioneering basic research and groundbreaking and original innovations." said the report of the nineteenth National Congress of the Communist Party of China. To build an innovation-oriented country at a faster pace, scientific and technological innovation plays a key role by providing strategic support for improving productive forces and overall national strength, therefore must be placed at the core of national development. With the national innovation-driven development strategy, China's investment in scientific and technological research and development has grown rapidly, and in 2015, the investment amount ranked second in the world. According

D. Li (✉) · Z. Zhang
National Natural Science Foundation of China, Beijing, China

J. Ma
Shenzhen Research Institute, City University of Hong Kong, Kowloon Tong, Hong Kong, China

C. Yao · W. Chen
IRIS Systems (Shenzhen) Limited, Shenzhen, China

© Publishing House of Electronics Industry 2020
China's e-Science Blue Book 2018,
https://doi.org/10.1007/978-981-13-9390-7_12

to National Bureau of Statistics of China [1], the expenditure on national research and development (R&D) increased from 128.76 billion yuan in 2002 to 157.67 billion yuan in 2016, with the input intensity reaching 2.11%. This is still far behind the average level of 2.40% in Organization for Economic Cooperation and Development (OECD) countries, but it has exceeded the average level of 2.08% in the 15 European Union countries. Moreover, the input intensity has been steadily rising, and the gap with developed countries has been narrowing year by year. For more effective R&D investment, it is urgent to develop an informatization "booster" for scientific and technological innovation, and to design an innovation mode that is compatible with the changing circumstances.

In today's world, the new scientific and technological revolution with Internet, cloud computing, big data, artificial intelligence and other information technologies is changing with each passing day. It has rapidly changed people's daily life. Furthermore, it has greatly promoted the productivity of scientific research and brought about significant changes in scientific research activities and scientific research management paradigm. Informatization of scientific research has become the most active part in the field of information technology. "The National Informatization Development Strategy 2006–2020" issued by the General Office of the Central Committee and the State Council in 2006 puts forward that speeding up the pace of informatization in education and scientific research is one of the strategic priorities for promoting social informatization. In the same year, the State Council promulgated "the Outline of the National Medium- and Long-Term Science and Technology Development Plan (2006–2020)," which formulated the overall goal and deployment of China's science and technology development by 2020, and clearly puts forward the key points of scientific research informatization.

As an important part of the national scientific research funding system, NSFC plays an important role in promoting the integration of informatization with scientific research activities and scientific research management system, fully playing the active supporting role of informatization in scientific research funding and fund management and providing advanced technical means to play the guiding role of science fund. Since the establishment of the leading work group on informatization of NSFC in 1992, after years of development, remarkable achievements have been made in the management of science fund, the sharing of achievements, the portal website, infrastructure, office automation, system operation and maintenance and security assurance, etc. A complete process management system for science fund and an information sharing environment have been established, which realized the digitization and networking of NSFC's office affairs, provided project management and daily office information environment for fund managers, and provided a fund application and project evaluation network for scientific research workers and review experts. The platform strongly supports the smooth workflows of NSFC's various business and provides a good technical and informational means for NSFC to play its guiding and coordinating role.

1　Current Situation of Informatization for Scientific Research Management

The informatization for scientific research management has gone through three main stages: computer automation, Internet informatization and cloud service of intelligent scientific research management information system.

In the computer automation phase (before the year 2000), research institutions and researchers used computers to input information about proposals and projects but also had to submit and retain paper documents such as proposals and project contracts. The information collected by the computer can help scientific research managers track the progress of the project, but it also requires researchers to print a large number of paper documents related to the project.

In the stage of Internet informatization (2000–2015), the whole process of information management from application to project conclusion was realized mainly by using Internet and database technologies. The Internet-based scientific research management information system enables users to complete the daily management of scientific research projects whenever they are online, which greatly improves the efficiency of scientific research. The database-based decision-making support system can also help scientific research managers to improve the level of scientific research decision-making. For example, the National Science Foundation (NSF) built the Internet-based science fund project declaration system (FastLane) in 2000, which has been continuously upgraded to today's "Research.gov" system. The system supports the functions of submission, status inquiry, application review, project progress/conclusion management and project financial management and provides one-stop scientific research management services for research institutions and users of scientific research personnel. In 1998, the Innovation and Technology Commission of the Hong Kong Special Administrative Region government developed and launched the Internet technology information system, which realized the functions of innovation and technology projects from application to final project management and fine project fund management, and ensured the risk management of technology projects.

On the basis of learning from the development experience of advanced scientific research management information system at home and abroad, NSFC launched the scientific fund network information system in early 2000, realizing the informatization of scientific research management based on the Internet. Using the computer-aided decision support system, the level of scientific research management and decision-making is improved.

In the stage of cloud services for intelligent scientific research management information system (since 2015), with the development of open access [2] and artificial intelligence, cloud services, platform strategy, deep learning, and new technologies such as intelligent recommendation algorithm are quickly applied to the scientific research management, which creates a new era of cloud services for scientific research management information system, and accurate scientific research management is realized. For instance, the UK proposed the E-Science plan to use the new generation of Internet technology to establish a new scientific research mode, so as to promote scientific and technological cooperation and resource sharing between university researchers and enterprise innovators and improve the efficiency of scientific research and innovation. The European company Elsevier has recently proposed a function of portraits of researchers, extracting information about their fields from their work to complete the selection and assignment of reviewers in scientific papers and peer reviews. Clarivate, a company in the US, uses the WoS technology database to find research hotspots and predict future research directions and achievements of researchers.

The development of new data-intensive scientific research paradigm makes the management of scientific research achievements an important part of scientific research and academic exchanges. Countries around the world have widely promoted the open sharing of scientific research results produced by public funding. Most national research funding agencies have introduced open-access policies, regarded as an important basis to promote innovation and development of the global community.

In March 2015, NSF published titled "Today's Data, Tomorrow's Discoveries" written plan and asked peer-reviewed journals and review meeting minutes to be accessed into database which was compatible with public access and available to download, aiming to speed up the public access to scientific publications and digital scientific data such as the results supported by funds. The European Union proposed "Horizon 2020" plan for open access, requiring all research papers of projects to be stored within the prescribed period of time to an open-access repository, aiming to achieve a single, opening European online research space, and researchers can make use of advanced, ubiquitous and reliable network and computing services and can realize seamlessly and open access to the E-Science environment and global data resources. The Australian Government launched the construction of the Australian National Data Service (ANDS) in 2008, aiming to comprehensively integrate the national data resources and realize long-term storage and sharing of data. The British research council revised the open-access policy guidelines in 2013, requiring the publication of funded open-access papers under the CC_BY agreement to improve the efficiency of research results. Established in 2012, the Global Research Council (GRC) aims to propose scientific development programs acceptable to the international scientific and technological community, promote more international scientific and technological cooperation and promote the sharing of high-quality scientific research data, methods and technologies. In 2013, the GRC conference adopted the "open-access action plan of scientific papers," which encouraged and called for open access to publicly funded research results, and also proposed a series of measures to promote and support the publication of research results in open-access journals and displayed them in open-access knowledge bases.

China is also actively adapting to the new trend of informatization development, accelerating the informatization of scientific research, and providing a powerful boost to the national capacity for scientific and technological innovation. In October 2015, the public service platform of the national science and technology management information system constructed by the information center of the Ministry of Science and Technology was opened for trial operation. The platform included management service module for various professional institutions, evaluation institutions, evaluation experts, researchers and the social public, mainly providing project management, information publicity, project application and related resources service functions. In addition, the Ministry of Science and Technology has taken the lead in building a national infrastructure platform, using modern information technology and other means to integrate scientific and technological resources and promote the development and sharing of scientific and technological resources. The Chinese Academy of Sciences (CAS) has established "technology cloud," "management cloud" and "education cloud" environment that can be shared openly and seamlessly by all the institute's researchers, and strongly supports scientific and technological innovation activities.

2 Historical Course of Informatization of National Natural Science Fund Management

Since leading group for information technology advancement was founded in 1992, after 20 years construction, the NSFC informatization has made remarkable achievements, including gradually realized the application submit and receipt based on Internet, project peer review and panel review, project approved management, development and cultivation management, project funds management, expert information completion and so on. Through the use of advanced information technology, the precise management for science fund was realized, providing better information services for the majority of experts and institution, and providing a good technical and information means for NSFC to play its guiding and coordinating role.

Learning the experience of scientific research management informatization in developed countries and complying with the requirements of scientific research management informatization, at the end of 1997, NSFC started to build scientific fund project management information system (MIS), and in March 1998, the system was preliminarily put into use, which was geared to the needs of NSFC staff for carrying out the project management business.

Since 2000, NSFC has started the development process of scientific research informatization based on the Internet. In early 2000, the Internet-based Science Information System (ISIS) was launched, realizing one-stop technology management services for the first time such as submission, review, progress and conclusion management based on Internet, which greatly improved the efficiency of scientific fund management. In 2003, because of SARS, the project assignment and review based on ISIS were realized ahead of schedule, which is of milestone significance in the process of scientific research informatization of NSFC. Since 2007, the ISIS has added a function of collection and open access for project results. When using the ISIS system to fill out the progress/conclusion report, the project leader requires to supply the results' link to the original literature database through DOI to ensure the authenticity and standardization of project results data.

Since the construction of ISIS, peer review expert database has been built as part of its functions. In the peer review expert database, expert information contains the basic information of experts, personal profile, research field, research results, project status and other information. The maintenance and update of expert information and the supplement of new experts are carried out directly by various discipline departments. To ensure the accuracy and effectiveness of the expert information, the maintenance, update and supplement of the expert information shall be carried out from January to March before the start of the evaluation of scientific fund project each year, and this work has become a major work process of the management of the national natural scientific fund. Up to now, there are more than 180,000 experts from home and abroad in the peer review expert database, which has become a valuable resource of NSFC and indispensable support for the management of scientific fund projects. Based on reliable scientific research projects, information of achievements and increasingly perfect expert information, since 2011, the research of experts assigned by computer assistance has been started, and the research results have been gradually realized in the ISIS system. With the continuous development of big data and artificial intelligence technology, more and more application project experts can be assigned through computer assistance to ensure that project review can be conducted more openly, fairly, justly and efficiently. At present, more than 3000 registered universities, research institutions and more than 1 million users of scientific research personnel have been served by the ISIS system, which has achieved the whole process information-based management of scientific fund projects from project application, project review, progress/conclusion management and result sharing.

In order to promote scientific fund project achievements and promote the transformation of scientific and technological achievements, in 2006, NSFC builds and opened the National Natural Science Fund project information sharing service system named "the scientific fund sharing service network (http://www.npd.nsfc.gov.cn)." The scientific fund sharing service network released basic information and academic research results obtained from 10 years period of projects before 2006 at first, including published papers, published books and conference papers. In 2012, the basic information and relevant achievement information of the final projects from 2008 to 2011 were further updated. In order to further fully share resources and better serve the whole society, a new edition of "scientific fund sharing service network" was released in 2013. The new version website supplemented the abstracts of released final projects and improved the system functions. In July 2014, according to the *National Natural Science Foundation Regulations* and *Several opinions of the State Council on improving and strengthening scientific research projects and funding management of the central government* (issued by the State Council [2014] No.11), the project achievements and full texts of conclusion report were released on the scientific fund sharing service network constantly. In 2016, according to the notice issued by the general office of the State Council on promoting scientific and technological achievements transfer and transformation action (issued by the General Office of the State Council [2016] No. 28) which required to construct the national scientific and technological achievements information system, NSFC further perfected the scientific fund sharing service network functions by integrating existing resources and avoiding redundant construction, such as adding the outstanding results show, hot word retrieval and connectivity interface. The further perfected system has become a comprehensive, integrated service platform supplying supported project, completed project and scientific research results information retrieval and service for the social public, enterprises and scientific researchers.

In May 2014, the Global Research Council meeting was held in Beijing, NSFC issued "NSFC Policy Statement on Open Access (OA) to the research publications of its funded projects." The statement demands research papers generated from projects fully or partially funded by NSFC, when submitted and published in academic journals, the authors of the papers should deposit the final manuscripts, which have been peer-reviewed and accepted by the journals to the NSFC repository with an embargo period of not more than 12 months. The announcement marks the release of the open-access policy, which fully embodies NSFC's responsibility and efforts in promoting open access, benefitting society and innovation-driven development. The policy statement will greatly promote the construction and development of open-access repositories, promote the open innovation resources and ability of the whole society and support the construction of an innovative country, also shows that our country in the world made great contribution to science and technology information open access.

On May 20, 2015, "NSFC basic research repository" (http://or.nsfc.gov.cn/) formally completed and opened online, at the same time, NSFC implementing rules for the basic research knowledge base of open-access policy also released, marking the Chinese open-access knowledge base platform for basic research results officially released.

For further carrying out the government information disclosure regulations, improving the transparency of science fund work and fully playing NSFC information service for the masses of the people, in November 2008, NSFC formulated and issued the measures for the administration of NSFC information disclosure, compiled the guides for NSFC information disclosure and annually released information public work report.

3 National Natural Science Foundation Management Informatization Achievements

3.1 National Natural Science Foundation Precise Management Platform—Internet-Based Science Information System (ISIS)

Since the online release of the Internet-based Science Information System (ISIS) in 2000, NSFC has been promoting the upgrade of the scientific fund project management system from traditional "experiential" to "digital." Digital management of scientific research support scientific fund project management throughout the life cycle, including how to retrieve, collect and verify project results from the scientific literature database, how to conduct comprehensive analysis of funded projects and results data and how to use artificial intelligence and machine learning technology to assist fund project management, etc. At the same time, NSFC conscientiously implemented *the Guiding Opinions on Accelerating the Work of "Internet + Government Services"* (issued by the State Council [2016] No.55), insisting on overall planning, problem-oriented, coordinate development, open and innovative principles, optimize service processes, innovate service methods, promote data sharing, open and transparent services and fully implement "distribution service" reform measures, so that project managers and researchers can share the development results of "Internet + Government Services." NSFC has achieved many important breakthroughs, for example, the comprehensive digital transformation of all projects from paper documents to online filing and the one-stop service for project application. Moreover, NSFC is actively promoting electronic signing, and striving to realize the paperless of the project application as soon as possible. The main features of the information management of NSFC are detailed below, including project application management, project review management, project execution management and project results management.

(1) Project Application Management

Due to the large number of applications and the non-standardization of keywords of different disciplines, the project application has brought different forms of difficulties to the staff of the funding agency, the managers of host institutes and the project applicants. The formal review of a large number of applications has become more and more arduous, while the traditional project management system is basically an information isolated island, and the integration with external information sources is heterogeneous and costly, which makes it extremely difficult to authenticity review the applicant's resume and content. In order to standardize the management of scientific research projects, it is imperative for NSFC to regulate and improve efficiency for data collection of scientific research results. NSFC has realized the transformation from paper documents to comprehensive digital for project applications. It has also made important breakthroughs in one-stop project application service, subject domain ontology library construction, distributed parallel processing mechanism, etc. A total number of over 190,000 applications were well handled by the system before the deadlines.

Firstly, the applicant/participant research resumes are standardized in the process of one-stop application. Different government-funded institutions have certain requirements for the research background of applicants and key participants in the project application. The design of simple and easy-to-use research resumes will greatly improve project application efficiency. For researchers, it is possible to generate research resumes that meet the requirements of the funding agencies based on their own research results, thereby improving the efficiency of scientific research. For the funding agencies, when examining the results of the application, they can directly link to the scientific literature database to ensure the authenticity and standardization of the scientific research results filled by the applicant.

Secondly, the proposal's keyword filling [3] is standardized through the construction of the subject domain ontology library. When the applicant fills the proposal's keywords and the reviewer selects the research field, the same concept often appears with different keywords. Therefore, it is necessary to construct a subject domain ontology library to standardize the keywords used in scientific research. The keyword standardization is carried out by word frequency analysis and social voting. On the one hand, the word frequency statistical analysis is used to extract the keywords in the subject area. On the other hand, principle investigators are invited to vote for the keywords for their familiar subjects. Afterward, keywords chosen from these two tunnels are combined to standardize the keyword library. The subject domain ontology library is the basis for project application and reviewer assignment. The established keywords database can be used to recommend relevant keywords when applicants fill out the application and reviewers update their resumes, thereby improving the matching between the application and the reviewer, which in turn improves the quality of the project review process.

Finally, with the rapid growth of the number of applications, how to maintain system efficiency during application period becomes more and more important. The Internet-based Science Information System has set up a distributed parallel processing mechanism to ensure good stability and high efficiency during the project application period, especially in the final stage of application submission, to ensure that the project application is safe and efficient.

(2) Project Review Management

The science fund project review requires the selection of several peer experts for each application to conduct project peer reviews. How to effectively use computer-assisted assignment system to avoid conflicts of interest and improve the quality of reviews is an issue we have been always studying. NSFC has made continuous improvements in the phase of project evaluation, making use of information technology to ensure openness, fairness and justness of project evaluation. The main measures include preventing duplicate applications through the similarity inspection system, comprehensive and accurate portraits of review experts using scientific data and intelligent grouping of applicants and recommenders using computer-aided assignment system [4].

The evaluation principles are "relying on experts, promoting democracy, funding excellent research and advocating fairness and justness." Peer review is the key step to ensure "excellence," "democracy" and "justness." Whether an application project can be selected fairly during the review process depends on whether the review experts can be appropriately selected for the project, and the computer assignment system provides auxiliary decision support for selecting the appropriate experts. The assisted assignment system makes full use of the application information (such as: funding category, application code, key words, etc.), applicant information (such as: basic personal information, research background, etc.) and reviewer information (such as: basic personal information, application code/field/key word, research background, etc.), accords to the big data research and analysis framework, extracts scientific research characteristics such as scientific research quality, relevance and cooperation, eliminates the reviewers whose relevance is not in line with the quality of the application and assigns the reviewers with similarity [5].

In order to improve the efficiency of scientific fund management, NSFC has done a lot of work on the computer-aided assignment system. First of all, NSFC has established a standardized "application code—research direction—key word" system to standardize the research field of the application. Secondly, through years of accumulation and improvement, NSFC has built an expert database with relatively complete information and comprehensive coverage, which has laid a good foundation for computer-aided assignment. After portraying the project application and qualified reviewers, the system manages to establish a scientific knowledge map for the research community. Supported by the theory of big data analysis for Chinese research, the intelligent matching algorithm from the three dimensions of scientific research quality, cooperation and relevance is designed utilizing the mainstream technologies of text mining, network analysis. Following the optimized reviewer assignment policy as well as the consideration of interest avoidance and workload limitation, the system assists the assignment of subject directors by quickly generating a candidate set of reviewers for each project application.

(3) Project Execution Management

Due to the wide variety of natural science fund programs, the wide range of research topics and the scattered directions, a unified management model [6] with strong operability and standardization is needed. After years of efforts, NSFC has strengthened the annual progress report and final report management of funded projects through the informatization, such as checking the labeling of funded projects.

The digital science fund project management scheme attaches importance to the whole process of project management, including the project annual progress report, final report and the standardized management of project funds. The standardization of the annual progress report emphasizes the collection and summary for project's staged achievements, which reduces manual entry errors, and sets project task reminders to ensure that principle investigators complete tasks and provide accurate and reliable data before the project deadline. Combined with the big data research and analysis framework, one can quantify analysis of scientific research input and output and provide decision-making basis for project acceptance.

The use of the achievements online system to automatically collect and standardize the results of scientific research projects can avoid or reduce the occurrence of irregularities, incomplete information and inaccurate project marks in the submission of annual progress reports and final reports by principle investigators, which promote scientific research integrity in scientific research project management.

(4) Project Achievement Management

In the process of project achievement management, managers can use the computer to statistically analyze the project funding and results output according to the subject/this funding plan/the host institution and conduct a performance evaluation of the project and decide whether the project is continuously funded. NSFC has made continuous improvement in statistical analysis and comparative analysis using information technology and has achieved remarkable results.

In the statistical analysis of project achievements, the results data is displayed in digital and graphical form, which provides effective decision support and analysis tools for fund development planning. It mainly includes scientific research competitiveness analysis, scientific research field summary analysis and scientific research cooperation and sharing analysis. Scientific research competitiveness analysis refers to the use of scientific research analysis framework, based on existing scientific research results data, to provide an analysis report on the scientific research competitiveness (quantity, quality and influence) of regions, universities, research institutes and individuals. The summary of scientific research field refers to the use of scientific research analysis framework, based on the existing scientific research results data, to summarize and analyze the research field and provide a summary analysis report in the field of scientific research. Scientific research cooperation analysis refers to the use of scientific research analysis framework, based on the existing scientific research results data, to summarize and analyze the cooperation among a certain region, universities, institutes, individuals and other regions and provide scientific research cooperation and sharing analysis reports.

In the comparative analysis of project achievements, in order to further strengthen the strategic planning of the natural science fund, to compare and analyze the objectives at different levels, we can use the scientific research analysis framework to conduct comparative analysis of scientific research results data and implement benchmarking management for selected subjects/funding plans/ institutions/regions. On the one hand, it helps us to clarify the strategic positioning of development and set development goals so that all institutions can develop rapidly in a benign competitive environment; on the other hand, the funding direction and funding focus of the natural science fund can be adjusted.

Digital scientific research management has also greatly promoted the development of open access and knowledge sharing. The scientific research achievements collected in the Internet-based Science Information System will be regularly published annually to the scientific fund sharing service network and the basic research knowledge base, which are to provide knowledge sharing and open-access services.

3.2 NSFC Sharing Service and Open-Access Platform

Through information sharing, researchers are provided with research results and scientific knowledge retrieval services, to promote and popularize scientific knowledge to the public, to meet the sharing needs of different social groups for basic research resources. It is of vital importance to greatly enhance China's scientific and technological innovation capabilities and accelerate the construction of innovative countries.

The Science Fund Sharing Service Network uses the accumulated information on the achievements of basic research in natural sciences and the relevant information of the NSFC, such as policies and management regulations, to provide researchers with a set of scientific research sharing services which is of benefit to information acquisition and cooperation. The system strives to create a comprehensive and integrated scientific research information sharing and service system platform [7]. On the one hand, it provides a portal for community to inquire and browse natural science research information at any time and also provides intelligent information retrieval tools for researchers to understand and master the research trends and research results of peers in the field; on the other hand, for the management department, it can provide comprehensive statistical analysis data of basic research results, fully demonstrate the important role and value of scientific research results for social development and scientific research and provide effective support for solving problems such as repeated research and multiple applications. Up to now, the relevant information of the funded projects from 1986 to 2016 has been published, and the achievements information of the completed projects in 2003–2016 has been published. Among them, there are 201,382 completion projects and 3,040,978 project achievements. At the same time, to further promote the sharing and linking information between the Science Fund Sharing Service Network and the National Science Reporting Service System from multiple angles, the project information and related achievements information currently published on the Science Fund Sharing Service Network will be submitted to the National Science and Technology Reporting Service System for opening.

The Science Fund Sharing Service Network currently provides three categories: funded projects, completed projects and achievements retrieval. For the completion project, it provides three search methods: "by application field," "by project type" and "by type of results," provides "by application area," "by project type," "by type of achievement" and provides five statistical methods of "by application area," "by project type," "by type of results," "by completed year" and "by geographic information." The application area covers eight scientific departments of the NSFC, and the project types include General Program, Young Scientists Fund, Fund for

Less Developed Regions, Key Program, Major Program, National Science Fund for Distinguished Young Scholars, Science Fund for Creative Research Groups and Joint Research Fund for Overseas Chinese Scholars and Scholars in Hong Kong and Macao, etc. The types of achievements include journal articles, conference papers, books, awards and patents, etc.

The NSFC's basic research knowledge base, as the infrastructure of China's academic research, collects and preserves the metadata and full text of research papers funded by the NSFC. At present, it has published 518,524 research papers on the results of funded projects. It contains 78,119 authors and 1792 institutions, covering 2000–2017, more than 27,300 journals and more than 19,600 conferences. The follow-up will incrementally release the metadata and full text of the research papers of the previous year's National Natural Science Fund project in November annually. Since the opening of the basic research knowledge base, the number of full-text downloads has reached 3.4 million. It will certainly promote the practical application of China's open access and promote the rapid transformation of knowledge generated by public investment into the ability of innovation and development of the whole society, thereby achieving the vision of "knowledge benefit society and innovation drive development."

4 Thoughts and Prospects on the Informatization of Scientific Research Management of NSFC

General Secretary XI Jinping stressed at the first meeting of the Central Leading Group on Cyber Security and Informatization that there can be no modernization without informatization, demonstrating the determination of the CPC and the country to promote the development of informatization. Without scientific research informatization, there will be no scientific research modernization. Scientific research informatization is the only way and key to the national "innovation-driven development strategy." As the most valuable core asset of the scientific research management informatization of NSFC, the dividend generated by the scientific research big data accumulated from the ISIS system over the years is still under-estimated. NSFC's research big data covers the whole process of scientific research, including researchers, research process, funding management, research environ-ment and research outputs, etc. It is the hard-won strategic resource of NSFC and the data basis of all operational systems. At present, the NSFC's research big data using is no longer a simple data warehouse, but through the depth of the data processing and mining, extract the knowledge service for managers' decision analysis and principle investigators' demand, and through continuous learning eventually translate knowledge into wisdom services, full release of NSFC data bonuses, give full play to the scientific research management strategic advantage of the whole process of science fund. Based on the current situation of scientific research informatization of NSFC and the new requirements of the big data era, the following thoughts are given to the future scientific research management informatization [8].

First of all, NSFC is one of the earliest government funding agencies that participated in international organizations to advocate open access of scientific and technological resources. The sharing of scientific research information and resources can greatly promote the development of scientific and technological innovation in China. The big data era needs a safe scientific research resource sharing mechanism to support the construction of scientific research management informatization in the future. The future scientific research resources sharing mechanism should consider adopting collaborative innovation sharing mode, encouraging all funding agencies and research institutions to open public API interfaces and use cloud services and using blockchain technology to guarantee the security of information transmission and resource sharing, so as to realize the cloud storage and on-demand sharing of scientific research information. As an important part of the national research funding system, NSFC will be active to carry out the government affairs information system integration, access to the national unification exchange platform as planned, so as to realize data resources sharing based on network and achieve cross-department business synergies.

Secondly, the informatization of scientific research management in the future should meet the needs of the integration of production, education and research and serve the development of scientific research, innovation and entrepreneurship in China. The ultimate goal of scientific research management informatization is to serve the innovation of the scientific research community and the whole society and support sustainable economic development. The informatization of scientific research management is to solve the problems of asymmetric information, low conversion rate of scientific and technological achievements and difficult cooperation between industry, universities and research institutes. Therefore, the future scientific research management informatization needs to open up the information barrier between universities, research institutes, enterprises and personnel, realize the seamless connection between supply and demand in the process of scientific research and innovation and form the collaborative innovation ecological environment of the integration of production, education and research.

Finally, the cultivation of scientific research management informatization talents is the most important factor to realize professional and efficient scientific research management. Especially, the big data era needs scientific research management informatization talents, and the talents' cultivation need to pay attention to three aspects: (1) pay attention to scientific management theory and practice of comprehensive quality training, as scientific research management information talent is high-quality management personnel, who can take on the important responsibility of universities, scientific research units, the funding agencies and so on; (2) all funding agencies, universities and research institutions should design information-based research management schemes suitable for their own characteristics according to the characteristics of their regions and different research fields;

(3) strengthen the talent cultivation in big data, artificial intelligence and other high-tech information field, as scientific research management information needs to walk on the leading edge of information technology, needs to use the latest Internet +, such as: social networks and mobile applications, data/artificial intelligence, cloud computing (SMAC: social, mobile, analytics and cloud technologies), to solve the problem of scientific research management.

Looking ahead, NSFC will design and provide more intelligent and personalized scientific research management services from the perspective of researchers and users of scientific research management. In the future informatization road, we will continue to combine the actual demand of scientific research management and carry out the informatization iteration.

The NSFC's informatization work during the 13th National Five-year Plan will be based on the national information development strategy, guided by NSFC's 13th five-year strategic plan, promoted by Internet + technology management, regulated by the *National Natural Science Foundation Regulations*, following the "overall planning, promote sharing, expand application, improve function, specification and efficient, safe and reliable" principle, and basing on the existing informatization construction results to further advance information systems from data service to knowledge service, from serving users to serving society and build an information environment that is people-based, service-oriented and intelligent. NSFC will comprehensively deepen the active supporting role of information system in the management of science funds, create an information service environment featuring transparency, openness, efficiency and intensity, rich functions, smooth interaction, safety and reliability and sustainable development and promote the rapid and steady development of China's science funds.

References

1. Statistical bulletin of national science and technology investment. http://www.stats.gov.cn/tjsj/tjgb/rdpcgb/qgkjjftrtjgb/
2. Jianjun Li, Mingshe Zhang, Dong Li, Wei Zhang, Jin Wang. Construction Scheme of NSFC Open Access Library. International Journal of Security and Its Applications. 2015, 9(8): 243–252.
3. Jiang, Hongbing; Yang, Chen; Ma, Jian; Silva, Thushari; Chen, Huaping, A Social Voting Approach for Scientific Domain Vocabularies Construction, Scientometrics, 108(2), pp 803–820, 2016/8.
4. Jian Ma; Wei Xu; Yong-hong Sun; Turban Efraim; Shouyang Wang; Ou Liu, An Ontology-Based Text-Mining Method to Cluster Proposals for Research Project Selection, IEEE Transactions on Systems, Man, and Cybernetics - Part A: Systems and Humans, 42(3), pp 784–790, 2012/5.
5. Silva T.; Guo Z.; Ma J.; Jiang H.; Chen H., A social network-empowered research analytics framework for project selection, Decision Support Systems, 55(4), pp 957–968, 2013/11.
6. Jian Ma, Wei Liu, Wei Xu, Hongbing Jiang. Smarter Research: Management Made Wise – A Lean Research Project Management Model. Bulletin of National Natural Science Foundation of China, 2011, 25(6): 331–334 + 347.

7. Yan Wang, Jianjun Li, Dongpeng Wang, How to Further Develop the "National Natural Science Foundation of China Sharing Service System": A Comparison with Research. Gov. Bulletin of National Natural Science Foundation of China, 2014, 28(3): 190–194.

8. Jianjun Li, Laiyun Qing, Prospect of Informatization Development in National Natural Science Foundation of China during 13th Five-Year Plan. Bulletin of National Natural Science Foundation of China, 2017, 31(2): 170–175.

Dong Li Senior Engineer, Deputy Director of Information Center of National Natural Science Foundation of China. Major in Computational Mathematics and Computer Software. She has engaged in the related research and construction of Scientific Fund Project Management and Information System Management for 20 years. In 2001, she started to build the National Natural Scientific Fund Information Management System and is currently in charge of Scientific Fund Information Management and Open Sharing Management.

Development and Prospect on National Center for Philosophy and Social Sciences Documentation

Lan Wang

Abstract This paper reviews the practice of Chinese Academy of Social Sciences in e-Social Science, analyzes the challenges faced by the information of philosophy and social sciences, and puts forward the strategies to speed up the construction of National Center for Philosophy and Social Sciences Documentation.

Keywords E-social sciences · National Center for Philosophy and Social Sciences Documentation/NCPSSD · Chinese Academy of Social Sciences/CASS

In recent years, with the rapid development of information technology, especially, the wide application of Internet, big data and artificial intelligence technologies is reshaping human society. The research methods of philosophy social sciences which take human society as the research object have gradually changed. And the new requirements have been put forward for e-Social Sciences. Chinese Academy of Social Sciences (CASS) actively responds to the new goal of accelerating the construction of philosophy and social sciences with Chinese characteristics and new think tank with Chinese characteristics, and explores and thinks about e-Social Sciences.

1 Practical Exploration on Information Construction of CASS

1.1 Improving Mechanisms and Systems and Strengthening Top-Level Design

To integrate the information resources of the whole academy and focus on the construction of the digital academy of social sciences, CASS has implemented the reform of information mechanism and system since 2013. With the continuous

L. Wang (✉)
Chinese Academy of Social Library, Beijing, China

© Publishing House of Electronics Industry 2020
China's e-Science Blue Book 2018,
https://doi.org/10.1007/978-981-13-9390-7_13

progress of the reform, the responsibilities of relevant institutions have been clarified. And the library of CASS has become the main responsible unit for the information construction. It clarifies the main principles of the information construction of the academy, that is, separation of management and construction, integration of resources, and nonduplication of construction. The main content of the information construction of the academy is "One Database, One Network and One Platform." "One Database" is philosophy and social sciences massive database. "One Network" is Chinese Social Science Network and the network infrastructure of supporting for the information construction of the academy; "One Platform" is the comprehensive integration laboratory and comprehensive management platform.

1.2 Promoting "One Database, One Network and One Platform"

Massive database of philosophy and social sciences of CASS has yielded initial results. The massive database construction project (phase I) officially started in April 2015. Scientific research achievements database, collection literature database, social investigation database, and ancient books database, which are the subdatabases of massive database, were basically completed and put into service at the end of 2017. National Social Sciences Database (NSSD) has become the largest open-access platform of philosophy and social sciences in China and won the title of "top 20 cases of integrated innovation of national newspapers and media" by State Administration of Press, Publication, Radio, Film, and Television of The People's Republic of China (SAPPRFT).

We insist on academic standard; follow the path of channelization and topicalization to upgrade Chinese Social Sciences Network continuously. There are four sections, respectively information, disciplines, integration and interaction, 54 channels, and more than 1300 columns. To strengthen the construction of academic network transmission, become the domestic and foreign influential philosophical social science academic portal, we start BBS, blog, microblog, open the mobile client, and make multimedia academic communication platform.

Network infrastructure is constantly optimized to provide researchers with good network services. The 21 disciplines of CASS are centered on the main part of the academy, which are connected to the Internet through SDH special line in the form of 1G bandwidth BGP autonomous domain to achieve load balance of the main link and gigabit to the desktop. We already established real-name Internet access, E-mail system, and other basic application systems, and a unified independent management of the academy computer system. The cloud platform of CASS (phase I) has been completed, providing solid hardware and software support for the construction of massive databases.

On the basis of the completion of the preliminary demonstration, "the integrated laboratory" and "the integrated management platform" have started, and continued to support the characteristic laboratories to carry out research by using new technologies. Based on public opinion monitoring and online public opinion survey, "China public opinion survey laboratory" is committed to building an index system of public opinion in China, providing social public opinion "barometer" and comprehensive solutions of public opinion and communication for all sectors of society. "Urban information integration and dynamic simulation laboratory" makes full use of modern information technology and the emerging technologies, such as the database, geographic information system, mobile Internet, and large data analysis to collect, sort, and analysis for urban complex systems of information. To provide powerful technical support of scientific city decision and management, they integrate with city economy, society, and spatial information, and combine statistical data and spatial data to simulate and monitor urban system.

1.3 Striving to Develop NCPSSD

In order to implement the spirit of General Secretary Xi Jinping's important speech at the symposium on philosophy and social sciences in September 2016, CASS began to develop the NCPSSD. The development of NCPSSD has been nourished by e-Social Sciences experience and achievements of "One Database, One Network, One Platform." The Center's portal started to provide online services on December 30, 2016.

At present, the portal of the NCPSSD has four sections. They are News and Bulletin, Resources, Topics, and Service, with over 11 million online data, including 1000 kinds of Chinese academic journals, 8000 kinds of foreign language of open-access journals. It realizes one-stop documentation retrieval, browsing, free download, and other functions, and provides users with nonprofit academy information services.

Up to September 30, 2017, the number of registered individual users surpassed 350,000, while the number of institutional users reached nearly 300. And there had been 75,000,000 times of clicks, nearly 8,000,000 times of resource search, and over 9,000,000 times of download. During the celebration of the fortieth anniversary of the founding of CASS, Liu Yandong, deputy prime minister, inspected the NCPSSD and spoke highly of it on May 17, 2017.

2 New Challenges in E-Social Science

2.1 New Technologies and Ideas that are Driving Change in Research Methods

With the development of information technology, the information form that the institute of philosophy and social sciences relies on is not only a traditional documentation but also includes the text, image, audio, video, and all kinds of data. Those data come from a variety of sources and increase quickly, and become "big data." Rapid increase of total research materials and diversification of their forms bring huge challenges to the philosophy and social sciences scholars. Big data is far beyond the processing category of traditional artificial search、collection and the reading ability. So, research data must be processed by using the platform of the computer system.

Especially with the development of artificial intelligence technology, natural language processing technology is constantly mature. Many computer processing patterns and analysis methods have been introduced into the field of philosophy and social sciences. Research topics, computer simulation, and demonstration based on complex operations and analysis, as well as business prediction based on facts and evidence and case evidence reasoning have been widely raised. It fundamentally changes the acquisition, annotation, comparison, sampling, interpretation, and expression of the knowledge. The research method characterized by text mining, complex network, and large-scale data analysis has been gradually accepted. New research forms have been formed, such as computational sociology, digital humanities, and so on.

In 2009, David Lazer et. proposed the idea of "computational social science" in a paper of *Science*. In 2014, American sociology held a seminar on "new computational sociology." Ray m. Chang discussed the transformation of social science paradigm that big data brought. He thought that big data brought more convenient collection technology of data and that social science was combined with computer science and information science. It is changing to the direction of "computational social science" and "network social science." In the fields of linguistics, literature, history, literature, ethnology, and other humanities, digital humanities have achieved remarkable result. And two major digital humanities research alliances have been formed; the international union of digital humanities institutions and the network of digital humanities centers

Innovation in philosophy and social sciences more and more depends on the supported of large-scale data platform and large-scale research and analysis platform. To realize knowledge objectification and computerization of big data, we use the platform to data modeling, descript, organize, save, access, analysis, and reuse.

2.2 Construction of the System of Philosophy and Social Sciences with Chinese Characteristic and New Think Tanks with Chinese Characteristics

To meet the needs of national development in the new era, the government puts forward new target that constructs of philosophy and social sciences with Chinese characteristic and new think tanks with Chinese characteristics.

The General Office of the Communist Party of China (CPC) and the General Office of the State Council issued *opinions on strengthening the construction of new-model think-tanks with Chinese characteristics* on January 20, 2015. The CPC central has proposed to promote scientific and democratic decision-making, to improve the modernization of the country's governance system and capacity, to enhance the country's soft power. Construction of new-model think tanks with Chinese characteristics is a major and urgent task. It clearly pointed out that the new think tanks should have multilevel platform of academic exchange and achievement transformation channels, as well as fully functional information collection and analysis system.

General Secretary Xi Jinping delivered an important speech at the seminar on philosophy and social sciences and profoundly answered a series of fundamental questions concerning the long-term development of China's philosophy and social sciences on May 17, 2016. The speech was a programmatic document guiding the work of philosophy and social sciences. The speech clearly pointed out that the construction of philosophy and social sciences with Chinese characteristics will be a systematic project and an extremely heavy task. We must strengthen the top-level design and coordinate efforts from all sectors to advance. We must strengthen the construction of philosophy and social science documents, networks, databases, and other infrastructures and information, must accelerate the construction of the NCPSSD, and build convenient and shared philosophy and social science research platform. This is the goal of accelerating the construction of philosophy and social sciences with Chinese characteristic, and the new requirement of strengthening philosophy and social sciences documents resources protection, and strengthening the application of information technology in the research of philosophy and social science.

The CPC central further issued *the opinions on accelerating the construction of philosophy and social sciences with Chinese characteristics*, which puts forward new requirements on how to accelerate the construction of philosophy and social sciences with Chinese characteristics on May 16, 2017.

The Nineteenth National Congress of CPC was held successfully, General Secretary Xi Jinping clearly puts forward that build a moderately prosperous society in all respects, open a new journey all-round construction of socialism modernization country on October 18, 2017. We should have cultural self-confidence to promote the prosperity of socialist culture prosperity and accelerate to construct philosophy and social sciences with Chinese characteristic and new think tanks with Chinese characteristics.

The construction of philosophy and social sciences with Chinese characteristic and the new think tanks with Chinese characteristics has become a national development strategy. To construct philosophy and social sciences with Chinese characteristics and new think tanks with Chinese characteristics, we must accelerate the information of philosophy and social sciences.

3   Speeding up the Construction of NCPSSD

The construction of philosophy and social sciences with Chinese characteristics and new think tanks with Chinese characteristics needs the support of solid documentation resources and information technology. In order to build comprehensive reasonable and effective system of documentation resources guarantee and information service of philosophy and social sciences, it is necessary to strengthen the top-level design and coordinate the efforts of all aspects to improve. NCPSSD that General Secretary Xi Jinping puts forward offers a new strategy to solve this problem.

Construction of NCPSSD can promote the integration, unified found, ubiquitous acquisition, long-term preservation, and depth use of the documentation, and research data. It develops openness and integration of national philosophy and social science research to realize integrated communication.

At present, the construction of NCPSSD is just beginning. Next, the social science community should unite to common discuss, build, and share. With the orientation of public welfare, openness, cooperation, and authority, we should cooperate with each other; improve the mechanism to promote the construction of NCPSSD continuously.

Firstly, drawing on the mature experience of relevant alliances and institutions, we will explore the efficient mechanism and system of management and operation. So, NCPSSD will play a truly joint and cooperative role in the field of e-Social Sciences.

Secondly, we will advocate coconstructing and sharing and establish unified standards. By means of network collection, digital processing, commercial procurement, and other manners, we will integrate various types of data resources to form an open, public welfare, and academic authoritative "data pool" with complete types and reliable quality. It provides data support for philosophical and social science research.

Thirdly, by using new technologies such as big data, data mining, and artificial intelligence, we will develop various information data processing and analysis tools to establish convenient and easy-to-use information platform of philosophy and social science research. It will provide technical support for the research of philosophy and social science.

Finally, adopting the concept of cloud computing and cloud service, we will construct the e-Social Sciences cloud platform to provide specialized IT infrastructure of social sciences, having mass storage, disaster preparedness, and large-scale computing functions. It will serve information construction of national philosophy and social science research institution and realize management and service mode innovation of national philosophy and social sciences in IT infrastructure.

Lan Wang Director of Library of the Chinese Academy of Social Sciences and National Center for Philosophy and Social Sciences Documentation.

Situation and Prospect of Security of the Science and Technology Cloud of China

Chun Long, Peng Gao, Yuhao Fu and Wei Wan

Abstract The science and technology cloud of China is an important infrastructure for the rapid development of scientific research with information technology, the insurance of this cloud is an important part of security project in CAS 13th Five-Year Program of Information, it aims at providing multi-dimensional and high precision security insurance system for cloud computing environment with integration of variety security technologies. This chapter presents an overview of security technologies in insurance of science and technology cloud of China. Its characteristics on security technology in cloud computing environment and the future development direction are briefly discussed.

Keywords Cloud computing · Science and technology cloud · Security technology

1 Introduction

Cloud computing technology is the evolution of the next-generation computing mode. With the further development of engineering practice, it has brought fundamental revolution to the scientific research mode and information construction. In cloud computing environment, it is necessary for the security to combine link security, application security, data security, active defense and other multiple levels, multiple types of security technology to achieve integrated security protection. For all kinds of application systems which are supported by virtualization technology platform, the user data interaction, whole cycle security of system operation and management behavior have been realized by using cross-domain link encryption, centralized control, two-factor and other security mechanism combination model. Finally, the integration of variety security technologies mode has been formed.

C. Long (✉) · P. Gao · Y. Fu · W. Wan
Computer Network Information Center, Chinese Academy of Sciences, Beijing, China

© Publishing House of Electronics Industry 2020
China's e-Science Blue Book 2018,
https://doi.org/10.1007/978-981-13-9390-7_14

The science and technology cloud of China is based on technology architecture implementation of proprietary cloud platform, and it is more complex than general computing environment. Security protection needs depth integration and adaptation, different levels, different granularity of security detection and protection in access environment, virtual resources, user environment and so on. In non-cloud environment, the system migration is required to be fast, safe and integrate technical support. This chapter presents a comprehensive introduction to the engineering and technical practice of the science and technology cloud of China security protection system from the three aspects: the security of cloud computing support environment, the security of cloud environment and deep security.

2 Security System Architecture

It is important to comprehensive guarantee the science and technology cloud of China network environment safety, construct security infrastructure continuously, improve security service support ability, and realize sharing and optimized allocation in fundamental resources of network security. On the basis of the achievements in network security during 12th Five-Year period, the security service capabilities of network intrusion detection, traffic cleaning, access control, data auditing and user protection in virtualized cloud environment will be improved comprehensively by constructing the integration of variety security technologies mode. Multi-tenant isolation, virtual integration, resource management and monitoring of multiple virtualization security technology are used to guarantee multi-type user information independently to run in isolated and safe environment, which can reduce the safety risk and realize the whole cycle control [1, 2] to the security function of the science and technology cloud of China. The security system of the science and technology cloud of China is shown in Fig. 1.

For the proprietary cloud computing environment of the science and technology cloud of China and the support facilities of network space, the whole cycle of multidimensional security environment is constructed. The cloud access security

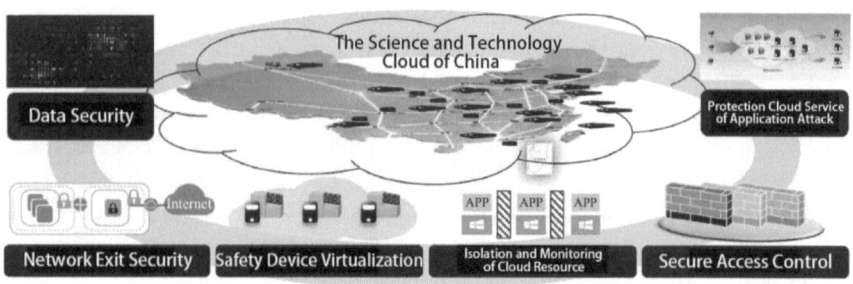

Fig. 1 The security system of science and technology cloud of China

environment is provided by the integration of access control, attack defense, traffic detection, login control, active evaluation, source design and other control measures, which can ensure all kinds of cloud business applications and satisfy whole actual business environment safety protection requirements of the science and technology cloud of China.

For the construction of cloud internal security protection, the main tasks are to control cloud platform horizontal and longitudinal flow, implement the monitoring, management, auditing, extensible, focus on the network data security protection of the cloud platform, and then realize the access platform between users and users and security protection in the interconnection of cloud. In the internal environment, business securities for tenants and access security or users are guaranteed, and access securities of hosts and virtual machines are ensured from east to west. Deploying virtualization software security protection measures for each tenant are required.

In terms of user side security services, cloud services for application safety and active safety defense and other series of security service are provided with a unified access environment, the environment of copying, distribution and processing are constructed. The related management and interface of data acquisition is provided for the business system construction module. The output data is regarded as all-round data support for situational awareness platform and other analysis and display systems.

On the basis of satisfying the safety requirements of the current cloud computing environment safety, the construction of the security system of the science and technology cloud of China also possesses a certain technology and capacity extension ability, which makes the safety protection measures to be expanded and upgraded to meet the requirement.

3 Realization of Security Technology

3.1 Cloud Computing Support Environment Security

For the network space facilities supported by cloud computing, we use the combinations of virtual and real security technology which satisfy all kinds of security requirements of cloud computing environment to provide independent and on-demand security protection. Integrated protection is performed from physical layer to application layer in cloud computing environment. This process can accurately identify data flows at each layer. In order to complete 2–7 layers network data security protection, access control, intrusion defense, worm defense, leak protection, web application security protection, risk assessment, terminal security, code auditing need to be implemented. Thus, the cloud computing environment can be protected from external networks in real-time, such as worms, viruses, Trojans, vulnerabilities, web attack, application attack and other threat behaviors. The cloud

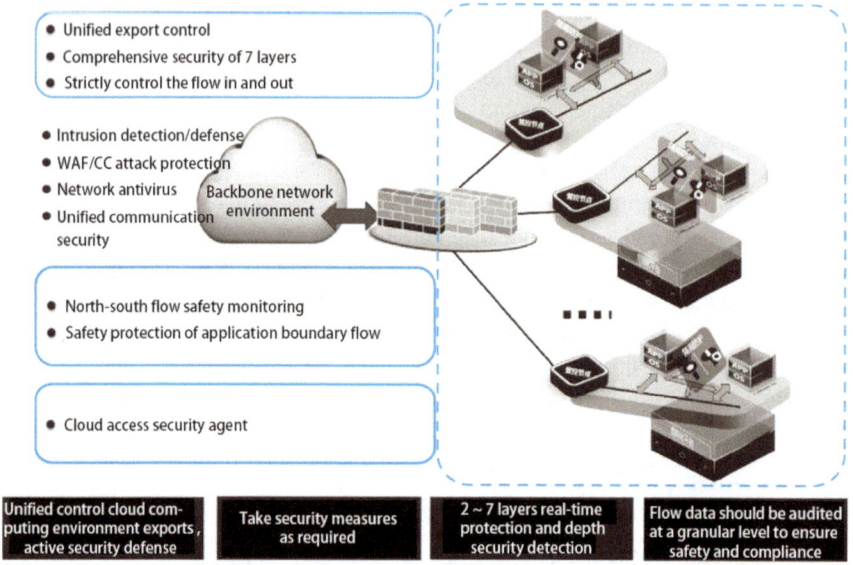

Fig. 2 The security protection architecture of cloud computing support environment

computing environment for safe operation of network and business system will be established. The security protection architecture of cloud computing support environment is shown in Fig. 2.

1. **Network Virus Protection**

Network level virus protection is carried out for the cloud computing environment. Viruses, Trojans and zombie clients are investigated and killed, which can prevent the virus invasion from invading the cloud platform. In backbone of science and technology cloud of China, HTTP, FTP, SMTP and POP3 protocol traffic are filtered and killed, such as virus in the compressed file (zip, rar, 7z, etc.), real-time update virus samples. Currently, the virus signature library already reached millions of levels, which can prevent the spread of the virus timely and effectively.

2. **Detection and Defense of Intrusion**

System and business application in cloud environment will become the focus of attack. Therefore we need to establish comprehensive defense means to against application layer attacks, and avoid the weakness in all kinds of application security protection in the cloud platform. For the host and the business application system, we carry on the security hole detection and protection, such as the backdoor hole, operating system vulnerabilities, FTP tool holes, loopholes in the database, and so on. These operations are provided with real-time leak detection and protection ability. In addition, real-time protections against exploits for the operating system and applications are implemented. The operating system and applications include

HTTP service (Apache, IIS, etc.), FTP services, Mail services, OpenSSH, Mysql, Oracle, safety protection, back door program prevention, protocol vulnerability protection, exploits protection, network sharing service protection, shellcode prevention and spyware prevention. For specific attacks, such as SQL injection attacks, cross-site scripting attack, CSRF cross-site request forgery, website scan, file contains vulnerabilities attack, a directory traversal vulnerability and risk weak passwords, we apply specific protective security and defense in depth through the series of security measures, combine with the application of attack rules of static and dynamic defense mechanism based on attack process, and realize the two-way content detection. Thus, a wide range of application security threats against protection can be provided.

Intrusion detection operations are performed based on import and export traffic in cloud environment, such as testing Trojan, malicious code, spyware, backdoor, and so on. When an internal host in the cloud computing environment is infected with viruses and trojans, it will attempt to communicate with an external network. These traffics are accurately identified, blocked and logged according to the strategy. It has the good safety protection effect to accurately locate the problem terminal, block its network traffic, and eliminate illegal malicious.

3. **Cloud Access Security Agent**

The cloud access security agent provides unified access environment for application security of cloud services, active safety defense and other series of security services. For all kinds of application systems, we construct the unified security control platform. Through link encryption, centralized login, dual factors and other security mechanisms, the security agent ensures management of traffic link encryption, unified control and operation audit. Users' multi-dimensional securities for their business system management and the security and unified business system management behaviors can be realized, which meet the related information security requirements.

3.2 *Security in the Cloud Environment*

In the virtualization environment of cloud computing, network traffics in virtual hosts are applied to security protections and controls of different services. In the virtualized environment, integrated security protection for north-south and east-west traffic is formed by virtualization security technologies, such as virtual firewalls, virtual intrusion prevention, and virtual antivirus. The security management mode of the cloud internal environment is shown in Fig. 3.

As for the security and application service protection of layer 2 to 7 of cloud internal environment, virtualized agentless mode is adopted to divide the fine field

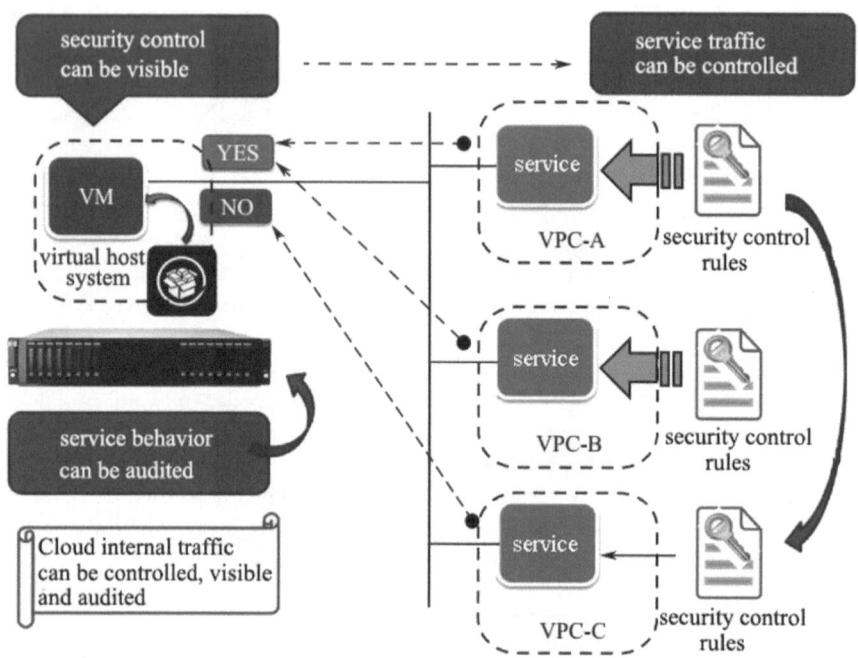

Fig. 3 Security management mode of the cloud internal environment

and visualize access control, such as security domain, access control, virus prevention, intrusion prevention, vulnerability detection, web security, data leakage protection. With security software resource pool, the security protection of on-demand allocation and flexible deployment can be achieved.

In the cloud computing environment, the boundaries of each tenant are separated and security isolation controls are provided for them. Considering the features of cloud, each tenant is divided as a separate virtual area, on top of this, different security level security domains are formed. Each domain is isolated by virtual firewall to avoid security risks caused by data exchange between unsafe level services.

Virtual firewall is deployed as a virtual host at the boundaries of the tenant virtual network and the cloud network, it provides security protection and isolation for east-west traffic of tenant application system, including security isolation between tenants, to prevent illegal personnel from conducting security attacks within the cloud platform. Tenants apply external security protection and isolation to prevent illegal personnel from conducting security attacks from the external network and the Internet.

3.3 Deep Security Protection

1. Security Detection

The capability to deeply detect the security threats in the cloud environment must be further improved. Traffic traction, distribution, detection, and re-injection security detection environment can be built to implement security protection, such as traffic cleaning, deep intrusion detection, and data flow auditing. According to important service system, service traffic of key terminals, and log characteristics, security attacks such as 0-DAY vulnerability discovery and APT can be deeply detected through fine-grained traffic division.

2. Risk Assessment

Based on platform-level vulnerability scanning analysis and test verification system, the science and technology cloud of China protection provides comprehensive security risk analysis and risk assessment with 1 + N distributed cluster architecture. With parallel and batch scheduling of work tasks, it provides information system backstage, website system, middleware and other special vulnerability detection, and adopts multi-mode injection, cross-site attack, counterfeiting cross-site request and other multi-environment technology verification methods. Further, a combination of threat warning and active defense mechanism is established to provide effective technical means for systemic risk assessment.

3. Service Audit

The security audit platform based on the cloud computing environment can audit the published content in the cloud, and discover the illegal and violation information in the service system traffic timely. And for the application protocols whose contents can be distinguished, such as e-mail, HTTP, FTP, it can check the transmission content, find out and record the violation information timely, and propose warnings.

Raw communication traffic will be recorded if attacks or violation information exists. Within it, logs can be recorded to provide traffic logs and behavior logs. Traffic logs provide information about traffic and behavior logs provide user activity logs, user behavior information (login IP, access resources, time, authentication method), user activity, user/user group traffic rankings and queries, user/user group flow rate trends and queries.

4. Code Audit

The active mode self-service code audit security service is provided for each service application system in the cloud environment. The security audit runs through the life cycle of the service system, and detects the security vulnerabilities with the white-box test mode in the software system code. For the coding phase, test phase, and delivery acceptance phase of the system development process, security audit detection is performed on the system source code of each phase. And the source code security problem is analyzed, detected and verified by using a data flow analysis engine, a semantic analysis engine and a control flow analysis engine to provide real status information on the security quality of the service system.

4 Summary and Outlook

From the perspective of supporting environment, cloud environment and deep detection, the science and technology cloud of China security protection establishes a security system of the cloud platform in terms of platform security, tenant security, service system security, service data security, and service access security, to ensure the service system of tenants on the cloud platform operate efficiently, reliably, stably and safely from the perspective of platform stability and service stability.

For the virtual environment of the science and technology cloud, a flexible, autonomous, on-demand, and visual integration security protection platform is provided, and advanced technologies and products are integrated in various ways. Each security protection function is deployed in the form of virtualization software and exists in the security protection platform in a virtualized form, which automatically adapts to VMware, KVM, XEN and the service environment of the domestic virtualization platform. It implements security detection in the manner of routing, one-arm, and vSwitch deployment. After being security detected and protected by device, the data traffic reaches and secure the service system.

In the future, it is necessary to further strengthen the capabilities of security protection in the cloud computing environment, and to strengthen the capabilities of the network security early warning, monitoring, and to emergency response in the important application environments of science and technology cloud. It is also necessary to strengthen the granularity of the deep detection of abnormal traffic information in the science and technology cloud computing environment, to enhance the global early warning capability for attacks performed by cloud platform user, to secure network information security for the science and technology cloud users, and to further enhance important information security support capabilities including content auditing, website security, and link security protection. Improving the capability of the cloud computing environment traffic distribution platform will provide raw traffic and output on-demand for applications such as security situational awareness. The detection of unknown attack modes needs to integrate security function linkage, and combine with big data mining and intelligent AI pattern deep recognition technology. The controllable management of security threats can be realized through the automation of security cloud platform. Based on key technologies such as protocol oblivious security service chain orchestration, the security assurance environment will be comprehensively improved to provide a secure link protection and security environment of access management for computing and data platforms, and to comprehensively enhance security assurance capabilities in cloud computing environments.

References

1. Office of the Leading Group of Informationization Work of the Chinese Academy of Sciences. Technology cloud overall implementation plan, 2012.
2. Office of the Leading Group of Informationization Work of the Chinese Academy of Sciences. Network security system construction project implementation plan, 2016.

Chun Long is a communist, he is a doctor of engineering and senior engineer. Long is also the director of Cyberspace Security Technology and Application Development Department of Computer Network Information Center, CAS. During his work, he hosts a number of national projects, including the research information infrastructure construction and application demonstration project based on CNGI, National next generation Internet security project, National development and reform commission information security special project and so on. In addition, he is in charge of many institutional scientific projects, such as Chinese Academy of Sciences security and service project, safety special project, etc. He is engaged in the researches of computer network and information security, secure data mining and analysis and next generation internet architecture. Recent years, he has published more than 10 technical specifications, journals and academic papers and obtained multiple patents and software copyrights.

Achievements of Informatization
in Interdisciplinary

Construction of Chinese Agricultural Science and Technology Cloud

Huoguo Zheng, Haiyan Hu, Le Tian and Yanchao Liu

Abstract The agricultural science and technology cloud of is an important project in Chinese Academy of Agricultural Sciences (CAAS) informatization planning from 2015 to 2017. The cloud aims at realizing the integration of network, computing, storage and application of scattered in various scientific research resources, and enhance the ability of innovation of science and technology of CAAS. This chapter presents the ideas and techniques of construction of the cloud, introduces the scheme of network, virtualization, storage and backup, security. Finally, the future development direction of the cloud is discussed briefly.

Keywords Cloud computing · IT infrastructure · Research network

1 Introduction

Through the network, cloud computing centrally manages and dynamically distributes distributed computing, storage, and software resources, enabling information technology capabilities to be delivered on demand like water and electricity. It has rapid flexibility, scalability, resource pooling, and extensive network access. Features such as multi-tenancy are major innovations in the information technology service model.

Cloud computing is the general trend of information infrastructure in the future, and it is also an important foundation for big data analysis and processing. In the domestic research institutions, the Computer Network Center of the Chinese Academy of Sciences has passed the "12th Five-Year" informationization project of the Chinese Academy of Sciences, and built the first research cloud platform in China, covering infrastructure cloud services, data storage and management cloud service environment, massive scientific data analysis and application. Demonstration, scientific data integration and sharing services, in the future, the

H. Zheng (✉) · H. Hu · L. Tian · Y. Liu
Agricultural Information Research Institute of CAAS, Key Laboratory of Agricultural Big Data, Ministry of Agriculture and Rural Affairs, Beijing, China

© Publishing House of Electronics Industry 2020
China's e-Science Blue Book 2018,
https://doi.org/10.1007/978-981-13-9390-7_15

Academy's systems for the hospital and the public will be migrated to the cloud platform.

For CAAS, building a private cloud platform that used to integrate the existing information system and realize the overall deployment the information systems not only could effectively solve the problem of decentralized construction, high operating costs and security risks, but also centralize research data resources, which could lay the foundation for scientific research management based on big data.

In order to enhance the ability of technological innovation, the CAAS Informatization Development Plan (2015–2017) clearly puts forward the task of "building the cloud computing platform that used to provide access to computing resources, and realize the effective mining, integration and sharing of decentralized resources" [1]. In 2017, the CAAS agricultural technology cloud platform was launched, and it is planned to build an information infrastructure environment based on cloud computing technology.

2 Demand Analysis of Agricultural Science and Technology Cloud

According to incomplete statistics, as of November 2017, there are about 500 application systems running in 12 institutes in Zhongguancun District of CAAS, involving more than 500 servers. Just CAAS Agricultural Information Research Institute has nearly 300 physical servers those purchased varies from 2007 to 2016, and more than 80% of servers used for more than 5 years. Through preliminary research, it is found that there are two major problems in the existing information infrastructure: on the one hand, most servers just deployed only one application system or one database system, and the usage rate of its CPU and memory is less than 10%, which caused a large waste of computer room's space, electricity and investment. On the other hand, with the development of scientific research information, more and more application systems are continuously developed and deployed online, the problem of lacking IT equipment is still existing, even if buy more new equipment. And more, the long running time from the procurement of equipment to a working system seriously affects the progress requirements of system development.

In response to the actual needs of CAAS, combined with the characteristics of cloud computing technology, the construction of agricultural science and technology cloud (hereinafter referred to as "Agricultural Science Cloud") will focus on the following issues:

1. Increase the utilization of IT equipment such as servers. Compared with the traditional IT architecture, virtualization technology can significantly improve the utilization of resources such as servers and storage. In the cloud computing environment, resource-intensive applications can run on the bare-metal architecture of the virtualization host. This architecture provides advanced memory

management capabilities and supports vertical scaling of physical and virtual machines. Users can also double the consolidation rate when running virtual machines and maximize resource utilization on CPU, memory, and hard disk space.

2. Effectively reduce server and storage costs. As processors become more powerful and the suite expands to accommodate the most demanding enterprise workload conditions, more and more applications can run in the cloud computing environment. According to industry experience, the resource pool model's rate of resource utilization can be more than three times higher than the traditional model, and the total IT cost can be reduced by 30%.

3. Shorten the time for business online, and respond to business needs in a timely manner. When new applications need to be deployed, the platform can be deployed quickly. After the deployment is complete, the storage and network need no additional configuration, and can be quickly adjusted to meet the needs of new IT applications.

4. Improve business system continuity. Cloud computing can increase the basic level of availability for almost any application, ensuring that service level agreements are met. Cloud computing can significantly reduce planned downtime, prevent unplanned downtime, and quickly achieve failure recovery by eliminating planned downtime from common maintenance operations, also could provide higher availability independent of hardware, operating system, and applications; And more, by using virtual machine's ability of automatical restarts, cloud computing could achieve the quick recover for the operating system and server.

5. Improve data integration capabilities. For any organization's management, after fully understanding the importance of data, it needs to have enough people and ability to integrate, build and improve the data management infrastructure. With the cloud platform, each unit can integrate various application systems and data, which is beneficial for data scientists to analyze and mine them to generate additional value.

3 Design Ideas and Technical Architecture

3.1 Design Ideas

The construction of the Agricultural Science Cloud will follow the principles of advancement, integration, sharing, reliability, scalability and manageability in its overall architecture.

1. Advancement: The overall architecture design has certain advanced and forward-looking, also in line with the direction and trend of information technology development, which could meet future needs.

2. Integration: Through the cloud platform, the information infrastructure and information resources are integrated, which could maximize the resources' ability, also achieve the unified deployment and management.
3. Sharing: Through the use of virtualization technology, network resources, computing resources, storage resources, and management resources are shared. Through the "virtual" resource pool, resources can be dynamically allocated and adjusted. And for managers, they can optimize or improve the system architecture just by simply configuration and adjustment on the management interface.
4. Reliability: With the adoption of mature and stable technologies and products, the high reliability and high availability of the information infrastructure are realized, and also could protect the key business data necessarily. So the overall system has the ability to prevent equipment failure.
5. Scalability: With the continuous development of scientific research management and the continuous improvement of management level, the requirements for the system will also change. Therefore, the system should have good scalability and can be easily extended horizontally and vertically.
6. Economic: The system must be based on mature and stable products, fully consider the utilization and continuation of existing equipment, reduce implementation costs, operational risks, and subsequent operational and maintenance costs and human resource costs of the system.
7. Manageability: The device needs to support all functions that can be centralized to facilitate routine maintenance and troubleshooting, and improve the efficiency of operation and maintenance.

3.2 Technology Architecture

From the perspective of cloud computing, the Agricultural Science Cloud is a private cloud, and it's different from the science and technology cloud of the Chinese Academy of Sciences, the later adopts a similar public cloud construction idea. In addition to providing services for the institutes under the Academy of Sciences, the science and technology cloud also provides Cloud computing services to other research units outside the Academy of Sciences [2].

According to the design philosophy of the private cloud, the Agricultural Science Cloud adopts versatile and mature technologies and solutions for overall architecture design [3]. The overall technical architecture mainly provides services for users from three aspects: infrastructure virtualization, on-demand service, and application system ubiquitous access [4]. Infrastructure virtualization and on-demand service are the basic features of cloud computing,and the research institutes and business teams do not need to pay attention to the infrastructure of the data center computer room, network, computing, storage, etc. when using the Agricultural Science Cloud. It is only necessary to propose the actual demand for the above resources, and the resource allocation mechanism of the cloud platform can meet its requirements. Ubiquitous access means that with the development of intelligent mobile terminals and wireless networks, various applications and services deployed on the Agricultural Science Cloud should be extended to all aspects of scientific research, so that researchers can obtain various services such as calculations and data anytime and anywhere [5].

The overall technical architecture of the Agricultural Science Cloud is shown in Fig. 1. The cloud IT infrastructure layer and service layer are the key to the overall architecture of the Agricultural Science Cloud. Through the virtualization technology, the Agricultural Science Cloud realizes the virtualization of computing, storage, network, security and other resources, and resource pools and automated resource scheduling layers provide interfaces for service layer that supports the operation of various application systems [6].

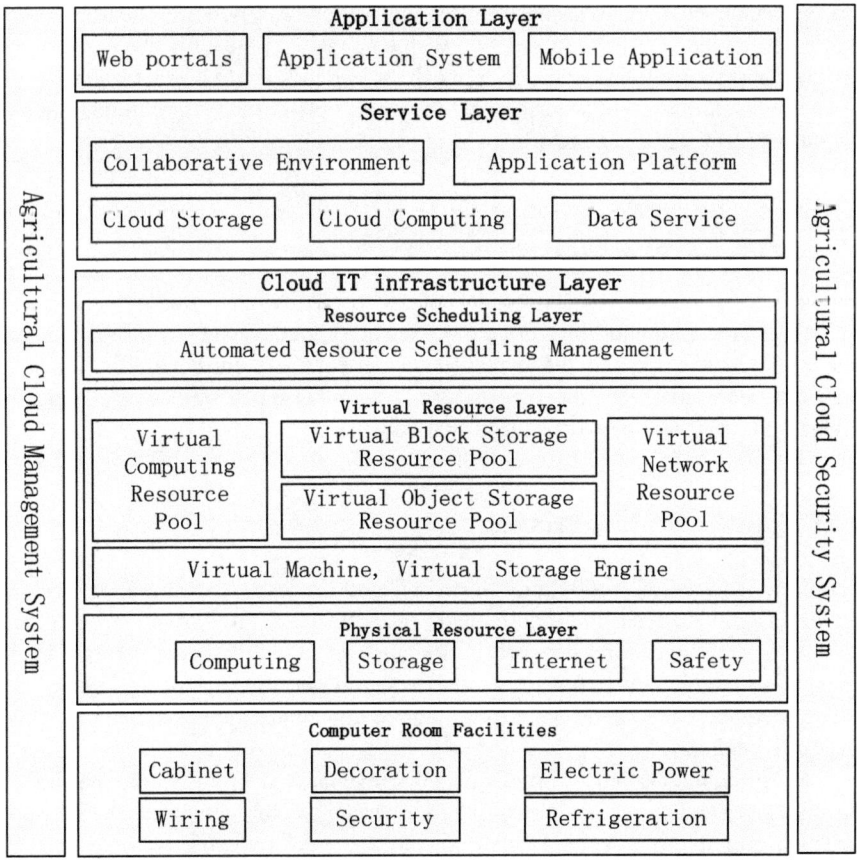

Fig. 1 Agricultural science and technology cloud overall technical architecture

With the continuous development of scientific research methods and means, agricultural remote sensing data, literature data, genomic data and other scientific research data are growing with geometric index. Agricultural research with big data thinking has become increasingly mainstream, and agricultural research is entering the era of big data. Various subordinate units of CAAS are building various types of big data processing platforms. In the future, it needs to integrate the computing platforms in accordance with the architecture of high-performance computing cloud services, and to realize the "cloud computing" platform in a broader sense.

4 Infrastructure of Agricultural Science and Technology Cloud Platform

4.1 Cloud Platform Infrastructure Overall Design

On the principles of phased construction, the Agricultural Science Cloud platform will be promoted step by step and also with emphasis. In the early stage, we focus on the construction of infrastructure, such as network, virtualized resource pool (computing, storage), and security. And the overall plan of the Agricultural Science Cloud platform infrastructure is shown in Fig. 2.

4.2 Network Planning

In the basic network part, the Agricultural Science Cloud use a typical separation of the business network and the management network like most cloud computing platform. And similar to the construction of computing resources, it's logically divided into physical network and virtual network.

4.2.1 Physical Network Planning

The basic network of Agricultural Science Cloud carries CAAS important service traffic. In order to meet the requirements of large-speed fast forwarding of cloud computing, this backbone network is designed to be a full-scale 10 Mbps interconnection. At the same time, in order to avoid service termination caused by network equipment's single point of failure, all key devices are designed with dual device redundancy.

The Agricultural Science Cloud's network layer adopts a two-tier architecture of the core layer and the access layer. The two-tier architecture simplifies the network structure and improves data forwarding efficiency by reducing the intermediate aggregation layer. And considering the lack of scalability of the Layer 2

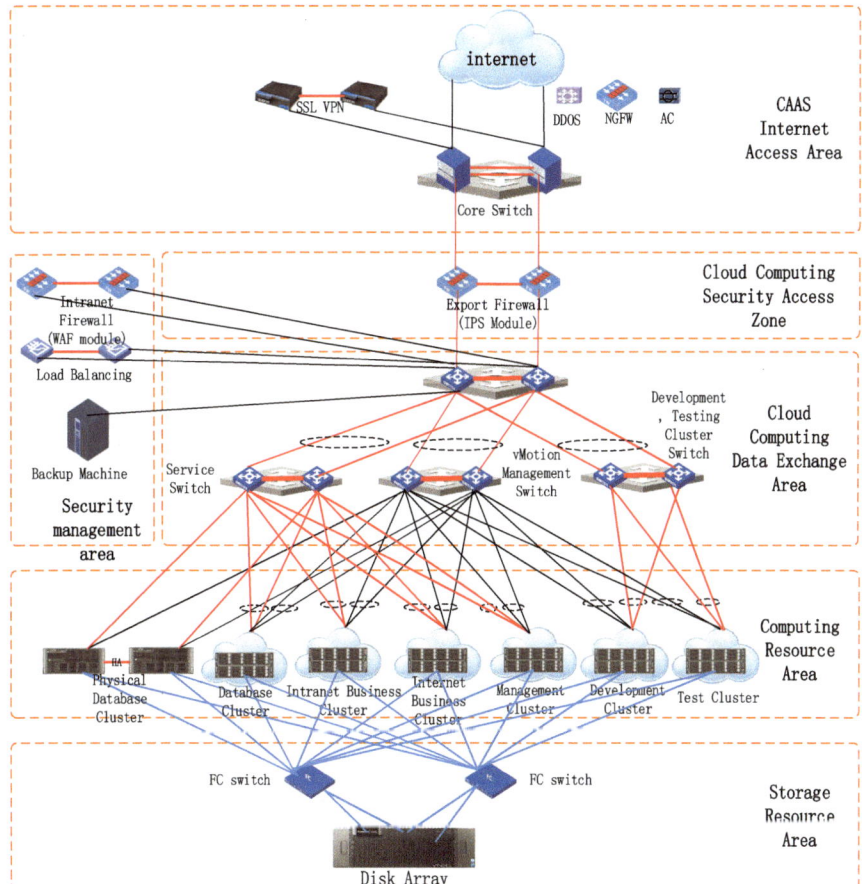

Fig. 2 Cloud platform infrastructure platform architecture

architecture design, the high-end box switch is used in the core switching layer to support the expansion of network.

Because the OSPF routing protocol is adopted in the current network, so this time, we use same protocol for the interconnection between the network and the existing network. The OSPF's running range is shown in Fig. 3.

4.2.2 Virtual Network Planning

The virtual network needs to be configured with a management network, a vMotion network, and a service network. According to the practices, the Management Network and the vMotion Network are best to be configured on the virtual standard switch for management and troubleshooting. The service network is configured on

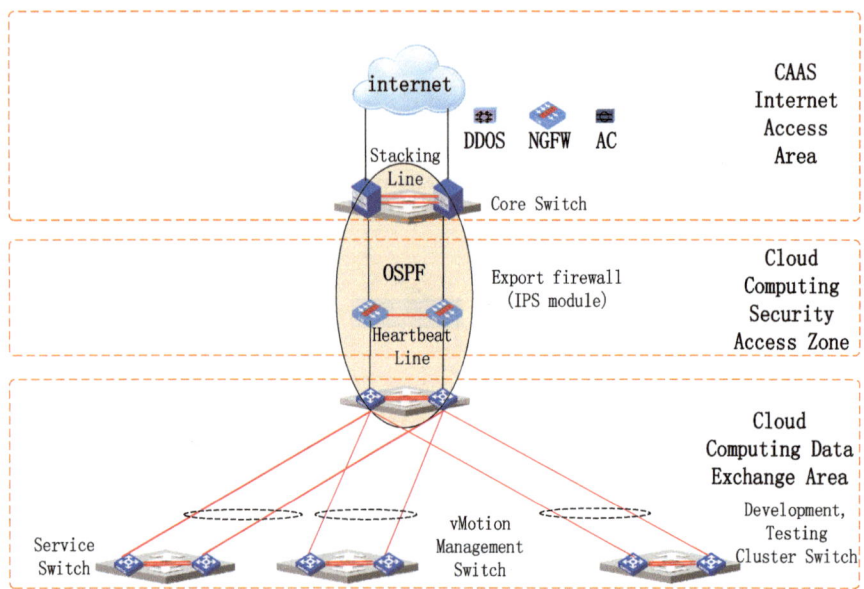

Fig. 3 OSPF docking

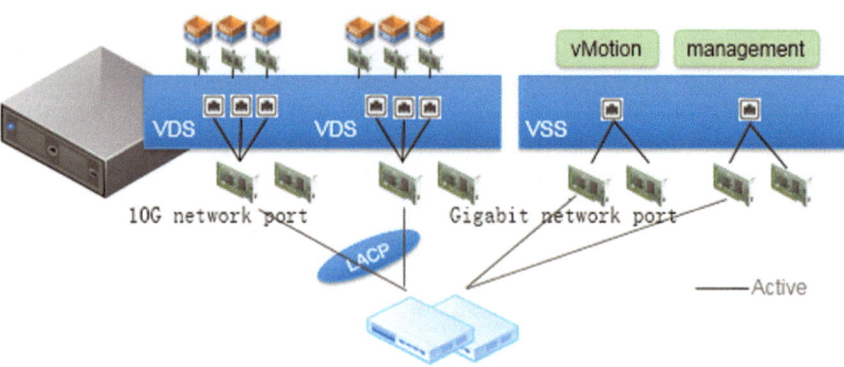

Fig. 4 Virtual network relationship

the virtual distributed switch of the dvSwitch0 to facilitate configuration change and unified management. See Fig. 4 for details.

The Management Network and vMotion Network are configured on the virtual standard switch vSwitch0, and the vSwitch0 uplink allocates two Gigabit NICs. The Management Network and vMotion Network traffic are configured in a dual-master mode for redundancy. The service network is configured on the virtual distributed switch dvSwitch0, and the dvSwitch0 uplink allocates two 10 Gigabit

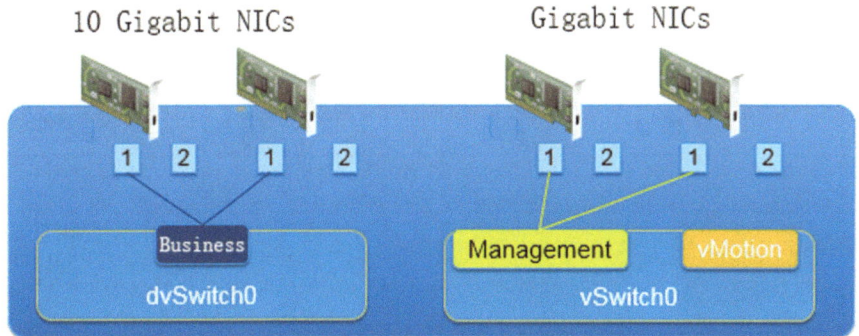

Fig. 5 Correspondence between virtual network and physical network

NICs, and the LACP binding is configured to provide maximum bandwidth and achieve dual- master mode redundancy (Fig. 5).

4.3 Virtualized Cluster Planning

4.3.1 Planning of Virtualized Computing Resource Pool

Virtualize the server with virtualization software, and configure the virtual infrastructure suite software on each server to split the powerful computing resources into virtual machines for different applications on a single physical server entity, then manage the resource pools according to different business, which could make full use of server resources. And for each virtual server, it is equivalent to a traditional single physical server in terms of function, performance and operation.

However, the server for the core database service still runs on the physical hardware platform of the enterprise server, ensuring the stability of the core data, and above these we also use third-party independent dual-machine software to ensure high availability of the data system.

4.3.2 Server Cluster Planning

For the different type of business needs of CAAS, the Agricultural Science Cloud virtualization platform divides the servers into 7 clusters, which are database clusters (including physical data clusters and virtualized database clusters), intranet business clusters, internet business clusters, management clusters, development cluster resource pools, and test clusters. Among the seven clusters, except the physical database cluster, all are implemented through virtualization technology. These cluster resource pools are divided into different service areas to achieve

Table 1 Design of each cluster function

No.	Cluster name	Main function
1	Physical database cluster	The institution's database of important business systems (such as collaborative office systems and mail systems) is deployed on clusters built by physical machines
2	Database cluster (virtual machine)	Other business systems' application system and database is separated, and the database is deployed centrally on virtualized database clusters
3	Intranet business cluster	Used to deploy business systems that are not released to the Internet and are only used internally by the organization
4	Internet business cluster	Used to deploy business systems for Internet publishing
5	Management cluster	Used to monitor cloud platform networks, physical servers, virtual servers, security management, etc.
6	Development cluster	Used in the internal application system development environment, after the system development is completed and is migrated to the test cluster for testing
7	Test cluster	Used for pre-line testing of the internal application system of the unit. After the test is completed, the application system will be migrated to the intranet service cluster or the Internet service cluster

security isolation, which facilitates unified management, simplifies resource allocation, and privilege setting (Table 1).

4.4 Storage and Backup Planning

In the cloud computing environment, different tenants and different business applications have different requirements for storage, which need the underlying storage devices have flexible policy's support, and allow users to perform storage policy configuration on demand.

4.4.1 Storage Planning

According to the Agricultural Science Cloud infrastructure plan, the hybrid storage architecture will be built in the future, including traditional SAN and NAS storage, and will also introduce software-defined storage to take full advantage of the characteristics of various storage devices to meet a variety of business needs. In the initial stage of project construction, considering the factors such as the number of business systems, platform scale and budget, it is planned to configure a set of fiber storage array products and configure redundant fiber switches to form a standard SAN centralized storage architecture. The package files that is produced by the virtual architecture suite of the machine are stored on the storage array. Through the shared SAN storage architecture, the advantages of the virtual architecture can be maximized, and also make the running virtual machine can be migrated online, fault tolerant (FT), dynamic resource management, and resource plug-and-play.

As the core storage of the data center, the storage product will adopt a dual redundant SAN network architecture, provide NAS functions, and support multiple connection protocols. In terms of selection, the storage device needs to support different types of disk intermixing, including SSD hard disks, SAS disks and SATA disks, and a layering technology that can significantly improve efficiency and save costs.

4.4.2 Backup Planning

As with traditional business architectures that require a data backup system, the Agricultural Science Cloud platform is essential for data backup while optimizing the virtualization system. The need to back up data includes not only application data and user data, but also system configuration, operating system, and so on. Because of the nature of the virtualized infrastructure, all of the virtual machine's data is encapsulated in several separate files, so virtualization greatly simplifies the process of backing up and restoring all of the system's data.

Considering the traditional backup software + tape library/optical disk library/ disk array backup mode, involving many software and hardware products, technology diversification, backup system deployment is complex, poor maintenance; and correspondingly, the backup machine has Simple deployment, cost-effective, simple maintenance, easy to use and other advantages. For the Agricultural Science Cloud platform business system, it is planned to adopt the backup integrated machine solution (Fig. 6) to realize the backup of Oracle, MySQL and other databases, also Linux, Windows and other system files, as well as virtual machine systems, which could improve the security of system data.

As shown in the figure above, a set of backup machines are connected to the Agricultural Science Cloud platform to centrally back up existing business system

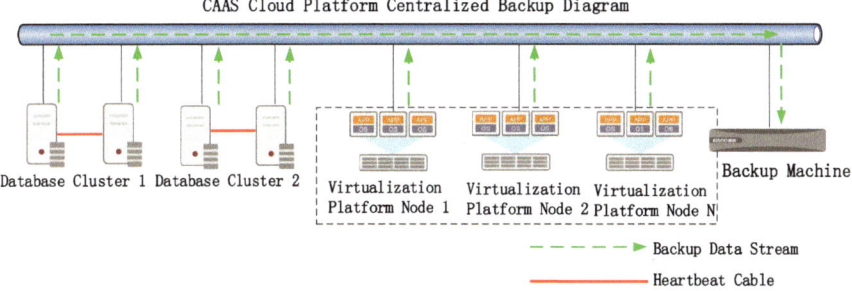

Fig. 6 Based on the centralized backup solution of the backup machine

data. For the database environment, you need to install a backup client on each server, the backup system comes with the client software under Linux and Windows. After the client installation is completed, under some certain strategies, the backup machine console issue backup commands to implement the backup of data. For the VMware virtualization environment, the backup appliance can extract platform data through the interface with the virtualization platform's own API to achieve centralized backup of the clientless agent.

4.5 Safety Planning

Cloud computing technology revolutionizes traditional IT infrastructure, applications, data, and IT operations management. For security management, the cloud computing architecture also brings new information security risks [7], especially to meet the cybersecurity law and new level protection regulations implemented in 2017.

According to the information system security protection requirements, the cloud platform security protection is launched around five levels: network protection, application protection, host protection, management protection, and virtualization platform protection. This chapter focuses on network security, virtual host security, and application security.

4.5.1 Network Security

In Fig. 2, two firewalls (with IPS modules) deployed in the cloud platform network exit are deployed in a three-tier master/slave mode to filter traffic from outside the cloud platform to provide Internet access security protection for the cloud computing network. Two firewalls are connected to the core switch in the Layer 3 master/slave mode to implement intranet security protection. At the same time, different security domains are classified according to different services. Different security domains are isolated based on VLAN. When the service system needs to communicate with each other, the traffic is directed to the internal network firewall for filtering, and then the traffic is forwarded to the destination service area. At the same time, two SSL VPN devices are connected to the core switch to provide secure encrypted tunnels for Internet users to ensure network communication security.

4.5.2 Host and Application Security

With the development of cloud computing and the application of virtualization technology, server boundaries have become increasingly blurred. Traditional firewalls have been unable to support the protection of network boundaries in

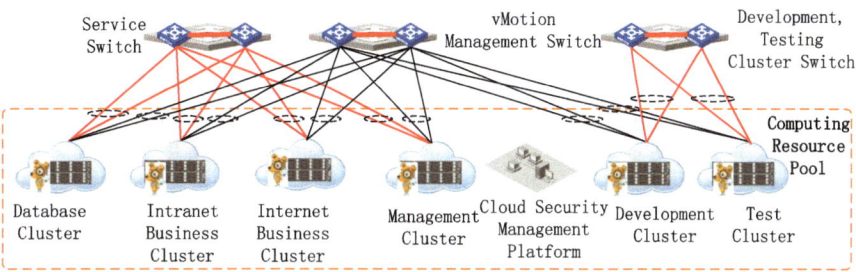

Fig. 7 Software-based server and application system protection

virtualized environments. In addition to traditional attacks against servers, new targets have emerged. The attack method of the virtualized server cannot effectively isolate and protect resources in a multi-tenant virtualized environment.

In order to solve the problems faced by traditional security protection in the cloud computing environment, a software-only security solution for servers and application systems is deployed in the Agricultural Science Cloud environment, and security agent software is installed on each virtualized server to provide security monitoring and protection for network layer, system layer and the application layer (as shown in Fig. 7). By deploying a private cloud security management platform in the management cluster, we could achieve rapid analysis and mining based on the collected logs, also accurately locating attack events and attack track and early alert.

At the same time, as shown in Fig. 2, the WAF module is added to the internal network firewall of the security management zone, and the HTTP and HTTPS service traffic is directed to the WAF device through the traffic diversion policy on the core switch, which provide application layer's security protection for the portal and application system.

5 Summary and Outlook

In order to embrace the new generation of information technology such as cloud computing, big data, and the Internet of Things, CAAS is actively planning to adopt a cloud computing key technology such as virtualization and distributed storage to build an agricultural technology cloud platform and create a new generation of information technology for the whole institute. The infrastructure platform realizes centralized management of IT hardware resources, shortens the deployment and upgrade time of business systems, and effectively reduces the TCO of information construction, also make application systems get rid of the constraints of physical hardware, and lay the foundation for the integration and sharing of agricultural scientific research resources [8].

However, we must be soberly aware that the CAAS cloud platform is still in its infancy, and there is still a large gap between the existing network or computing power and the rapid demand growth of researchers, there still has a long way to go from infrastructure (IaaS), platform (PaaS) to software (SaaS). With the deepening of research on agricultural remote sensing, bioinformatics, agricultural internet of things, and intelligent equipment, information technology is increasingly infiltrating the agricultural research process. In the future, demand for high-performance computing, big data analysis, and knowledge services will continue to emerge and new ones will be proposed. With higher requirements, more and more platforms and systems will appear on the Agricultural Science Cloud. It is hoped that CAAS could seize the rare historical opportunity and deepen the integration and sharing of scientific research resources. After the completion of the cloud platform of CAAS, it will provide various information services for 34 research institutes and more than 300 innovation teams.

References

1. CAAS Informatization Work Leading Group Office. CAAS Informatization Development Plan (2014–2025). 2014.
2. Kai Nan, Jianhui Li, Hong Wu, et al. Status and prospects of science and technology cloud of Chinese Academy of Sciences. Blue Book of China Scientific Research Informatization. Beijing: Science Press, 2015.
3. Jiongjiong Gu. Cloud computing architecture technology and practice. Beijing: Tsinghua University Press, 2016.
4. Wanyun Lei. Cloud computing technology, platform and application case. Beijing: Tsinghua University Press, 2011.
5. Michael J. Kavis. Architecting the Cloud: Design Decisions for Cloud computing Service Models. Wiley, 2016.
6. Yili Gong, Lian He, Chuang Hu. Cloud computing concept, technology and architecture. Beijing: China Machine Press, 2016.
7. Chi Chen, Jing Yu. Cloud computing security system. Beijing: Science Press, 2014.
8. Yongwei Wu, Zhongyuan Qin, Zhenyu Li, et al. Cloud computing and distributed systems: from parallel processing to the Internet of Things. Beijing: China Machine Press, 2016.

Huoguo Zheng, Ph.D., associate researcher, master tutor, mainly engaged in agricultural information technology, network technology research. He is currently the deputy director of the CAAS Agricultural Information Research Network Center. He has hosted and participated in more than 20 scientific research projects such as the National High-tech R&D Program of China (863 Program), the National Key Technology Support Program, and the Ministry of Agriculture special. He has won 3 provincial awards and published more than 30 chapters.

Key Technology and Application Platform of Agricultural Economic Spatial Information Service—Chinese Agricultural Economic Electronic Map

Shengping Liu, Yeping Zhu, Yue E and Xiuming Guo

Abstract Taking agricultural economic data management, analysis and decision service as the main line, Key Technology and Application Platform of Agricultural Economic Spatial Information Service (Chinese Agricultural Economic Electronic Map) established the largest country rural economic basic database, carried out long-term interdisciplinary application and research, provided excellent information services for the agricultural decision making departments through three aspects which included data management, information analysis and decision service. Establishment of Chinese Agricultural Economic Electronic Map played an active role in the development of rural and country informatization in China, and stimulated the economic, social and environmental development. At the same time, our achievement carried out widely academic cooperation with colleagues both at home and abroad, and was awarded the prize of Scientific and Technological Advancement by Beijing and Chinese Academy of Agricultural Sciences successively.

Keywords Spatial information service · Electronic map · Agricultural economy · Intelligence service

The Intelligent Agriculture department from agricultural information institute of CAAS started collecting and managing basic rural economy information about and has created the biggest county-level basic rural economy information database in China. Following data management, analysis and service, long-term cross-disciplinary and cross-department technological development and application was developed. The work was carried out from data management, information analysis and decision service, and the research achievement "the key technology and application platform for rural economy spatial information service (Chinese rural economy electronic map)" was achieved, which supported effective information service for agricultural policy-making and production departments, research and

S. Liu (✉) · Y. Zhu · Y. E · X. Guo
Institute of Agriculture Information, Chinese Academy of Agriculture Sciences/Key Laboratory of Agro-information Services Technology, Ministry of Agriculture, Beijing, China

© Publishing House of Electronics Industry 2020
China's e-Science Blue Book 2018,
https://doi.org/10.1007/978-981-13-9390-7_16

development institutions, promoted harmonious development for agriculture, society, economy and environment and pushed forward agricultural and rural informalization in China. The achievements were shared at home and abroad, and wined 2012 first prize award of CAAS science & technology progress and 2013s prize of Beijing science and technology progress.

1 Background

China's Ministry of Agriculture started creating county-level basic rural economy information database from 1980s, and has owned a batch of valuable information resources which have become an important reference for production management and macro decision-making for agricultural policy-making and production departments, research and development institutions. However, quite a few difficulties still exist in agricultural economic information resource management and utilization. Firstly, agricultural economic data collection and management is not normative, which appears in the single standard absence of statistical index, difficulties in managing the dispersive data, and difficulties in data sharing and exchange. It seriously affected the information utilization efficiency, and agricultural economic data management rules are cried for to achieve centralized and normative management; Secondly, agricultural economic data analysis was lack of technologies and systems, and was weak in automation, which seriously influenced the deep mining and analysis for agricultural economic data, so the technologies and systems development for agricultural economic data was urgent [1]; Thirdly, the absence of exhibition visualizations for decision-making analysis service based on agricultural economic information, the simplex terminal for service and the difficulties in collaboration among many decision-making analysis services seriously influenced the decision-making service effect [2], and agricultural economic information spatial service platform is urgently needed to build (Chinese agricultural economic electronic map data service platform).

Aiming at the above issues, Intelligent Agriculture department from agricultural information institute of CAAS carried out some special researches on macro decision analysis and application software development for agricultural economic data from 1990s in some projects such as "agricultural information system application software research and development (863 Program)", "agricultural economic management and policies and laws information system (the ninth five-year plan for scientific and technological breakthroughs)", "agricultural macroscopic management and decision-making information technologies research (key scientific and technological breakthrough in the tenth five-year plan)", "agricultural resource utilization and management technologies research and application (Support of science and technology in the 11th five-year plan)"; Meanwhile, along with space information technologies development and widespread use in other areas, department of market and economic information from ministry of agriculture launched the "Chinese agricultural economic electronic map" project to research how to apply the spatial information technology to agricultural economic information management and service.

Based on the above projects, a batch of software has been created for practical application, and the goals to vitalize and excavate agricultural economic basic resource and support portable information technology service tools for agricultural macro decision making and analysis were achieved. It played an important role in management for agricultural resources information, utilization promotion for area agricultural resources, capacity improving in analysis and control for agricultural emergency, it supplied effective and exact information service for relative departments, and it will promote the coordinated development of agriculture, society, economic and environment to realize sustainable agriculture and agriculture informatization in China.

2 Key Technical Framework

The Key Technology and Application Platform for Agricultural Economic Spatial Information Service is a complete technical system for Chinese agricultural economic statistics on the basis of normalized data management, with the key function of systematic data analysis and for the purpose of visualized data decision services. To fundamentally solve problems in current agricultural economic statistic information such as nonstandard management, insufficient analysis technology and unvisualized display, the project focuses on data management, data analysis and decision-making service to perform research [3] and has made innovative breakthroughs in a number of core items, including construction of China agricultural economic database, analysis model of agricultural economic information, analysis of regional industrial advantages, optimized layout of agricultural production, analysis of agricultural economic spatial information, specific analysis of agricultural normal disasters and unexpected incidents, dynamic analysis of agricultural product market prices, analysis of income gaps of countryside residents, Agent-based agricultural economic information analysis and group decision-making, construction of China agricultural economic electronic map decision-making service platform. The technical framework of the project is shown in Fig. 1.

After creating Chinese largest agricultural economic basic material database, the Key Technology and Application Platform for Agricultural Economic Spatial Information Service (Chinese Agricultural Economic Electronic Map) has performed research on data standardization and intelligent processing technology for long time periods, continuous changes of administrative divisions and inconsistent index units; for long-term shortage of agricultural economic information analysis tools, especially shortage of highly flexible and adaptable information technical means and software to meet industrial needs, has integrated an application database, GIS, GPS, modeling and intelligent technology, and performed technical system research on a series of core technologies with systematic data analysis and assisted

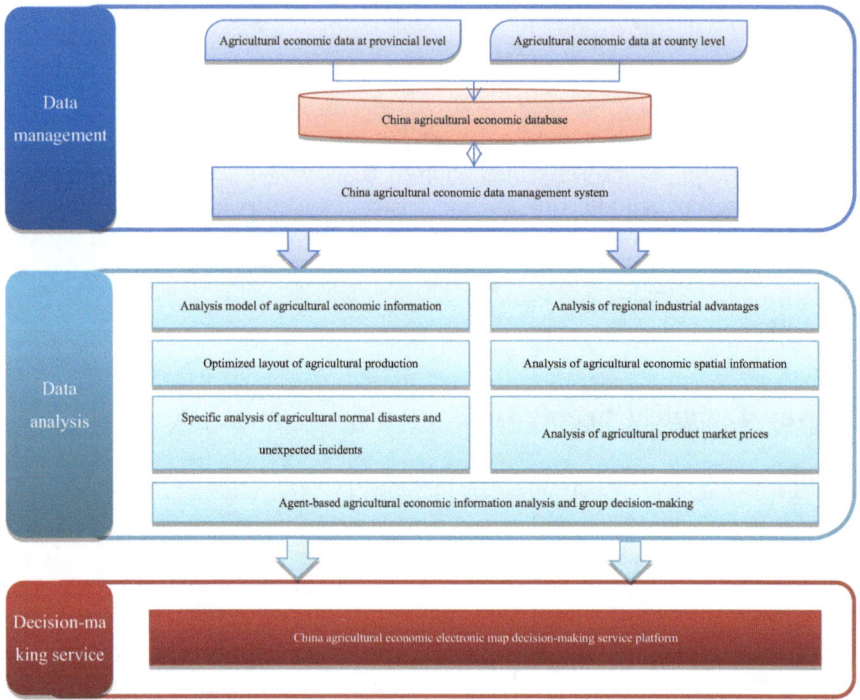

Fig. 1 Technical framework of the project

decision-making technology as its key, visualized decision-making service as its goal, the core technologies including database management, agricultural economic analysis and prediction model, Agent-based agricultural economic information analysis and group decision-making and construction of China agricultural economic electronic map decision-making service platform. Main innovative breakthroughs include:

2.1 Construction of Chinese Agricultural Economic Database

Construction of Chinese agricultural economic database is an important work of scientific management and effective utilization of agricultural economic data, it is the foundation of data sharing, data analysis and decision service. For a long time, there are many problems in the agricultural economic statistical data at all levels in China, such as the lack of unified standards of collection and storage, lack of unified management and difficulty in sharing. This study has formulated the standard specification for agricultural economic data acquisition and storage, and on this basis, it has built a Chinese agricultural economic database, which contains 9 categories and more than 500 indicators from 1980 to 2010. This study has developed a Chinese agricultural economic data management system, it has laid the foundation for centralized management and efficient utilization of agricultural economic data in China.

2.1.1 Norm Setting of Agricultural Economic Data

Chinese agricultural economic data are large and scattered, inconsistent statistics, and difficult to share. Aiming at these problems, this study has carried out the research on the standards of agricultural economic data acquisition in China, and formulated the rules of agricultural economic data acquisition and storage referring to the relevant national standards, in order to meet the data needs of GIS data service platform. The unified coding of agricultural economic statistical indicators in different counties in China has been carried out, and the unified coding of the names of all provinces, prefectures (cities) and counties (districts) in the database of basic rural economic data in different counties has been formulated. In this study, 468 metadata and data dictionaries of agricultural economic statistical indicators at county level have been developed, which laid a foundation for the standardized management of agricultural economic data.

2.1.2 Construction of Chinese Agricultural Economic Database

Chinese agricultural economy database is formed by abstracting, transforming and modelling various agricultural economic data by using metadata and metadata mapping technology, heterogeneous data integration technology, data cleaning technology, large-scale heterogeneous data extraction, conversion and loading technology, etc. This database covers agricultural economic statistics of provinces and counties from 1980 to 2010, including 9 categories of statistical indicators, such as natural conditions, macro-economy, crop production, animal husbandry production, fishery production and agricultural production conditions, with a total attribute data of more than 300 M. With the establishment of the database, the centralized processing of the storage, management and application of agricultural economic data has been realized, which provides a technical basis for the integration, sharing, analysis and service of various agricultural economic data at different granularities.

2.1.3 Development of Chinese Agricultural Economic Data Management System

Through the in-depth analysis of the needs of different application level users for agricultural economic data, the Chinese agricultural economic data management system has been established for the users of agricultural management at the national, provincial, County (city) and township levels, and the cleaning, transformation and integration of agricultural economic data under various granularity conditions have been realized; in order to improve the objectivity of agricultural economic statistics, this study uses GIS technology to develop thematic maps on three categories of existing data—the distribution of agricultural dominant industries, regional distribution of characteristic agricultural products and the fixed wholesale market distribution of Ministry of Agriculture, this study has produced 39 fixed thematic maps and on this basis, a custom thematic map module has been developed to automatically generate thematic maps with arbitrary economic and statistical indicators; this study has developed a map display and analysis report template for various applications which can generate reports, charts and text reports automatically. It improves the macro-analysis and control ability of various departments for different applications, and provides technical tools for government departments to form analysis reports quickly through map representation of data.

2.2 Research on Agricultural Economic Data Analysis Technology

The inadequate use of agricultural economic data, the untimely analysis of data, and the lack of corresponding data analysis technology system can lead to the data analysis cannot be normalized. Aiming at these problems, this study focuses on the analysis and optimization of regional grain crops, normal disasters and emergencies in agriculture, income disparities among rural residents and market prices of agricultural products. It has built corresponding analysis models, and established four technical systems, which improve the utilization efficiency of agricultural economic data effectively.

2.2.1 Analysis on Regional Grain Crops and Optimal Layout Decision

Facing the need of agricultural macro-decision-making for the analysis of regional superiority industries, this chapter puts forward the design idea of using comparative advantage theory, model technology and computer technology to establish the analysis software system of main grain crops production superiority industry in counties of China, which fills the gap of comparative advantage analysis of agricultural products in different counties of China by using agricultural information technology at present. It is difficult for human beings to deal with such problems as lack of essential information data, unconformity of data collection standards and complex data structure. Aiming at these problems, this study has established an optimal layout model for wheat, corn, soybean and rice crops by choosing the natural conditions, production costs, material input, market supply and demand factors involved in the optimal layout decision of crop varieties and applying the theory and method of system engineering optimization [4]. Collecting, cleaning and transforming the above data on the area, variety, yield, purchase price and unit area employment of grain crops, the data warehouse for decision-making of distribution of all kinds of grain crops was constructed, a data warehouse for grain crop layout decision has been constructed, an assistant decision system for comparative advantage analysis of regional agricultural industry and layout optimization of crop varieties has been developed. Decision makers can generate optimal layout plan in real time according to the input factors with different constraints [5]; they can obtain the price trend of all kinds of crops and changes in the proportion of farmers' final income to grain crops [6], which provides a basis for evaluating decision results and effects quantitatively; the system can build new decision models based on different data for data analysis and decision making in multiple regions. The project results provide an easy-to-use and customizable intelligent decision-making tool for agricultural management and decision-making departments at all levels, and provide an important assistant means for regional crop layout and optimization of agricultural production structure.

2.2.2 Thematic Analysis of Agricultural Normal Disasters and Emergencies

In view of the increasing occurrence of agricultural natural disasters and agricultural production emergencies in China, this study selected drought, waterlogging, typhoon and other major disasters as the research object, and carried out the research on thematic analysis technology of agricultural normal disasters. According to the principle of dominant factors of disasters and the dominance of related crops, taking county-level administrative regions as units, according to the background of Regional Disasters (time, space), disaster characteristics (drought, flood, typhoon, etc.) and the production situation of related crops (planting area, yield, etc.), probability statistical analysis methods (including Analytic Hierarchy Process (AHP), Information Quantity Method, etc.) is adopted, five evaluation and analysis models of major agricultural disasters such as drought, waterlogging, typhoon and agricultural production emergencies were established, and the disaster status, development trend, overdue consequences, intervention measures and emergency decision support of emergencies could be provided according to the model. It has realized the omnidirectional, scientific and quantitative prediction of normal agricultural disasters. On the basis of disaster evaluation model, integrating data dynamic analysis technology and thematic customization technology, a thematic customization analysis tool for agricultural emergency disasters and emergencies is developed. When emergencies occur, thematic maps of agricultural information related to disaster areas can be quickly customized, and the areas involved in emergencies can be quickly identified. Provide disaster area, affected population, affected crop area, affected crop yield and other content, give disaster assessment, provide an efficient and convenient tool for agricultural managers to study and prevent disasters, and to quickly formulate disaster mitigation measures.

2.2.3 Analysis of Rural Residents' Income Gap

In view of the unbalanced economic development in the eastern, central and western regions of China in recent years and the continuous expansion of the income gap among farmers, this study uses the relevant theory of income difference and the calculation method of absolute difference index and relative difference index to study and analyze the income difference of rural residents in China from 1997 to 2010 [7], and the income gap analysis model of rural residents in China using absolute income difference and relative income difference is put forward, which is the first time to realize the whole country [8]. Combining the income gap analysis model technology and database technology, the national income gap analysis system of rural residents at county level has been developed [9], rapid and customized analysis of rural residents' income gap in 31 provinces and automatic generation of analysis reports provide an efficient and convenient tool for agricultural managers to formulate macro-agricultural economic policies.

2.2.4 Dynamic Analysis of Agricultural Product Market Price

In recent years, abnormal fluctuation of agricultural product prices has occurred frequently in China. How to quickly and accurately grasp the fluctuation of agricultural product prices and analyze and forecast the trend of agricultural product prices has become a difficult problem for the managers of agricultural products market. This project is aimed at the designated wholesale markets of the Ministry of Agriculture collected by the Department of Market and Economic Information of the Ministry of Agriculture in 30 provinces (municipalities and districts) of China. It carries out the analysis and monitoring of the price data of agricultural products, and puts forward the dynamic analysis method of the prices of agricultural products in the main wholesale markets. Using GIS technology and database technology, the price analysis system of China's agricultural economy market was studied and established. The digital management of price information of agricultural products in 12 key wholesale markets and more than 500 fixed-point wholesale markets was realized. And the use of thematic maps, curves, charts and other forms to achieve an intuitive display of price fluctuations in agricultural products. At the same time, the price changes of agricultural products are analyzed and modeled in three dimensions: wholesale market, agricultural products varieties and time. Three kinds of price fluctuation analysis models of agricultural products are established. The dynamic analysis of price changes of multi-wholesale market single-day single-variety, single-wholesale market multiple-day single-variety and single-wholesale market single-day multiple-variety is realized. Combining with automatic report generation technology, the dynamic automatic generation of analysis report is realized, which provides a powerful technical means for agricultural market managers to grasp the information of agricultural price fluctuation in time.

2.2.5 Agricultural Economic Analysis and Group Decision Based on Agent

Aiming at the common problems existing in the existing agricultural economic information analysis and decision-making aided system, such as too much manual intervention, no cooperation between systems, not satisfying distributed computing and not realizing group decision-making support, the idea of using agent-oriented method to construct decision-making support system is put forward, and an agent-based agricultural economy information analysis and assistant decision-making methods and technical solutions is designed, which decompose decision-making tasks into multiple independent agents, greatly simplifying the solving process of complex decision-making. At the same time, the same knowledge (such as predictive mathematical model) is established as Agent, which is used to solve different problems (such as predictive problems with different indicators) and greatly improves the reusability of Agent [10]. An agent-based distributed group decision-making environment is constructed, which effectively utilizes the dispersed computing resources and achieves effective collaboration among different decision-making service systems and ensures the scientificity of decision-making results to the greatest extent. An agent-based prototype of agricultural economic information analysis and assistant decision-making system is established, and the design idea of the system is applied to the specific content of agricultural economic data analysis (Fig. 2).

2.3 Construction of China's Agricultural Economic Electronic Map Decision Service Platform

Aiming at the problems of narrow service user range, decentralized service function and single service terminal existing in various types of agricultural economic analysis service systems, this chapter puts forward rapid processing and analysis method of spatial and temporal information of agricultural production by using the integrated application of Agent technology, GIS technology and database technology based on the analysis of the service demand for agricultural economic data analysis by users at different application levels, which has changed the traditional agricultural economic data service's uncertainties in time and accuracy, which uses tabular statistics to complete data information collation and analysis. A multi-level, multi-terminal, comprehensive and extensible electronic map service platform for

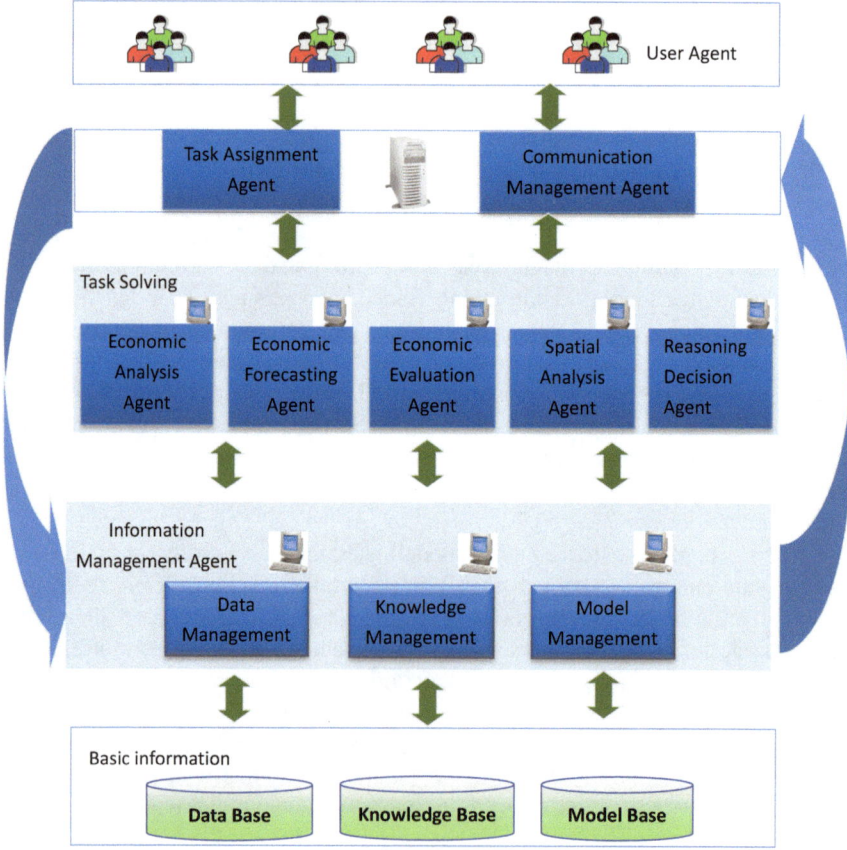

Fig. 2 Schematic diagram of agricultural economic information analysis and auxiliary decision-making method based on agent

China's agricultural economy has been developed [11]. The platform architecture is shown in the following figure (Fig. 3).

2.3.1 Level Data Management Services for All Provinces, Counties (Cities) and Townships

In order to meet the needs of data upload, download and sharing of agricultural management users at the national, provincial, County (city) and township levels, this study designed a multi-level control data management system, which realized the multi-level data management of data filling data-auditing and data-query. The data submission mechanism and norms suitable for the use of electronic map service platform are formulated, and the data level auditing mechanism is adopted to

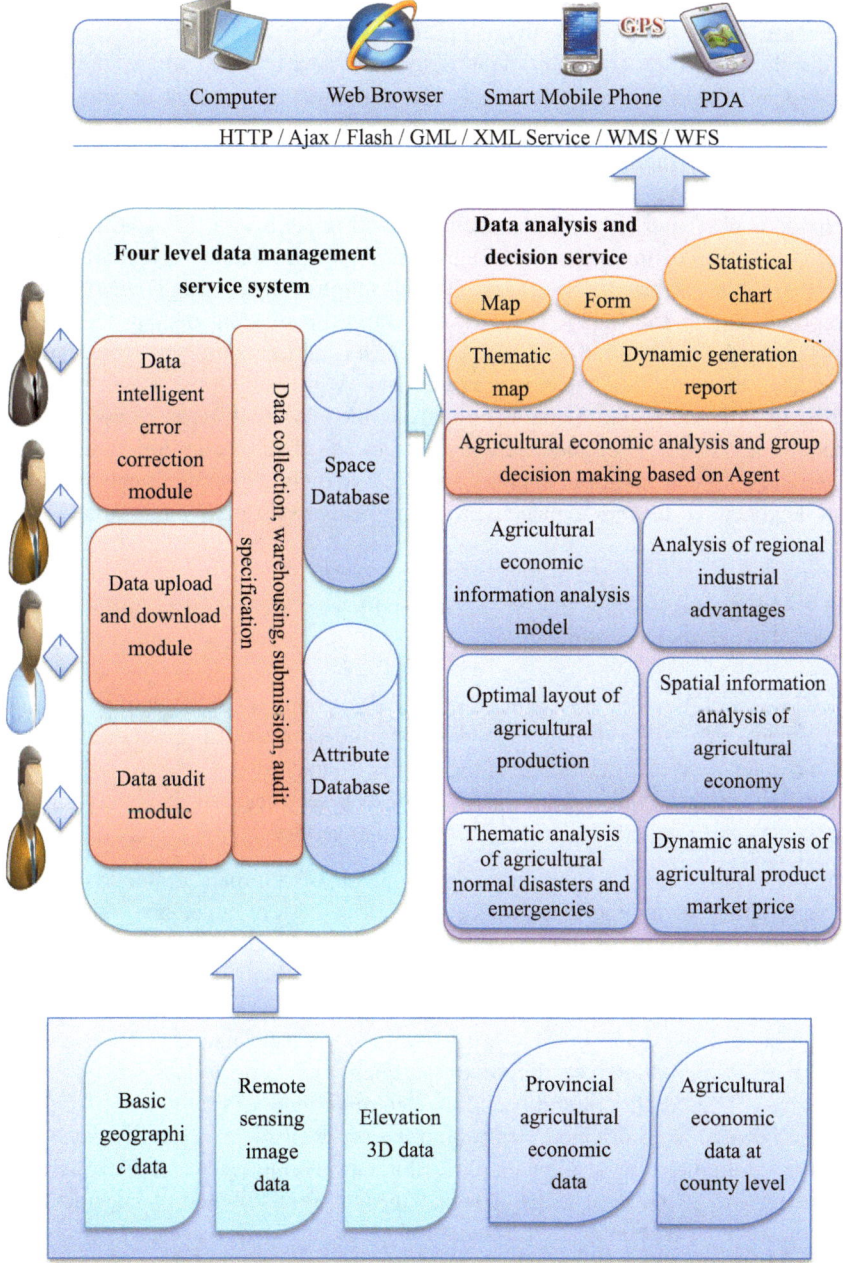

Fig. 3 Structure of decision making service platform for China's agricultural economic electronic map

realize the standardization of data reporting at all levels of the country →
Province → County (city) → township; the intelligent data error correction model
is developed, which combines with years of historical data, adopts preliminary
processing, linear analysis and interval adjustment. Intelligent error correction is
carried out on the data submitted by the whole method to reduce the manual error
rate in the data submission stage. Meanwhile, the method of change curve and
probability analysis is used to make a deep judgment on the data, confirm the
suspicious data, and ensure the authenticity and reliability of the data; In order to
meet the needs of the Ministry of agriculture and provincial agricultural departments. In this chapter, a convenient, fast and intuitive data management method is
proposed, which can query, statistics and multi-dimensional display the existing
data using different types of data such as years, regions and types, etc. and can
complete the mastery of the existing state data. At the same time, according to the
needs of different departments, data analysis models of different departments are
established. Through online analytical processing and data mining technology,
historical data and real-time data are analyzed, and corresponding analysis reports
can be generated to help management departments make macro-decisions.

2.3.2 Map Services for PC, Web, Smart Phones, PDA and Other Diversified Terminals

Faced with computer, Web, smart phone, PDA and other diversified terminal
equipment, this study designs and develops the corresponding agricultural economic map service system according to the characteristics of various terminal
equipment in the acquisition and display of map services, which greatly enriches
the means and methods for users to obtain map services.

1. A computer-oriented agricultural economic electronic map service system. The
   agricultural economic map service stand-alone system developed only for personal computer made full use of the advantages of strong computer computing
   ability and good display effect to display intuitively analysis functions of
   superior industry analysis, emergency disaster analysis and price analysis of
   agricultural products in the form of various thematic maps, Chart charts and
   reports, which effectively improves the quality of map services.
2. WebGIS-based agricultural economy electronic map service system. Aiming at
   the need of Web-oriented electronic map service, this chapter designs a Web
   service architecture of China's agricultural economic electronic map based on
   SOA, studies the technology of mass information distributed management, and
   constructs a WebGIS-based agricultural electronic map service system, which
   provides a comprehensive, deep-seated and easy-to-operate Information analysis
   service for departments and bureaus of the Ministry of Agriculture and related
   subordinate departments.
3. Agricultural economic information spatial service system based on GPS.
   Aiming at mobile terminals such as smart phones and PDA, integrating GPS,

PDA and GIS technology, EVC 4.2 is used to develop the background data, using SQL CE3.1, making full use of the powerful functions of smart phones and PDA to process the real-time data received by GPS, and then combining electronic maps to achieve rapid positioning and display of basic agricultural economic information [12]. It mainly develops a small embedded GIS platform to display and manage map data, realizes the query of agricultural basic economic information based on GIS, realizes the query of agricultural basic economic information based on text and map by combining SQL CE database and ADO technology. Combining embedded GIS and basic agricultural economic data, the distribution map display of basic agricultural information is realized, which effectively improves the convenience of map service.

2.3.3 Extended Interface of Agricultural Economic Data Management Service Platform

To solve the problem of "data gap" and "data barrier", this chapter puts forward a data service interface method of agricultural economic electronic map, designs and develops the extended interface of agricultural economic data management service platform, realizes the universality of China's agricultural economic electronic map service platform, and ensures extensibility in data and function. A metadata management system suitable for the whole agricultural industry has been developed. Users can generate corresponding data dictionaries according to specific business to realize the import and validation of user business data. An automatic association model between attribute data and spatial elements has been established to realize the automatic association and verification between user input data and map elements. A map service generation module is developed, which combines the service functions of data display, report generation and thematic map generation in the platform to generate data services, map services and analysis services for specific business data for users, and effectively solves the extended application problems of the platform in other agricultural industries.

3 Platform Function

3.1 *Operation Example of Computer Oriented Agricultural Economic Electronic Map Service System*

1. Data analysis function

   The data analysis function can provide data analysis service which combines data display, icon analysis and map data analysis for different economic indicators in different regions and years.

2. Thematic analysis function

   Thematic analysis function can build corresponding thematic maps according to the selected provinces, counties, cities, economic indicators and other contents, and provide thematic data analysis services.

3. Feature customization function

   Thematic customization function can quickly customize disaster thematic regional analysis and emergency management plan according to different disaster situations, providing emergency management and disaster rapid response services for the Ministry of Agriculture and local agricultural authorities in response to emergencies.

4. Analysis report auto-generation function

   The automatic generation function of analysis report can customize the function report template according to the needs of different business types. It can automatically generate disaster impact analysis, disaster summary analysis, dynamic analysis of market prices of agricultural products, etc. according to the selected content (Figs. 4, 5).

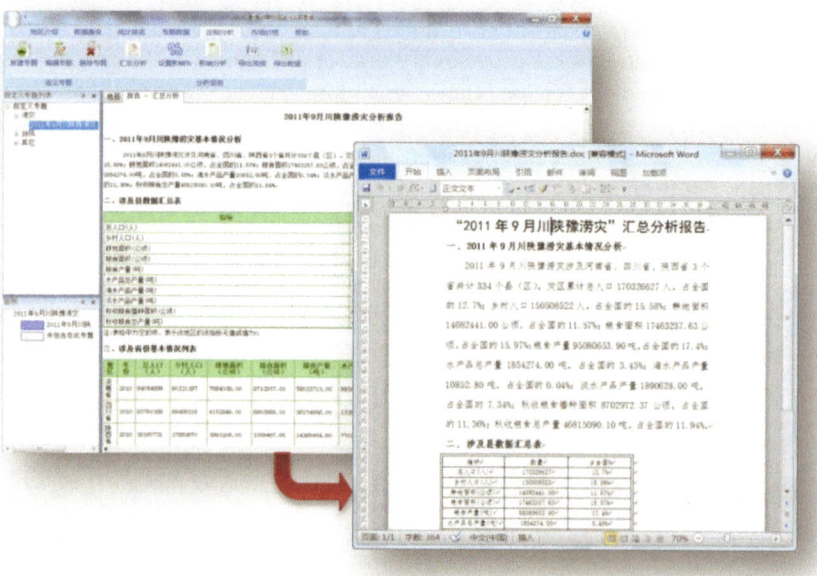

Fig. 4 Analysis report on normal disasters and emergencies

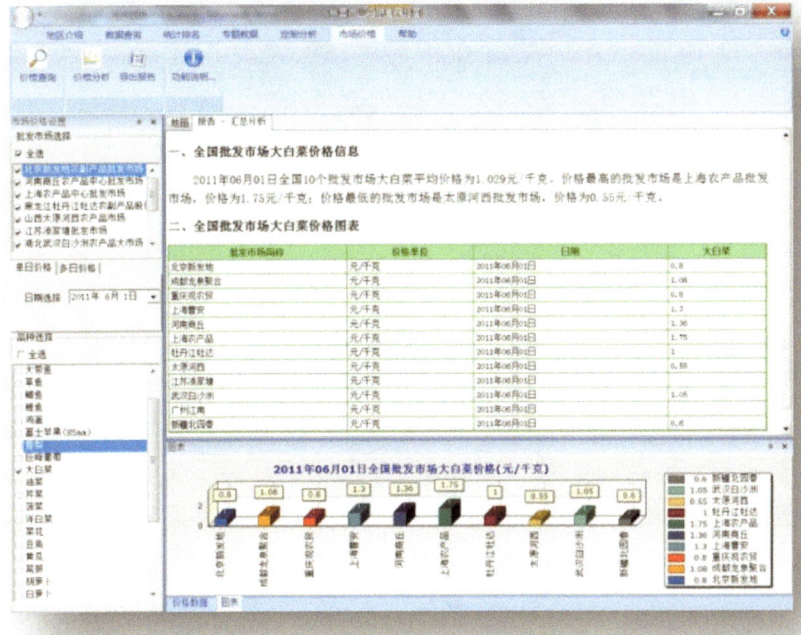

Fig. 5 Dynamic analysis report on market price of agricultural products

3.2 Operation Example of Agricultural Economic Electronic Map Service System Based on WebGIS

1. Data browsing function
   Realize various data browsing and query of different economic indicators in different provinces, counties and cities in WebGIS (Fig. 6).
2. Thematic analysis function
   Provide network version of the map thematic analysis and thematic service functions.

3.3 Operation Example of Agricultural Economic Information Space Service System Based on GPS

1. Map browsing function
   Realize map loading, zooming, reducing, roaming, fenestration scaling, location query.
2. Information query
   Implement the property data query in the property library (Fig. 7).

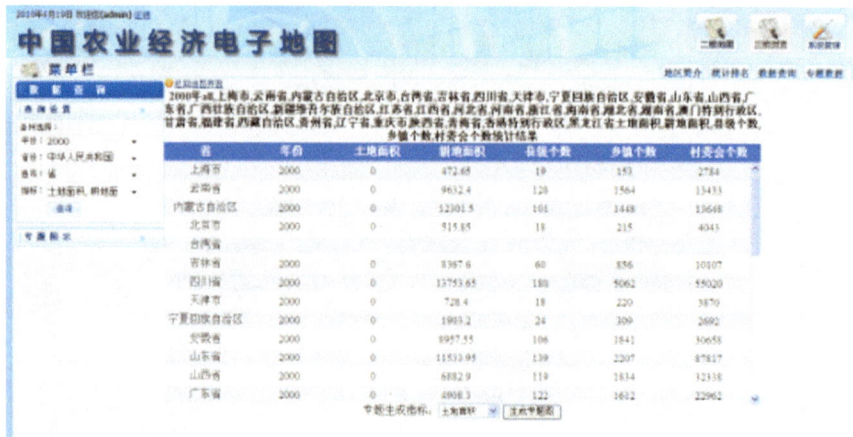

Fig. 6 Oriented web of data browsing services

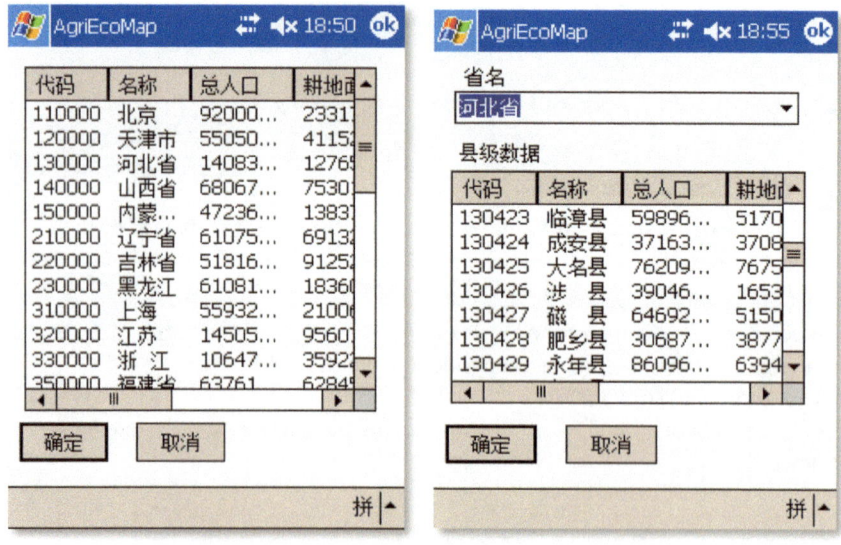

Fig. 7 Information query oriented to mobile terminal

3. Map click query

   Click the left button on the map to query the agricultural economic data in the area where the mouse is located. The map of the clicked area will be highlighted first, and then the agricultural economic data of the area will be displayed.

4. Data distribution map

   Displays distribution diagram or histogram of the query data.

4   Technical Innovation

In order to solve prominent problems in the current agricultural economic statistic information such as nonstandard management, insufficient analysis technology and unvisualized display, the project performed long research on construction of China agricultural economic database, analysis model of agricultural economic information, analysis of regional industrial advantages, optimized layout of agricultural production, analysis of agricultural economic spatial information, specific analysis of agricultural normal disasters and unexpected incidents, dynamic analysis of agricultural product market prices, analysis of income gaps of countryside residents, Agent-based agricultural economic information analysis and group decision-making, construction of China agricultural economic electronic map data service platform, and has achieved a series of innovative breakthroughs to supply firm technical support for developing modern agriculture.

1. In terms of data management technology, the project has developed Chinese Agricultural Economic Database that covers the period of 1980–2010, and contains over 500 province-level and county-level agricultural economic indexes in 9 categories. As Chinese agricultural economic data are massive but scattered, inconsistent in statistical standards, difficult to share and integrate, the project has performed research on collecting standards of Chinese agricultural economic data, and has established regulations on agricultural economic data. China agricultural economic database is developed by applying metadata and metadata mapping technology, heterogeneous data integrating technology, data cleaning technology, large-scale heterogeneous data Extract-Transform-Load technology to achieve centralized storage, management and application of agricultural economic data. By developing China agricultural economic data management system, the project team has achieved efficient audit, transformation and integration of agricultural economic data under various granularities; by applying GIS technology, has made and studied three categories of specialty maps on existing belts of agricultural advantageous production belts, distribution of specialty agricultural product regions and distribution of wholesale markets authorized by the Ministry of Agriculture, and has developed customized thematic mapping modules to automatically produce thematic maps for any statistic index; has developed map exhibition and automatically produced

statements, statistic charts and text reports for various applications to effectively improve macroscopic analysis and control capabilities of different departments.

2. In terms of data analysis technology, the project has broken through a series of key technologies on agricultural economic data analysis. For issues that are hard to solve by human such as information without key elements, inconsistent collection standards and complex data structures, the project has applied optimized theory and method of system engineering to develop optimized layout models for wheat, corn, soybean and rice, decision-making data libraries for various crops, assisted decision-making systems for analyzing regional comparative advantages of agricultural industries and optimized layouts of crop varieties. For agricultural natural disasters and unexpected agricultural productive incidents that frequently occur in China, the project has performed normal disaster analysis technology research, established thematic analysis models for normal agricultural disasters, and achieved scientific and quantitative prediction analysis for normal disasters; the project has integrated dynamic data analysis technology and customized thematic technology to develop analysis tools for emergent disasters; under the background that income gaps of Chinese farmers are widening, the project has proposed an income gap analysis model and developed an income gap analysis system for rural resident income by using absolute income gaps and relative income gaps to rapidly and automatically acquire, analyze and produce analysis reports on rural resident income gap indexes for over 2800 counties and districts in China; for abnormal price fluctuation of agricultural products, the project has developed China agricultural economic market price analysis system to dynamically monitor and analyze changes of agricultural product prices.

3. Multi-level, multi-terminal, comprehensive and scalable China agricultural economic electronic map service platform. For issues existing in agricultural economic information analysis and assisted decision-making systems such as excessive human intervention, inefficient inter-system collaboration, insufficient capabilities for distributed computation and group decision-making support, the project has developed and applied a decision supporting system oriented to intelligent entities and Agent-based methods and technical solutions for agricultural economic information analysis and assisted decision-making; developed Agent-based distributed group decision-making environments; developed an Agent-based prototype for agricultural economic information analysis and assisted decision-making system. For narrow scope of customer services, decentralized service functions, monotonous service terminals existing in current agricultural economic analysis service systems, the project has integrated Agent, GIS, WebGIS, database technologies to develop the multi-level, multi-terminal, comprehensive and scalable China agricultural economic electronic map service platform to achieve a multi-layer control and management system for data filing, data review and data query. For acquiring authentic and reliable data and in-depth data judgment, the project has applied preliminary processing, linear analysis and interval adjustment methods to develop an intelligent data-error-rectification model, change curves and probability analysis methods. For various terminals

such as desktop, Web, PDA and smart phones, the project has developed SOA-based China agricultural economic electronic map service framework, and agricultural economic map service system to enrich means and methods for customers to receive map services. The project has applied advantageous industry analysis, emergent disaster analysis, and agricultural product price analysis for visual display of thematic maps, charts and reports. By establishing automatic correlation models between attribute data and spatial elements, generative modules for map services, the project has developed extension interfaces for the agricultural economic data management service platform.

The achievements made by the project in agricultural economic information normative management, systematic analysis and visual decision-making services can be extensively applied agricultural production, governmental agencies and administrative departments at various levels with extensive and promising prospects. The achievements have been extensively introduced on national and international occasions, and relevant software has been widely promoted and applied in the Ministry of Agriculture and 31 provincial agricultural departments, supplying effective technical support for agricultural management decision-making and handling emergent incidents, effectively improving regional agricultural resource utilization efficiency, optimizing agricultural production layouts, supplying efficient and precise information services to agricultural departments at various levels, promoting harmonious development between agriculture, society, economy and environment, and actively boosting Chinese information development in terms of agriculture and countryside fields.

In terms of data management technology, the project has effectively solved some problems in Chinese agricultural economic data such as massive and decentralized data, inconsistent statistic standards and share and integration obstacles. Its agricultural economic data regulations, China agricultural economic database, China agricultural economic data management system can be extensively applied in agricultural statistic agencies, governmental entities and administrative departments at all levels of China, and can be used as main methods and specifications for collecting Chinese agricultural economic statistic information, for further and widespread promotion and adoption in China.

In terms of data analysis technology, the project has developed regional agricultural advantageous industry analysis, agricultural normal disaster and emergent incident analysis, rural resident income gap analysis, agricultural product market price analysis, intelligent entity based agricultural economic information analysis and group decision-making systems can supply substantial assistance to regional crop layouts, optimization of agricultural production structures, prediction of agricultural product market prices and disaster prevention and mitigation. Therefore, it is an intelligent decision-making assistant tool that can be easily used and customized by agricultural producers and managers.

In terms of decision-making service technology, the project has developed Agent-based multi-level, multi-terminal, comprehensive and scalable China agricultural economic electronic map service platform for management decisions that can be used on various terminals such as desktop, PDA, smart phone, extensive interfaces for the agricultural economic data management service platform, WebGIS-based agricultural electronic map service platform will be an application system group that is the most efficient in Chinese agricultural information service fields, and be a great technical upgrade and leap in Chinese agricultural intelligent information services.

References

1. Mueller, T, N. Kitchen, D. Clay. Know Your Community: Precision Agriculture Systems. Crops, Soils, Agronomy News, 2013.58(12):22–22.
2. Zeng, Z, J. Cao, Z. Gu, et al. Dynamic Monitoring of Plant Cover and Soil Erosion Using Remote Sensing, Mathematical Modeling, Computer Simulation and GIS Techniques. American Journal of Plant Sciences, 2013.4(07):1466.
3. Yeping Zhu. Key Technologies Innovation Achievements of Agricultural Information Intelligent Service. Agricultural Products Market Weekly, 2013(46):8.
4. Bo Su, Lu Liu, Fangting Yang. Comparison and research of grain production forecasting with methods of GM(1,N)gray system and BPNN. Journal of China Agricultural University, 2006, 11(4):99–104.
5. Fei Fan, Debin Du, Heng Li, et al. Spatial-temporal characteristics of scientific and technological resources allocation efficiency in prefecture-level cities of China. Journal of Geography, 2013, 68(10):1331–1343.
6. Hongjie Bao, Deguang Liu. Study on Spatial Statistical Analysis of Gansu Province Regional—Ecomonic Growth Based on Geoda095i. Industrial Technology Economy, 2011, 30(9):54–59.
7. Fuyun Yang, Yeping Zhu, Yue E. Design and Realization of GIS-based Agricultural Economy Information Service System. Chinese Agricultural Science Bulletin, 2008.24 (5):429–433.
8. Xiaoyi Ma, Tao Pei. Exploratory Spatial Data Analysis of Regional Economic Disparities in Beijing during 2001–2007. Progress in Geographic Science, 2010, 29(12):1555–1561.
9. Yuming Wu. Estimation of Input–Output Elasticity of Regional Agricultural Production Elements in China. Rural Economy in China, 2010(6):25–37.
10. Hualin Xie. Spatial Inequalities Analysis of Regional Agricultural Economic in Poyang Lake Ring Based on ESDA Method. Study on Agricultural Modernization, 2010, 31(3):299–303.
11. Yue E, Jianbing Zhang, Yeping Zhu, et al. Agriculture Economy Information Service System Based on Embedded GIS. Computer Engineering, 2008, 34(23):269–271.
12. Zhou Hong. Using Mapinfo Data to Build Spatial Index Based on MAPX. Surveying and Mapping and Spatial Geographic Information, 2013, 36(3):118–121.

Shengping Liu, Ph.D., is the deputy director of the Intelligent Agricultural Technology Research Office of the Institute of Agricultural Information, Chinese Academy of Agricultural Sciences, also the first outstanding young person of the Agricultural Information Research Institute. His research interests include agricultural product quality and safety control technology, crop simulation model technology and agricultural GIS application. He proposed a framework of crop production management including field data acquisition, agricultural IOT application, crop model simulation, production plan formulation, production process monitoring and product quality control.

He has presided over, participated in and completed the National Science and Technology Support Program, 863 Program, the Ministry of Agriculture key projects, the National Natural Science Foundation project and other types of scientific research topics more than 10. He took a major role in the completion of wheat, corn planting management system, Chinese agricultural economic electronic map, bee product quality traceability system and other scientific research results, gained 9 institutional, provincial and ministerial scientific Research awards. He has published over 10 papers in international and domestic journals and conferences, gained 9 provincial and ministerial research awards, attained 1 technical invention letter of patent and 18 software copyright registrations.

Development and Prospect
of Identification Service Technology
for Advanced Manufacturing

Ye Tian, Jie Shen, Yuan Tao and Jia Liu

Abstract To develop the advanced manufacturing is the urgent needs of the economic globalization and improving the international competitiveness. Meanwhile, it is an inevitable choice to promote industrialization with information technology and the structural adjustment, optimization and upgrading of the traditional manufacture industry. The interconnection and interworking is the prerequisite to realize intelligent manufacturing, especially for the interconnection and correlation of cross-domain and multi-link applications. However, in the case of various heterogeneous identifiers coexisting, each application link becomes an information silo, furthermore, it can be predicted that different identification systems coexisting is normality in the long term. This chapter describes the core technology areas of advanced manufacturing. For example, supply chain management of raw material, customization, networked cooperative, remote operation and maintenance, and so on. It is focused on establishing the heterogeneous compatible, multi-level, distributed and peer-to-peer interworking identification service system architecture so as to provide reference for various innovative models in the field of advanced manufacturing.

Keywords Advanced manufacturing · Identification technology · Heterogeneous identification service

1 Introduction

In different procedures of advanced manufacturing, in order to realize the interconnection between machines, workshops, enterprises and people, and to meet the requirements of data interaction of machine-to-machine and thing-to-thing, it needs

Y. Tian (✉) · Y. Tao · J. Liu
Computer Network Information Center, Chinese Academy of Sciences, Beijing, China

Y. Tian
China Academy of Industrial Internet, Beijing, China

J. Shen
Computer Network Information Center, Chinese Academy of Sciences, Guangzhou, China

© Publishing House of Electronics Industry 2020   291
China's e-Science Blue Book 2018,
https://doi.org/10.1007/978-981-13-9390-7_17

to use the identification technologies to recognize the sensors, products, people, terminals, network nodes, and various business applications, and then, to perform collection, mapping, and conversion of the data through the services such as identification coding, identification assignment, identification registration, identification query, and identification discovery, so as to obtain corresponding addresses or related information [1]. The interconnection and interworking is the prerequisite to realize advanced manufacturing, especially for the interconnection and correlation of cross-domain and multi-link applications. While in the traditional manufacturing system, the type, coding format, mapping relationship of identifier and others are fairly simple, and it can be effectively managed based on assignment and query of closed-type identifier distributed across different business systems. However, with the deployment of advanced manufacturing or industrial Internet applications, all types of identification will be in an explosive growth stage, and massive multi-source heterogeneous identifications appear in the same scenario of advanced manufacturing, which will bring great challenges to advanced manufacturing.

Industrial Internet is an important support for advanced manufacturing. Compared with the IoT, it faces a large number of complex equipment and systems in the industrial environment, such as industrial manufacturing equipment, industrial control systems, complex network environments, and heterogeneous information. At the same time, due to the wide range of suppliers and the inconsistent production data structure of hardware and software equipment, the internal and external factories, as well as the upstream and downstream of the supply chain are not interoperable, which seriously affects the development of the industrial Internet. In the future, China plans to establish a fully functional identification resolution system for industrial Internet, and it is proposed to establish more than 20 public identification resolution service nodes, with a target of more than 3 billion registrations. It is an inevitable trend to achieve the interconnection and interworking in all procedures of advanced manufacturing by adopting the advanced manufacturing orientated identification service system, and providing identification service technologies such as heterogeneous identification resolution and discovery. The identification service system will certainly become the central nervous system and important infrastructure of advanced manufacturing.

In this chapter, we discuss the key role and development trend of the identification service system, and depict the architecture of identification service system for the advanced manufacturing. Moreover, we analyze some use cases, and explore the evolution of advanced manufacturing. This chapter is organized as followed. It firstly presents the identification service system for advanced manufacturing. Then the typical application links such as raw material supply chain management, personalized customization, networked collaboration, and remote operation and maintenance are described, which is to illustrate how the new mode of application is supported by heterogeneous identification service. Final section is the summary and forecast of the identification service system.

2 Advanced Manufacturing Orientated Identification Service System

The identification service system is the basic support of interconnection and interworking between various applications of advanced manufacturing. It is used in the entire process of coding, assignment, registration, resolution, discovery and other identification services, and it can realize the query, management and control of object's information. Based on this, identification service system provides the support services for various advanced manufacture applications. The identification service system is shown in Fig. 1 [2].

As increasingly prominent of the importance of identification technology in advanced manufacturing, it has become an essential part in the global application planning. Industry 4.0 is proposed by the German government. IIC (Industrial Internet Consortium) is established by GE, and AII (Alliance of Industrial Internet) is led by the Ministry of Industry and Information Technology of the People's Republic of China. These organizations have widely applied the identification technology to various industrial applications. By associating the identifiers to core resources of all procedures of advanced manufacturing, industrial manufacturing and product production can achieve efficient positioning, query, control and management. At present, the mainstream identification systems include GS1, EPC, Handle, OID, uID, Ecode, NIOT, etc., and the detailed illustrations are as follows:

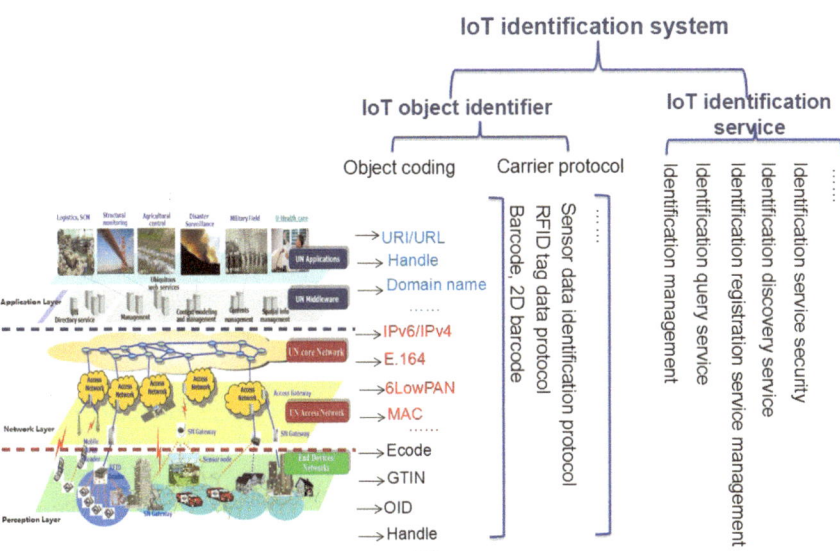

Fig. 1 Identification service system

1. GS1: GS1 is for commodity circulation, mainly including trade (GTIN, Global Trade Item Number), logistic units (SSCC, Serial Shipping Container Code), assets (GRAI, Global Returnable Asset Identifier), locations (GLN, Global Location Number), and services (Global Service Returnable Number, GSRN). The most widely used application for is GS1 is the barcode. In advanced manufacturing, GS1 is also applied to automotive production and supply chain management of automotive components.

2. EPC [3]: EPC has its origins at Auto-ID Center of the Massachusetts Institute of Technology (MIT) since 1996. It is a global standard for RFID-based identification. RFID electronic tags are widely used in all procedures of advance manufacture industry, and it is also one of the most widely used identification systems in this area. For example, the BRIDGE project in the European Union studies on how RFID and EPC networks can assist the related operations in industrial manufacture, and improve the traceability of individual product information by using globally unique identifiers [4].

3. ucode [5]: ucode is supported by the uID center (ubiquitous ID center), which is a non-profit organization based in Japan. It is mainly used for RFID.

4. Handle [6]: The Handle System was originally developed at CNRI (The Corporation for National Research Initiatives) in 1995.
   It is widely used in the field of digital book publishing, such as the DOI adopts the Handle identifier starting with 10. In China, the Handle system has been further promoted and applied to various application fields of the IoT, it is currently being applied to advanced manufacturing.

5. OID [7, 8]: OID is used to identify different data objects, and it has been widely applied to information security, health care, network management and other fields. In South Korea, OID has been used in many industrial enterprises. For example, the electric power company KEPCO applies OID to its IoT platform to identify the equipment and resources, which can uniquely identify all devices and gateways, and each resource information can be accurately located [9].

6. Ecode [10–12]: Ecode was proposed by GS1 China. Up to now, the Ecode-V0 segment of the NSI is assigned to the EPC, and the Ecode-V1 segment of the NSI is assigned to the barcode and sensor identifier. In the advanced manufacturing, Ecode is being applied to the entire process of product traceability and management, product precision marketing and so on.

The mainstream identification systems descried above are widely used in advanced manufacturing, different procedure require different types of identification systems, and each object may be corresponded to multiple different types of identifier. Furthermore, different countries, industries and organizations have various heterogeneous identification coding and service systems, how to achieve cross-domain and cross-industry interworking and avoid information silos has become a common problem.

Computer Network Information Center, Chinese Academy of Sciences is commissioned by the National Development and Reform Commission to lead the construction of the China Internet of Things Names Service Platform (NIOT), and propose the NIOT identification system [13]. The core technical route is to propose a compatible and extensible hierarchical tree structure based on the existing identification systems, and realizes the compatible and unified management for multiple heterogeneous identification systems, and further realizes compatibility and interoperability between various mainstream heterogeneous identification systems such as GS1, EPC, Handle, OID, Ecode. The heterogeneous identification service technology of core independent intellectual property rights has been accepted by oneM2 M and completed the standardization in CCSA (as shown in Fig. 2) [14]. Up to now, The NIOT platform has established many sub-nodes for different industries such as product traceability, building materials, environment monitor, data management, etc., The total amount of registered identifiers is over 79 billion, and the total number of time of identification resolution is over 8 billion. The independent intellectual property rights of IoT identification management service technology system is initially established, which also formed an open, shared and autonomously controllable IoT identification architecture.

The Internet/IoT identifies various devices and resources through the identifier. It is the essential and important link for the Internet/IoT to support various applications in advanced manufacturing. The phenomenon of identification isomerization is becoming increasingly fierce with the development of network technology and advance manufacture applications, and in terms of sufficient attention of strategic position. To Support a variety of identification systems, and achieve the

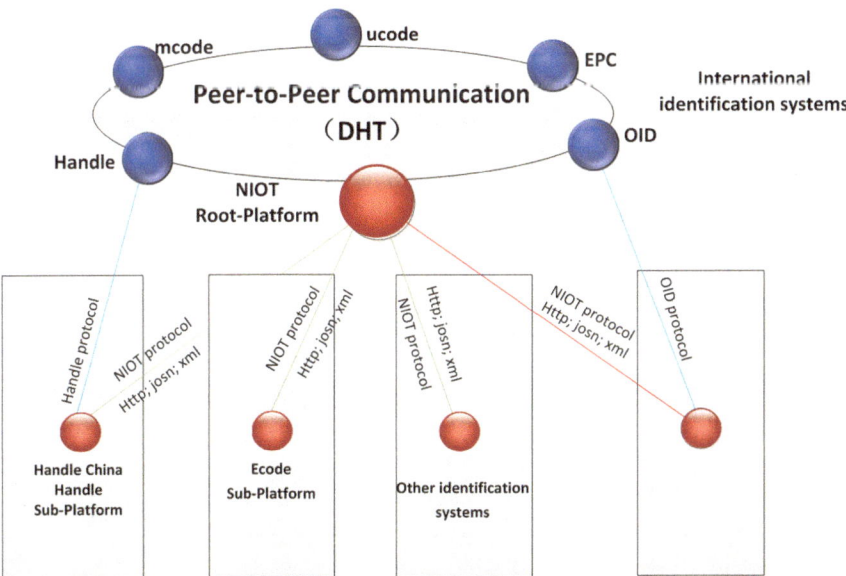

Fig. 2 Architecture of NIOT identification system

interconnection of multiple identification systems, and further support the integration and development of advanced manufacturing, is the inevitable choice for the development of advanced manufacturing.

3 Heterogeneous Identification Service Supports the New Mode of Application

The advanced manufacturing of countries is gradually moving towards the stage of application deployment. As an important carrier of massive heterogeneous industrial data integration and manufacture application innovation, the identification system is becoming the core of a new round of industry competition. At present, the heterogeneous identification service system has completed preliminary research and application deployment in advanced manufacturing. With the continuous breakthrough and experience accumulation on various heterogeneous identification service technologies, the heterogeneous identification service system will be deeply applied to the advanced manufacturing with innovative mode of application, and realize various aspects of exploration including supply chain management of raw material, personalized customization, networked collaboration, and remote operation and maintenance.

3.1 Supply Chain Management of Raw Material

Today's global competition for enterprises is increasingly manifested as competition between supply chains, and is no longer a competition between individual enterprises. The efficiency of supply chains directly affects the competitiveness of all enterprises within the chain. Applying the identification technology to the supply chain can improve the efficiency of the supply chain and enhance the competitiveness of the enterprise. Because the identification improves the information transmission in the supply chain, the automation, safety and efficiency of procedures such as warehousing, transportation, production, sales and after-sales service of the supply chain are improved.

With the continuous development of science and technology and logistics management, the use of innovative technology to improve the intelligent level of supply chain management of logistics has become the main focus of the industry. Modern logistics is a systematic project including procurement, storage, production, packaging, loading and unloading, transportation, processing, distribution, sales and service activities. It especially requires fast and accurate information acquisition and processing, nevertheless, the traditional identification technologies such as barcode have become more and more difficult to meet the needs of logistics informatization.

The different identification systems and the diversified carriers are more and more widely used in the exchange and transmission of logistics information, so as to promote the development of logistics modernization, however, the identification application of different systems also brings the challenge of heterogeneous identification resolution for enterprises. Through the application of the heterogeneous identification service system, enterprises can realize the resolution of information of different identification systems, so as to realize the data collection and real-time monitoring of each link of each product, and manage the logistics system. It can not only monitor and share the information of raw materials in the supply chain, but also analyze and predict and information of raw materials at all procedures of the logistics supply chain. By predicting the current stage of information of the raw materials, estimating the future trend or the probability of accidents, and timely taking remedial measure or early warning, the ability of enterprises to respond to the market can greatly improve and speed up.

The traceability system of full life cycle of the raw material is based on heterogeneous identification services, and distributed constructing in the participants of all procedures, and each participant can be traced the full life cycle of information of product, as shown in Fig. 3. It can realize the effective sharing of

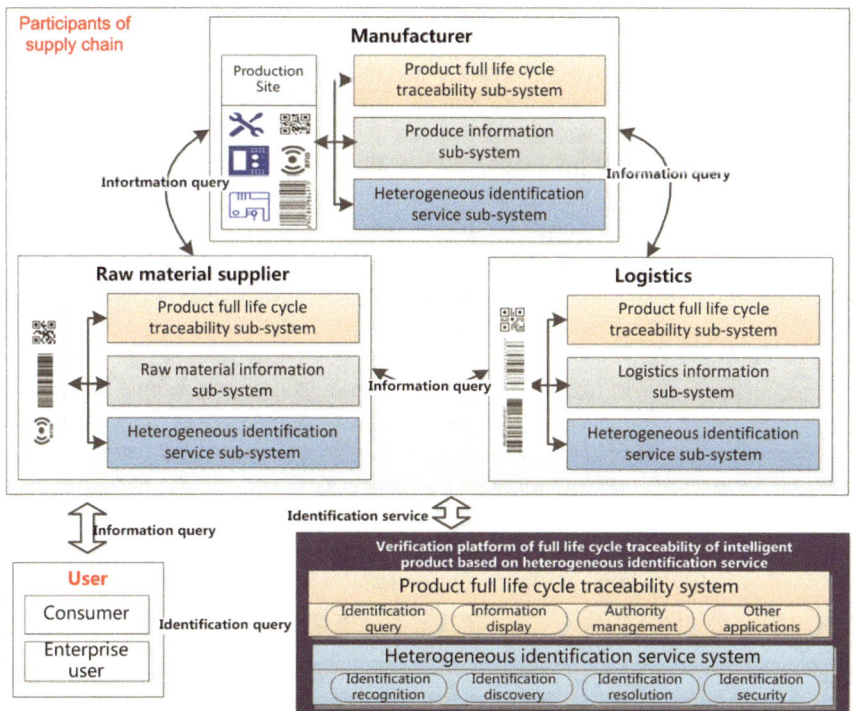

Fig. 3 Architecture for full life cycle traceability of supply chain of raw material

information between internal information systems, different enterprises and different links, and promote the interconnection and interworking of the whole industry chain. That is to say, the raw material is associated with the IoT identifier from the beginning of production, and the process information that generated by the raw materials in the supply chain is resolved and collected. Then the process information is transmitted to the processing center for inquiry by enterprises and consumers. The real-time monitoring of the supply chain process can realize timely identifying the problems in the supply chain process and improve service quality and consumer satisfaction effectively.

3.2 Personalized Customization

The transformation of standardized manufacturing into personalized customization becomes the main direction for industrial enterprises to complete intelligent transformation in the future. The heterogeneous identification service system is an important technical basis for locating and addressing devices in a personalized and customized scenario, and realizing the collection and query of device information. To realize the intelligent indexing and integration of relevant information of devices and products in the mass industrial field through information technology service such as identification resolution and discovery. Through identifying the personnel, machines, materials, products, equipment parameters, environmental parameters, quality conditions and other resources in the factory, the information corresponding to all resources can be interconnected in the production system. Furthermore, the integration and sharing of production information can be achieved in the industrial field, and the key data is provided to support the enterprise decision makers.

For example, the identification service system is used to upgrade large-scale manufacturing to mass customization. It achieves efficient production of customized batch orders and the construction of a full-process real-time interconnection visible interconnection factory system, through the cooperation of a large number of sensors, informatization tools of big data and robots. Using RFID as the hub of the intelligent library, it can realize the rapid out and in storage of personalized orders, improve the work efficiency, improve the quality of inventory operations, increase the throughput of distribution centers, effectively utilize storage resources, and improve the overall transparency of logistics. By deploying RFID and sensors, it can realize the interconnection and visualization of full process of customized products. In the process of material development, it can also be traced by QR code. The module manufacturer produces the personalized customized products according to the requirements, prints the QR code on the back of the customized module, and the module distribution and acceptance process which is verified by the scanning module QR code whether it meets the requirements of the product customization to achieve the error prevention and traceability.

In the clothing field, the unified identification resolution service system of personalized customization for the whole process can realize the information

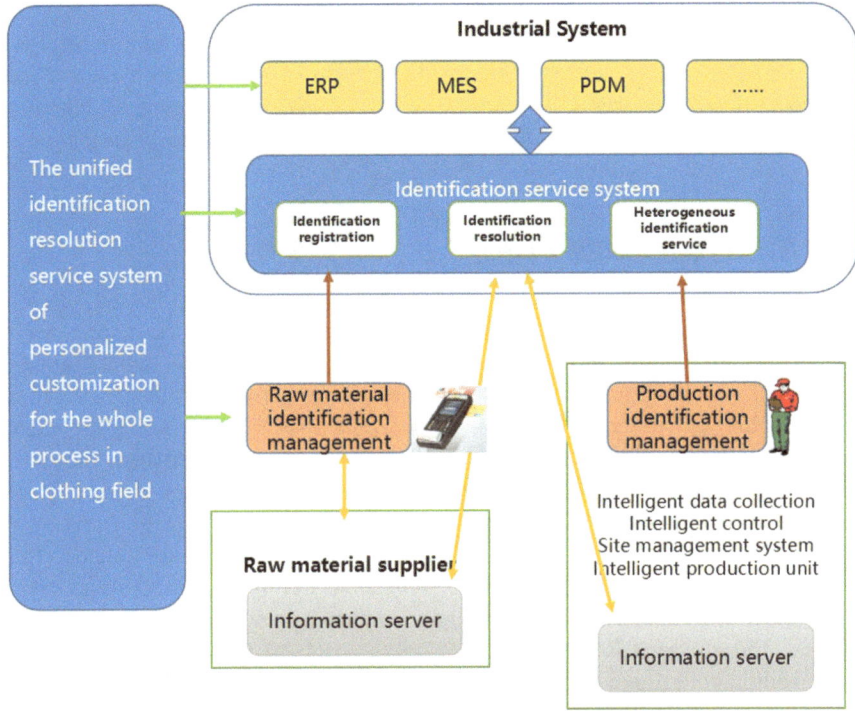

Fig. 4 Architecture of product personalized customization

exchange and supply chain interconnection without changing the existing identification systems by the heterogeneous identification service technology throughout
the whole process, it is described in Fig. 4. The heterogeneous identification service
system includes an identification registration service, a heterogeneous identification
service, an identification resolution service, an identification discovery service, etc.,
and it is compatible with various heterogeneous identification systems to provide a
unified identification service.

For example, the traditional clothing production process can be upgraded by
using the identification service technology, which constitutes an integrated information platform, with a customized order as the information flow and a full digital
operation system with the radio frequency chip as the carrier. The automatic, online,
real-time and comprehensive recording of each processing action of people, raw
materials and machines in the production process realizes automatic collection,
automatic transmission, automatic processing and automatic execution of data in all
procedures. Moreover, it realize sending the accurate data to the right person and
machine at the right time, which solves the problem of production uncertainty,
diversity and complexity brought about by personalized customization. Through a
personalized customization platform, consumers can choose fabrics, styles, and

personal accessories to participate in the design. By identifying the various parts of the raw materials and product design, it establishes the mapping relationship. By identifying and querying the devices based on identifiers, the cutting style and size of each part are determined. Based on product identifier, the product tracking, positioning, and automated production information query for product lines.

3.3 Networked Collaboration

Networked collaboration enables sharing information throughout the enterprises in the supply chain through the Internet, enables tight collaboration between the various links of the industry chain, and promotes a comprehensive interconnection of production, quality control and operational management systems. In the enterprise, in order to avoid information silos in the manufacturing process, the coordination between different production elements management is required, and therefore, the compatibility between various systems is required. Collaboration is not only the internal collaboration of the enterprise, but also it is the collaboration between the upstream and downstream organizations in the industry chain. Through network collaboration, the consumers and manufacturing companies work together on product design and R&D to meet the requirements of personalized customization. In addition, through network collaboration, the configuration of raw materials, equipment and other production resources, as well as the organization of dynamic manufacturing, which can meet differentiated market demand, and achieve horizontal industry integration in the value chain.

In the process industry (such as the gold mining industry), the networked collaborative platform applies the heterogeneous identification service technology to the established big data platform of gold mine industry. This is based on the establishment of a golden mine industry network collaborative model for many mining enterprises, as shown in Fig. 5. The interaction of production data is carried out based on the identification technology between each system and enterprise. The identification service is the core function module and needs to be networked collaborative with the relevant industrial production system. By building a big data management chain that integrates data collection, data integration, data storage, and data analysis. It highly integrates the heterogeneous identification data such as production, equipment, security, human resources, compensation, materials, technology, transportation, and energy. Moreover, it realizes the optimization and coordination of management functions of the mining industry. Through the establishment of the cloud service platform, it can establish a communication bridge for enterprises in all procedures of the mining industry, providing enterprises with cloud application services such as innovative design, intelligent manufacturing, supply chain collaboration and remote maintenance. Through the sharing and intercommunication the comprehensive information throughout various procedures in various industries, it can promote the coordinated development of various enterprises in the industry.

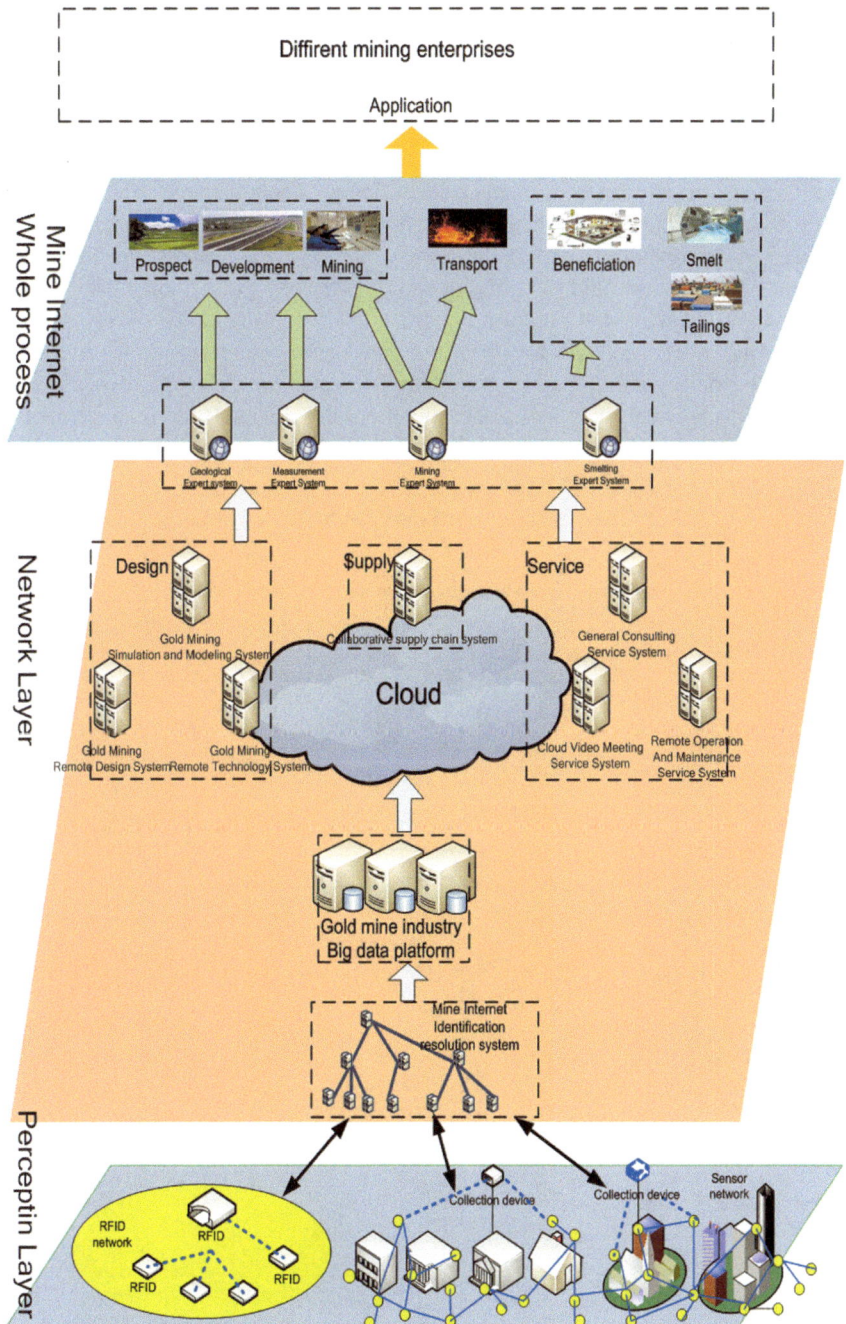

Fig. 5 Architecture for process industry networked collaborative

3.4 Remote Operation and Maintenance

At present, most enterprises lack the automated remote operation and maintenance management mode for product management, and lack comprehensive tracking records. It leads to complex operation, inefficiency, and waste of resources. The identification service can also manage identified products through big data, cloud platform and mobile internet technology, realize standardized information collection of remote operation and maintenance, and provide remote operation and maintenance hardware and security operation services of network. The remote operation and maintenance platform integrates the identification discovery service, provides the query service for the authorized user, implements the mapping between the object identifier and the multiple related identifiers, and applies the identification discovery service to the operation and maintenance management of

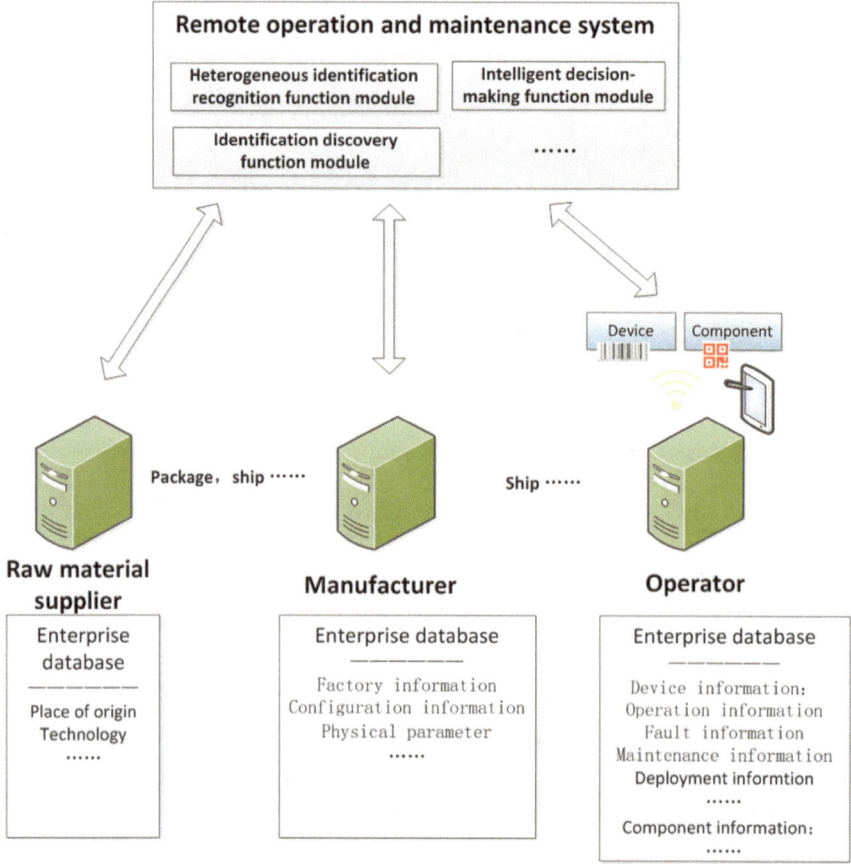

Fig. 6 Architecture of remote operation and maintenance platform

the enterprise, which makes the complex and diverse operation and maintenance management become simpler and more efficient, and realizes remote operation and maintenance management of the product life cycle, as shown in Fig. 6.

For example, in the field of home appliances, raw material suppliers, manufacturers, and operators need to maintain and manage their own information servers, and store each electrical appliance or component and related information in the information server based on identifying each individual electrical appliance and key components. The remote operation and maintenance platform provides identification discovery service to participants of supply chain to obtain full life cycle information. The function module of heterogeneous identification recognition can realize the identification of various identification systems. The identification discovery function module maintains the mapping relationship between the identifiers, and obtains the corresponding all related information server addresses, and identifiers associated enterprises, maintains and manages the industry data information efficiently.

In addition, based on the common needs of home appliance industry for the management of home appliance products, and various key functions such as device positioning, device management, fault alarm, automatic dispatch, and data statistics, it can collect various identification data of home appliances and realize efficient data interaction. Through analyzing big data, a solution for smart home appliance remote operation and maintenance service platform based on heterogeneous identification service is formed to explore the management requirements of home appliances in various scenarios. The platform is based on open source and has good extension.

4 Summary

The identification service system is important for promoting the integration of manufacturing, Internet and IoT. It plays an important role in advanced manufacturing and will also become the central nervous system and important infrastructure. However, in the advanced manufacturing application scenario, the long-term coexistence of various heterogeneous identification systems is the normalcy in the future, and the resulting in many problems to various procedures of intelligent manufacture industry. For example, at the production site, due to the diversity and closure of the various manufacturing control systems in the protocol and identification system aspects, mass production data cannot be fully used and analyzed. In the business management layer, due to the lack of unified planning during the construction of different software systems, the integration and interoperability between software data becomes extremely complicated and costly. In this context, the identification or data intercommunication between enterprises and business collaboration are only explored preliminarily in a few business-specific scenarios. In the future, advanced manufacturing urgently needs a unified heterogeneous identification service system for building basic conditions for intelligence. Therefore, the heterogeneous and compatible, multi-level, distributed and peer-to-peer identification service architecture has become the key point to realize the organic connection between the massive resources and advanced manufacturing applications.

At present, all countries and organizations have designed a variety of heterogeneous and incompatible coding rules based on their own interests. How to maintain the mapping relationship for various types of heterogeneous identification properly, and construct heterogeneous identification management and service infrastructure has become an urgent problem. In addition, China's advanced manufactured industry is on the early stage with weak foundation and capabilities, the comprehensive universal identification service system across industries and platforms has not yet been formed, and it is urgent to strengthen overall coordination. Moreover, many new technologies will be presented in the form of platforms and software, in addition to work out the relevant standards for the technology itself, the standards for the identification service system are urgently needed to be researched and formulated. The requirements of security for manufacturing industry are much larger than other traditional fields. How to find a balance in openness and security, and achieve common development is also an urgent problem to be solved. These problems have brought challenges to the research of heterogeneous identification service systems for advanced manufactured industry.

As we look to the future, the advanced manufacturing ecology is becoming a spotlight of industrial competition, and the opportunities are fleeting. Chinese Academy of Sciences will build a globally unified heterogeneous identification service system based on its own superior resources, technology and experience in the field of identification service. This system establishes a peer-to-peer identification resolution service architecture with mainstream identification systems such as GS1, EPC, Handle and OID, which achieves the compatibility and interworking with the international advanced manufacturing identification systems, furthermore, it will play an important role in various advanced manufacturing scenarios such as supply chain management, personalized customization, networked collaboration, remote monitoring of industrial equipment/product operation status, operation and maintenance, and full life cycle management, as well as to promote the industrialization, informatization, and intelligent development in the process of industrial transformation and upgrading. In addition, through exploring unified management and services of heterogeneous identification service systems in the application scenarios of advanced manufacturing field, it can provide valuable experience for integration of identification services and advanced manufacturing applications.

The new wave of the fourth industrial revolution is surging, and a new era of intelligent manufacturing is coming. In the future, the unified heterogeneous identification service system will become a solid foundation for building intelligent manufacturing ecosystem.

References

1. Industrial Internet platform whitepaper, Alliance of Industrial Internet (AII), September 2017.
2. Common identification management service platform for Internet of Things whitepaper, Computer Network Information Guangzhou Sub-center & Chinese Academy of Sciences (CNIC), Computer Network Information Center & Chinese Academy of Sciences, March 2017.
3. The EPCglobal Architecture Framework. EPCglobal, Standard Specification: Final Version 1.3, 2009.
4. BRIDGE. Serial Level Manufacturing Control, University of Cambridge, BT, SAP, June 2009.
5. 930-S101-1.A0.10/UID-00010-1.A0.10 Ubiquitous Code: ucode, Ubiquitous ID Center, August 2009.
6. RFC3650, Handle System Overview, IETF, November 2003.
7. ITU-T X.oid-res|ISO/IEC 29168. Information technology—Open Systems Interconnection - Object Identifier Resolution System. 2009.
8. China Object Identifier (OID) whitepaper, China Electronics Standardization Institute, July 2015.
9. TP-2017-0114-KEPCO_e-IoT_deployment, oneM2M, May 2017.
10. GB/T 31866-2015 Identification system for Internet of Things—Entity code, September 2015.
11. Xinxin Wu, Identification system for Internet of Things and Identifier Ecode, Xiamen Institute Standardization Xiamen Science & Technology, February 2016.
12. Juan Tian, New forces of China intelligent manufacturing, China Auto-ID, September 2017.
13. Common identification management service platform for Internet of Things Acceptance. http://www.cnic.cas.cn/kxcb/mtsm/201703/t20170315_4759414.html.
14. T/CCSA 209-2018 Technical requirement of service for heterogeneous identifier in Internet of Things, China Communications Standards Association (CCSA), July 2018.

Ye Tian received the Ph.D. degree in computer system structure from Institute of Computing Technology Chinese Academy of Sciences. He is currently an Associate Professor, the Deputy Chief Engineer, China Academy of Industrial Internet, and the concurrently tutor of master degree candidates at Computer Network Information Center, Chinese Academy of Sciences. He is a Senior Member of China Computer Federation and Youth Expert of Internet Society of China. He serves as Communication Member of CCF Task Force on Big Data. He is engaged in research on IoT technology and application, Industrial Internet, smart city, next generation Internet, network security and so on. He has authored more than 30 journal papers and international conference papers, applied for more than 30 invention patents (12 granted patents). He is the executive director of the project "China Internet of Thins names services platform" approved by The National Development and Reform Commission, and responsible for the architecture, technology research and development, platform deployment and industrial application.

Augmented Reality Technology for Assembly and Maintenance of Complex Equipments and Facilities

Haoyu Xu, Shijie Mao and Zhenqi Han

Abstract Augmented reality technology is becoming increasingly practical. This chapter first introduces the origin of this technology. Then, it pointed out that because AR can realize the integration of the atomic world and the bit world, this technology is bound to greatly improve the efficiency of human collaboration and has far-reaching significance. Next, the main technologies and applications of AR in the assembly and maintenance of complex equipments and facilities are discussed. The fourth part discusses the important role of this technology in the research informatization in China. The fifth part of this chapter mainly discusses the relationship between AR technology and the current popular technologies. It demonstrates the complementary relationships among them. At the end of this chapter, the problems and prospects of AR technology are described.

Keywords Augmented reality · Complex equipments and facilities · Maintenance · Assembly · Collaboration

1 AR: Originated from Complex Equipment Assembly and Maintenance Needs

Augmented Reality, or AR for abbreviation, is not a new emerging concept. As early as the late 1960s, scientists had explored augmented reality technology. More interestingly, the birth of the term "Augmented Reality" is related to aviation. The term is created by Boeing engineers.

In 1990, Boeing launched the development of the 777 aircraft. Because the 777 uses a fresh new group of avionics, flight control, landing gear and other design, it was required to install a large number of aircraft harness. These wiring harnesses,

H. Xu (✉) · Z. Han
Shanghai Advanced Research Institute, Chinese Academy of Sciences, Shanghai, China
e-mail: xuhy8@lenovo.com

H. Xu · S. Mao
Lenovo Research, Shanghai, China

© Publishing House of Electronics Industry 2020
China's e-Science Blue Book 2018,
https://doi.org/10.1007/978-981-13-9390-7_18

Fig. 1 Complex wiring harness inside the aircraft. *Source* https://www.fresher.ru/2013/03/27/
kak-delayut-superjet-100/?replytocom=371860

like the nervous system and energy transfer system on the human body, provide not
only the traditional power distribution function but also the information transmission of various systems. The installation position of wiring harness must be
accurate. It is necessary to ensure that all kinds of wiring harnesses do not interfere
with each other. Besides, it is necessary to avoid being interfered by external signals
or interfering with external signals (Fig. 1).

In order to install these complex harnesses, the manufacturer must check the
instruction manuals and perform the assembly job simultaneously. This working
mode is not only inefficient but also difficult to ensure the installation accuracy.

To solve the above problem, in 1990, two Boeing engineers, Thomas Caudell
and David Mizell, proposed a head-up see-through device [1]. The device generated
digital CAD drawings from the scenes captured by head cameras and automatically
extracted the digital parts that matched the current scene from the installation
instruction manuals. It generated the installation instruction virtual image of the
current operation and added it to the real vision. Workers can conduct various
wiring installation with the help of the virtual guidance appearing on the head-up
device. Therefore, it was easy to improve the harness installation efficiency while
reducing its error. So, they created the English phrase "Augmented Reality" to
describe the technology of automatically overlapping virtual content in real-world
situations based on what the user sees.

To realize AR technology, there are two key technologies that need to be broken
through, namely, intelligent recognition technology and tracking registration
technology. The former solves the problem of the location for superposing virtual
objects, and the latter solves the pose of superposition. Because the supporting

upstream and downstream industrial chain was not mature, and the display and tracking registration technology was not ideal, the final idea was a long way from practical application. But AR technology has already sprouted in the assembly demand of complex equipments.

2 AR Realizes the Fusion of Atomic World and Bit World, and Improves the Efficiency of Human Collaboration

2.1 AR Can Realize the Fusion of Atomic World and Bit World

The objective world of human existence has the atomic world and the bit world. Before the birth of digital electronic equipment, there was only one atomic world. The atomic world refers to the material world in which atoms are the basic units. It is the entity we can see and feel. After entering the digital age, human beings have the parallel bit world relative to the atomic world. Bit world refers to the virtual world with the bit as its basic unit. It is an invisible and intangible information world.

Before the maturity of augmented reality technology, although the bit world originated from the atomic world, the representation of the same object in these two worlds was disconnected. For example, when an aircraft engine in the atomic world appears in front of us, the three dimensional model of the engine, maintenance manual, operation data and so on are stored in the space of the bit world. The representations of this engine in both the atomic world and the bit world cannot be presented simultaneously to people in a convenient way (Fig. 2).

Fig. 2 AR realizes the fusion of atomic world and bit world

Augmented Reality (AR) can simultaneously present an object with its bit world and atomic world representations in a natural way through key modules such as object recognition, three-dimensional modeling, scene understanding, SLAM, and rendering in an appropriate way, and thus realize the fusion of atomic world and bit world. For example, the AR system understands that it is in a landing gear replacement scenario through scene understanding and object recognition when the airplane maintenance personnel takes the system walking into the landing gear replacement workshop. If there is a problem that cannot be solved by the on-site crew during the replacement process, with the help of the AR system, the problem labeling or repair process description on the three-dimensional model of the invalid parts are carried out by experts or intelligent machines in the background. At the same time, the annotated three-dimensional model can also be real-time presented in AR devices and assist maintenance to improve the efficiency. The SLAM technology can help the AR system understand the position and posture of the crew, and better understand the scene, so as to achieve more accurate superposition of virtual objects.

Because AR can integrate the atomic world and the bit world well, it can greatly improve the human collaboration ability and efficiency. Therefore, AR can further promote the development of human productivity.

Different objects, from small molecules to large planets, mutual collaboration is one of the decisive factors for their development.

2.2 Life Originates from Collaboration

German Nobel Prize winner and chemist Manfred Eigen studied the reproductive performance of life genetic material in collaboration with Peter Schuster of the University of Vienna [2]. Their results showed that the final winner in the life selection is not the most adaptable RNA sequence, but the most adaptable quasi-species. Quasi-species refer to a group of affinity RNA molecules that are slightly different from each other. RNA replication always has mutations. In the most adaptive quasi-species, if the neighbors of a mutated-molecule mutate, it is more likely to produce the original RNA replication factor. Therefore, adjacent RAN sequences can mutually coordinate through mutation.

In the process of evolution, the quasi-species with good collaboration in group operations eventually defeated the quasi-species with poor collaboration and the RNA sequences in a single operation, won the victory of evolution, and finally created the dawn of life.

2.3 Language: the Important Collaboration Tool for Human Beings Achieving Civilization

The significance of language to the birth of humankind is comparable to that of the first life. Only the most cooperative quasi-species resulted in the life birth after billions of years of evolution. With the mastery of language as an important collaborative tool, human beings can overcome other species in the process of evolution and gradually move toward civilization.

Language, the collaborative tool, is an important symbol that differentiates human beings from other species. In the very early stage, the intelligence level of human beings was not that different from that of some high-level animals. These slight intellectual differences did not distinguish humans from other higher animals as more advanced species. Elephants can find water sources that they had gone 40 years ago. Monkeys are able to move stones from a kilometer away to destroy nuts. But these animals are less intelligent than humans. This is because in the long evolutionary history, human beings have realized the collaboration among groups and accumulated a lot of knowledge through language, so that they can think more deeply and express more complex emotions.

Human beings can master language because they own some features at the same time. These features include a certain level of intelligence, a well-developed vocal organ, and enough life time. Smart dogs can achieve the intelligence level as an 8 years old child. But they have only elementary vocal organ such that they can only make simple barks. Parrot's vocal organ is very close to human's. But they have a short life span. The whole parrot species can't evolve into a systematic language. Neanderthals in Europe are human-like species extinct thirty thousand years ago. They had bigger brain capacity, stronger muscle strength than human beings. Because their larynx was very small, they can only make simple roar and cannot form complex language. When Neanderthals faced with our ancestors' aggression, although a single Neanderthal is stronger than human beings, they cannot effectively cooperate with each other due to lack of language. Finally, they were exterminated by mankind and completely disappeared in the long history.

2.4 AR Can Greatly Improve the Efficiency of Human Collaboration

The collaboration among RNA quasi-species eventually evolved into life. The collaboration using language enabled early humans to stand out in species competition. The emergence of AR technology will further enhance the efficiency of human collaboration and promote the further evolution of human society.

Without considering the time factor, language is a one-dimensional information representation and a sequence of pronunciation, intonation and semantics.

Language, the one-dimensional information representation, can already help mankind win in the species' competition on the earth and move toward civilization.

On the other hand, AR is a multi-dimensional information representation. AR realizes the integration of the atomic world and bit world, and these two worlds are at least three dimensions. The dimension of the new world integrating the bit and atomic world is far greater than 6 dimensions, and its impact on human beings is unimaginable at this stage. As we all know, language and imagination are two unique features that differentiate human beings from other species. According to the viewpoints of some scholars [3, 4], the imaginary community shapes our society, including a series of concepts such as state, nationality and company. Human beings have shown their rich imagination through various literary and artistic means, such as music, movies, and novels. But these forms of imagination cannot be displayed in real time in front of people, and they are disconnected from the real-world scenes. With AR technology at hand, human imagination can be displayed in real time, seamlessly integrated with the atomic world, and interact with a large number of users. It will inevitably bring further profound changes to the human society.

3 AR is Promising in the Field of Assembly and Maintenance of Complex Equipment and Facilities

AR technology is a kind of technology to enhance human collaboration ability. It has a great use in most areas where humans and robots are not good at. For example, in the aircraft manufacturing process described in the previous section, it is difficult for crews to have the accurate and complete memory of the large number of steps and winding relationships. Although it is very easy for robots to memorize this information, robots are still unable to complete the maintenance and repair work considering the limitations of current mechanical technologies, such as the fluctuation of the body, the flexibility of jumping and the kneading of the fingers, etc.

3.1 AR has a Broad Applicability

The scenario for AR application is that a job involves a large number of complex tasks, and these tasks themselves have strict operational sequence and quality requirements. Common scenarios include wiring harness and accessories installation in the manufacturing process, fault detection and troubleshooting assistance in maintenance process for high-end equipment and facilities.

Let's take a look at Boeing's example again. Although Boeing did not succeed in its attempt to assemble the harness in the 1990s, Boeing did not stop applying AR technology in the assembly of the harness. The birth of Google Glasses created conditions for the application of AR technology in the aviation industry.

Google eyeglasses are an AR glasses launched by Google in 2012. Google Glasses is equipped with a head-mounted micro-display on the upper right of the lens, which can project the required information on the display. The display effect is similar to that of a 25 in (635 mm) HD screen outside 2.4 m. Google Glasses only provided a basic hardware platform, so Boeing found Skylight, an application software developer, to develop AR applications that meet the harness assembly requirements. During the trail, a worker first received Google glasses and completed a device authentication by scanning a two-dimensional code. After the authentication was completed, the system would automatically push a wire harness assembly application to Google glasses. Then the worker went to the assembly site and scanned the two-dimensional code of a part on the assembly site with the camera on the glasses. The assembly instructions of the wiring harness of the part were automatically displayed on the glasses. The worker would follow the instructions step by step to carry out the wiring harness assembly (Fig. 3).

So what's the result of this project? Boeing Engineer DeStories said AR glasses help installers save at least 25% of assembly time and significantly reduce assembly error.

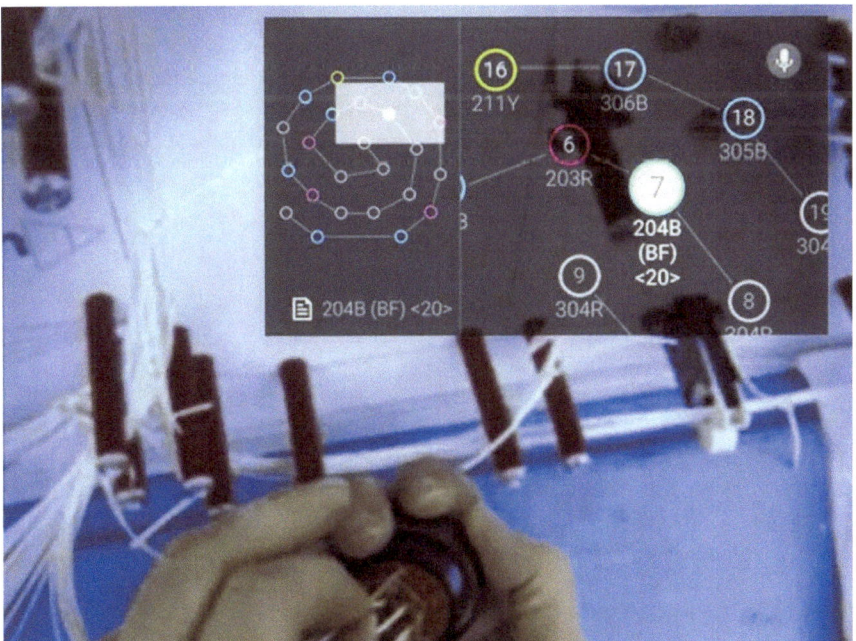

Fig. 3 Workers are wiring according to the instructions of Google Glasses. *Source* https://tecnoblog.net/198410/boeing-google-glass-avioes/

3.2   The First-Person View of AR is Especially Suitable for Remote Collaboration

Remote collaboration is not a new application. After the invention of telephone, the concept of remote assistance began to appear. However, it is very difficult to describe the situation of the scene completely by only relying on the voice description of the telephone. Several years after the birth of the Internet, people began to try to use remote video to help each other. Compared with voice, video can really demonstrate more content and richer details. But video cameras are usually located in a fixed place, so the images presented to remote collaborators and those seen by field workers are biased in perspective. This kind of bias will bring great inconvenience in some operations, such as some operations with precise azimuth requirements.

In contrast, the first-person view function provided by AR glasses is especially suitable for remote collaboration. After wearing AR smart glasses, the remote expert sees the same picture as the on-site person. This kind of situation is called as first-person view. With the help of first-person view, the clarification of the relative positions of different objects is omitted, thus eliminating unnecessary misunderstandings, which not only ensures the security of the results, but also enhances the efficiency of the work. Therefore, this function is warmly welcomed by users.

Skylight developed the remote expert collaboration function based on Google glasses. When on-site workers encountered problems that cannot be solved independently, they used Google Glasses to transmit real-time videos to experts in other places for help. Because the camera of AR glasses was in the middle of the person's two eyes, the videos seen by the experts were basically the same as those seen by the on-site workers. Thus, remote experts were able to give more accurate operation suggestions aiming for quick solution (Fig. 4).

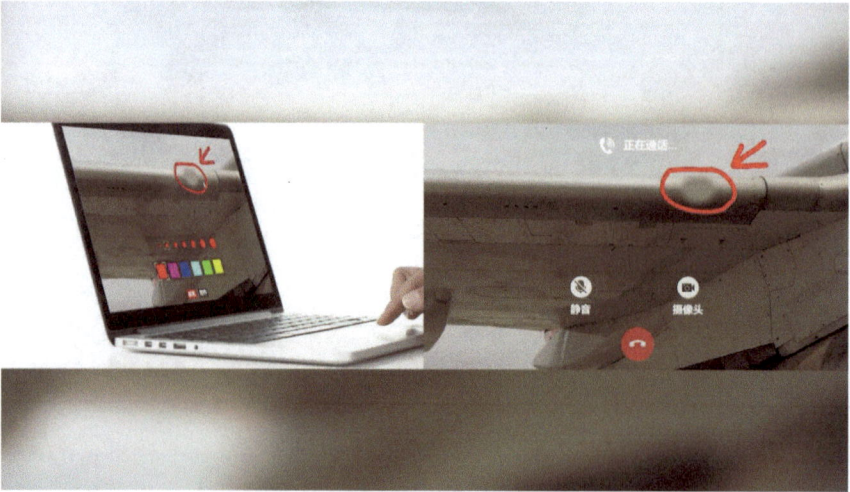

Fig. 4 First-person view for remote collaboration

3.3   AR has a Strong Extensibility

AR technology not only has wide applicability in the field of complex equipment and facilities, but also has strong extensibility in functions. In addition to the work steps to achieve complex task and the quality requirements, AR can also realize automatic defect detection and troubleshooting functions using man–machine interactive question–answering.

At present, automatic defect detection mainly relies on visual sensors, such as visible light and infrared, to complete signal acquisition. After the signal acquisition is completed, the signal can be processed locally or sent back to the backend servers for processing. When the on-site worker wearing the AR smart glasses walks into the workshop, the automatic defect detection program will start automatically. The AR system can intelligently detect the defects of related components involved in each maintenance task.

Due to the limitation of power consumption, size and other conditions, the types of sensors that can be integrated into AR glasses are still limited. For example, the present ultrasonic flaw detector has a wide range of applications in industry. It can quickly, conveniently, non-destructively and accurately detect, locate, evaluate, and diagnose various defects (cracks, porosities, inclusions, etc.) inside the workpieces. It is widely used in aerospace, aviation, power, shipbuilding, automobile, machinery manufacturing, steel structure, railway transportation, nuclear power and other industries. If the ultrasonic detection technology can be miniaturized to the level integrated into the glasses, AR glasses will greatly expand the scope and type of defects it can inspect in addition to visible and infrared visual inspection.

Man–machine question–answering troubleshooting system is especially suitable for some difficult situations when the remote experts are not available. By digitizing the maintenance manuals of complex facilities and equipment and the maintenance records in the history of enterprises, knowledge is extracted and constructed into a dynamic knowledge base. The dynamic knowledge base refers to the database that can update the new maintenance cases at any time. After the on-site workers ask questions in natural language, AR system automatically finds the appropriate answers based on the knowledge base and feeds back to them in the form of text, image, animation and even video.

Man–machine question–answering systems have at least two benefits. Firstly, it can promote the corporations to cherish the knowledge accumulation both outside and inside. They can still operate as usual if the senior staff leaves. Secondly, it can also reduce the demand for technical experts and liberate their productivity.

4 AR Technology Will Strongly Boost the Informatization of Scientific Research in China

As AR technology is not yet fully mature, its role in the process of scientific research informatization in China has not been fully reflected. However, AR technology realizes the integration of the atomic world and the bit world, thus improving the collaborative efficiency of scientific researchers, and will strongly promote the scientific research informatization in China.

AR technology is well suited to research activities that require extensive experience but the on-site conditions are not friendly. For example, in the process of field investigation or archeology, because of tough site conditions, experienced elderly researchers inevitably cannot carry out their work on the site.

Through the first-person view function provided by AR technology, senior experts in the laboratory can provide remote assistance to front-line staff in animation, labeling and other ways, thus speeding up the process of field research. Moreover, due to the special requirements for site protection of investigation and archeology, with the remote assistance of senior experts, there is an additional layer of protection for the site.

AR technology is also suitable for the assembly of complex high-end equipment that requires frequent manual inquiries. For example, the manufacturing of rockets or satellites has a large number of assembly steps and components. With the step-wise guidance provided by the AR equipment, the assembly work can be completed quickly and accurately. It reduces the possibility of errors in the process of assembling and querying manuals simultaneously.

AR technology is of great significance for the maintenance of large-scale scientific research facilities in China, such as the FAST Radio telescope, controlled nuclear fusion device, and Shanghai light source. With the help of AR technology, maintenance personnel have improved robustness in detecting defects and anomalies. In addition to real-time sensor monitoring and manual inspection, AR is added to improve the maintenance level of major scientific research devices.

Above are just some simple examples to illustrate the role of AR technology in scientific research informatization. As AR technology breaks through the barriers between the atomic world and the bit world, it will be more widely used in scientific research activities, thus playing a more powerful role in promoting scientific research informatization in China.

5 "Bit-Atom" System Will Realize the Improvement of Maintenance Efficiency

"Bit-Atom" system is developed by Shanghai Advanced Research Institute, Chinese Academy of Sciences and Lenovo. The system applies AR technology to the maintenance of high-end complex equipment and facilities, reduces the requirements of maintenance work for staff, and achieves the efficiency improvement. Based on AR equipment and AR service platform, it provides seven functions: intelligent object recognition, pose estimation, remote assistant, defect detection, human–computer interaction, process recording, etc.

Intelligent object recognition function is the bridge between the atomic world and the bit world. AR devices can intelligently identify the target objects that need to be assembled or maintained and automatically display the corresponding 3D model or guidance information. For example, in the inspection process, when the AR equipment is aligned with the parts that need to be repaired, the AR equipment can automatically identify the parts and intelligently prompt the inspection requirements of the parts. Therefore, AR devices can quickly distinguish different kinds of objects through intelligent object recognition technology, focus on the objects assembled or maintained, reduce interference, and ensure the efficiency of assembly or maintenance.

The pose estimation function lays the foundation for the intelligent operation step guidance in assembly and maintenance process. Pose estimation is to determine the location and posture of the AR device through SLAM technology or other location techniques. For example, through SLAM technology, AR devices can intelligently prompt the current work position and personnel posture, the completed work locations, and the work locations to be completed. When repairing parts, AR equipment provides intellectualized prompts for which parts to disassemble or install in the upcoming steps.

Remote assistant functions are aiming for helping to solve difficult problems. With the help of AR equipment, the on-site worker can transmit real-time video to the remote experts. Remote experts can add guidance information on the received video and meanwhile the annotated video is shared with the on-site worker simultaneously. They can carry out the repairment steps based on the guidance given by the remote experts.

Defect detection function intelligently detects any abnormal situations in the view of AR equipment. Suppose in a routine inspection, the Z handle of a hatch is not correctly reset. When the crew watches the cabin door through AR equipment, the prompt of "incorrect positioning of Z handle" is immediately popped up on the equipment, accompanied by an alarm sound. Defect detection enables operators to get rid of the work pressure brought about by high tension and make them feel the happy work.

"Bit-Atom" system uses the gesture or speech recognition technology to complete human–machine interaction. The operator only needs to put his finger in front of the AR device and gently close his forefinger and thumb to start the application. Operators can also use natural language "start XXX application" to start the application.

Process recording realizes controllable and traceable management in assembly or maintenance process. Through the abundant visual sensors on AR equipment, they record the whole operation process (similar to traffic recorder), provide video analysis for standardization of operation process, and provide data source for future assembly or maintenance knowledge base construction.

With the help of seven intelligent functions, the operators are thoroughly liberated from the complicated, tedious and professional assembly, and maintenance tasks. As a result, only the staff with preliminary experience can complete maintenance tasks with high quality in time.

6 AR is Complementary to Other Information Technologies

The development, maturity, and commercialization of AR technology are closely related to other upstream and downstream supporting technologies. This section mainly discusses some popular technologies and their relationship with AR.

6.1 5G

5G technology is a new generation of mobile communication technology being developed all over the world. Its technical goal is to achieve a single terminal peak throughput of 10 Gbps and end-to-end delay of 1 ms. Compared with the current 4G network, 5G network will provide super-high capacity, ultra-low delay, and high reliability.

The high capacity, ultra-low latency, and high reliability of 5G network will provide wireless transmission guarantee for AR application and further improve the user experience. The application of AR aims to share and interact at anytime and anywhere. In order to realize the functions of 3D image reproduction, real-time location and reconstruction, high-definition video detection, 3D HD multimedia content download, and so on, the throughput of several hundred megabits or even Gbps is required for a single terminal. AR scene recognition and reconstruction, image and gesture recognition, the delay from the occurrence of action to the presentation of the image in front of the user should not exceed 6 ms. The power consumption and battery endurance of AR mobile terminals also have higher requirements for energy saving design on the network side. If the power control and coverage have no special optimizations, excessive power consumption will lead to a significant increase in power and heat consumption of AR terminals, thus affecting the AR experience.

6.2 Artificial Intelligence

Augmented reality and artificial intelligence are two closely related popular technologies. AR can be used as an entry to AI, and the advances of AI technology promote the improvement of AR applications.

The so-called entry is a tool that people must use when experiencing or using a service. PC and mobile phones are entries for the Internet and mobile Internet, respectively. To be an entry, it must be a tool that people use frequently in daily life. AR equipment will also be used frequently in the field of assembly and maintenance of complex equipment and facilities mentioned in the previous section.

On the other hand, AI will enable AR technology to provide more fascinating functions. AR devices can realize object recognition with the AI enabling. Object recognition is the foundation of virtual objects superim posed objects. Defect detection needs computer vision technology, and man–machine question-and-answer troubleshooting needs various technologies in natural language processing.

6.3 Brain–Computer Interface

Brain–computer interface technology refers to the technology of realizing the interactivity between human brain and machine. Brain functions are roughly classified as perceiving the world (such as auditory, visual, and olfactory sense), limb movements (such as limb movements, eye movements, and speech), and other more endogenous functions (such as decision-making, memory, and emotion). Common brain–computer interface technology is widely used in the care of the elderly, the loss of sports people, and so on.

Brain–computer interface technology has not been used in AR, but this does not hinder our imagination on combination of these two techniques. There are two direct uses, corresponding to read-out and write-in functions, respectively. Read-out brain–computer interface technology refers to the operation of overlapped virtual content after reading and decoding nerve signals to realize human–computer interaction, such as virtual content clicking, amplifying, shrinking, closing, and moving position. Write-in BCI converts AR virtual content into signal coding and writes to human nervous system for visual perception. In this case, there is no need for the display device.

6.4 Robots

AR provides data acquisition and case preparation for robots to expand their applications. The target area of AR technology, as mentioned above, should be those most people or robots are not good at them. However, not good for today, does not mean it will be the case for tomorrow, especially for robots. With the advances of computing power, storage capacity, high-energy battery, flexible materials, and other technologies, robots are becoming more and more powerful.

Therefore, when using augmented reality technology, people are carrying out data accumulation for future robots to be competent for the same job. During engine maintenance, after recording the finger joints, wrist joints, and elbow joints of the worker in each step of the operation process, people can summarize the movement rule and force direction of the hand in complex operation, so as to provide basic data input for robots' operation.

7 Key Technologies and Challenges of AR

7.1 Tracking and Registration

Tracking and registration are two core technologies of AR to achieve virtual and real objects fusion. Tracking refers to determining the position and orientation of an object in the physical world. An object can be a real object to be superimposed on a virtual image, or it can refer to a part of the user's body, such as the head or hand. Registration is based on tracking, realizing the alignment of virtual objects and real objects, so as to achieve the goal of virtual and real objects fusion.

In order to meet the requirements of AR glasses, tracking technology must meet the mobility, portability and centimeter accuracy requirements. Therefore, hardware-based tracking technologies, such as Beidou/GPS, mechanical tracker, electromagnetic tracker, ultrasonic tracker, inertial tracker, and optical tracker, cannot meet the above requirements. Vision-based tracking technology provides a non-contact, accurate, and low-cost solution, which is becoming more and more practical in AR glasses. Especially with the increase of computing power of CPU and the sharp decrease of cost and size of 3D sensors, the SLAM has become the mainstream of AR eyeglasses tracking technology. SLAM technology usually uses 3D data collected by depth sensors (ToF, etc.) and binocular vision sensors, combined with inertial sensor data, to complete scene construction and tracking in the run time without the need to save scene information in advance. This technology meets the needs of mobility, portability, and centimeter accuracy requirements and has been applied in products such as Lenovo's Daystar.

The main error of registration comes from two aspects. The first and the most important source is the tracking error, and the second is the display distortion. There are two ways to improve tracking errors: one is to use more sensors and better algorithms, such as SLAM to reduce tracking errors; the other is to consider both rendering and tracking errors to achieve joint optimization. Therefore, with the maturity of SLAM and other related technologies, tracking and registration issues in AR have been basically solved.

7.2 Display

The display of AR glasses includes two parts: image source and optical system. The image sources currently include LCoS (used by Microsoft Hololens and Sony), mini-OLED (used by Apple), and DLP (used by Oculus). The focus of display technology nowadays is on the optical system, which is ODG reflective scheme, optical waveguide scheme, and optical field technology according to the maturity level from high to low.

The ODG reflection scheme is an improved version of the prism scheme. The ODG scheme adopts a monolithic half lens technology. The scheme removes other parts of the prism, leaving only the semi-inverse and semi-permeable film. In this way, it is easy to realize the thinning of the glass sheet sandwiching the film. The features of this scheme are low-cost and mature technology. The disadvantage of this scheme is that the lens size and space occupied are slightly larger than that of the other two schemes. The optical waveguide scheme uses the diffraction principle of light to realize image projection. Take the Hololens as an example. It uses holographic diffraction waveguide grating. The advantage of this display technology is that its lenses can be roughly the same thickness as ordinary lenses. Small field of view and high cost are two disadvantages. Optical field technology controls the direction of laser emission by changing the shape of fibers in three-dimensional space, especially the tangent direction at the fiber ports. This technology has the merits of large filed of view and high-definition display. However, the disadvantages of the optical field technology are also obvious. The computation is too large to realize on the AR glasses. Meanwhile, it is also extremely difficult to adjust the flutter mode of optical fiber dimension in real time in the way of data synchronization so as to change the output direction of light naturally.

ODG reflection scheme and optical waveguide scheme have commercial products. While optical field technology is still in the conceptual stage, the launch of products has no accurate timetable.

ODG reflective scheme is acceptable for industrial users despite its large size.

7.3 Adaptability in Complex Environments

At the present, AR glasses and technology developed by major companies aim for the application in the indoor environments with small space size. The environment of complex equipment and facilities usage includes outdoor open areas and large space workshops with a scale of several hundred meters.

Outdoor work scenarios, in addition to the requirements of electronic equipment such as three proof, suitability for high-and low-temperature, also pose new challenges to the core technologies of AR. The illumination conditions of outdoor working scenes are complex and changeable, which bring problems such as backlighting, reflection, and insufficient illumination. They will make object recognition and SLAM technology intractable.

The outdoor large space size challenges SLAM's effective distance. Nowadays, the maximum working distance of depth sensor is not more than 50 m, which is smaller than the size of hangar or workshop. In addition, in the apron, hangar or workshop, the huge noise emitted by the engine, and other equipment also causes technical obstacles to voice interaction.

Therefore, in the aspect of complex environment adaptability, the existing AR technology needs further feasibility evaluation and even technical tackling.

7.4   The Application Breakthrough of AR

People should develop the AR technology system on the basis of the existing mature technology to meet the application demands. As long as the "single breakthrough" of market pain points can be realized, there will be opportunities to promote further vertical and horizontal expansion of applications and also to promote the gradual maturity of technology.

The core technology that has not yet been resolved must be highly valued. Although there are still many difficulties in realizing optical field technology, by the large investment of Microsoft and Magic Leap, it is believed that in the near future, the technology will be expected to be commercialized. Light field technology is expected to become the preferred display scheme for major AR manufacturers as it is the most natural and comfortable human eyes. Besides, object recognition and SLAM are difficult problems in complex illumination conditions.

In conclusion, with the improvement of chip computing capability, miniaturization and low cost of various new sensors, rapid development of artificial intelligence technology, commercial use of 5G transmission technology, and even maturity of brain–computer interface technology, AR technology will be much widely applied in the assembly and maintenance of complex equipment and facilities beyond people's imagination.

References

1. Caudell TP, Mizell DW, Augmented reality: an application of heads-up display technology to manual manufacturing processes, Proceedings of the Twenty-Fifth Hawaii International Conference on System Sciences, 1992.
2. Martin A. Nowak, Super Cooperators: Altruism, Evolution, and Why We Need Each Other to Succeed (in Chinese), Zhejiang people's Publishing House, 2013.
3. Yuval Noah Harari, A brief history of humankind (in Chinese), CITIC Press, 2014.
4. Benedict Richard O'Gorman Anderson, Imagined Communities: Reflections on the Origin and Spread of Nationalism (in Chinese), Shanghai people's Publishing House, 2005.

Haoyu Xu is with Shanghai Advanced Research Institute, Chinese Academy of Sciences as an Associate Professor and Lab Director. He has published more than 20 academic papers and once received the best student paper in IEEE ICNS 2012. He has been authorized seven invention patents of China and one of them is PCT. He was once awarded the second prize of Science and Technology of Shanghai Municipality. His current research interests include the machine learning and its applications in AR and robots.

Zhenqi Han works in Shanghai Advanced Research Institute, Chinese Academy of Sciences as Research Assistant. His main research area is the application of machine learning in defect detection and AR. He has published a SCI-indexed paper, four EI-indexed papers, and filed five invention patents.

Study on Big Data-Supported Clinical Oncology

Ying Jin, Chaofeng Li, Zixian Wang, Ying Sun and Ruihua Xu

Abstract Population health is the most important issue for social well-being, which is closely related to national economic development and social progress, and a great concern of governments around the world. There are millions of new cancer cases in China every year, and the incidence is increasing year by year. Cancer has become a common disease and the leading cause of death in China, transcending cardiovascular and cerebrovascular diseases. With the rapid development of big data and cloud computing technologies, their application in basic medicine, clinical medicine, and public health is rapidly expanding. This chapter mainly introduces the importance of big data to the development of oncology; the key technologies for the development of oncology big data; the framework of big data platforms for oncology and how it would work for oncologic clinical practice and researches; and the current status and challenges facing cancer big data, as well as perspectives to its future development.

Keywords Big data · Oncology

1 Background

"Big data" refers to data sets which are large in volume, complex in structure. They are difficult to acquire, store, format, extract, curate, integrate, analyze, and visualize by traditional data-processing software. The International Business Machines Corporation (IBM) summarizes properties of big data with the five "V" acronym, namely, (1) Volume (the volume of data), (2) Variety (the number of types of data), (3) Velocity (the speed of data processing), (4) Value (the potential significance of the data), and (5) Veracity (the trustworthiness of the data). Biomedical big data is significantly different from that of other fields such as economy, social media, and environmental science. They are greatly diverse, high in generation rate, difficult to reuse, sometimes doubtful in quality, and constrained by regulatory authorities to

Y. Jin (✉) · C. Li · Z. Wang · Y. Sun · R. Xu
Sun Yat-sen University Cancer Center, Guangzhou, China

© Publishing House of Electronics Industry 2020
China's e-Science Blue Book 2018,
https://doi.org/10.1007/978-981-13-9390-7_19

share between clinical and basic research institutes. These properties pose unprecedented challenges to extract meaningful information from them, in all aspects of medical work.

Tumors are complex diseases with varying etiology, pathogenesis, and risk factors. Correspondingly, there is multitude of treatments: surgery, radiotherapy, chemotherapy, interventional therapy, traditional Chinese medicine, target therapy, immunotherapy, and the combination of different regimens. Randomized controlled trials (RCTs) are the current gold standard to evaluate the efficacy of cancer treatments. However, RCTs' restricted inclusion criteria often limit the population of interest to an insufficient size lacking representativeness, especially for rare cancer types. On the other hand, daily medical practice generates large volume of real clinical data, which if fully utilized could compensate the short comings of traditional RCTs. The current treatment for tumors in China is mainly based on guidelines and experts' experiences. Many of them are not well supported by high-level clinical evidence. Mining medical big data will empower biologists, clinicians, epidemiologists, and health care policymakers to make data-supported decisions and such decisions will ultimately benefit patients and the entire society. Several recent influential studies have pointed out a few important aspects particularly suitable for big data application: understanding disease mechanism, population-level evaluation of disease prevention and treatment systems, diagnosis and treatment decision-making, and new therapy development.

2 Big Data Promotes Development of Oncology

2.1 The Significance of Big Data in Explaining the Carcinogenesis and Improving Tumor Prevention

Similar as to the development of tumors which is a multistep process, the risk factors for tumors are multifaceted and multileveled. The useful information can be found in the medical history, genomic and laboratory tests, personal health records. Traditional analysis of tumor risk factors mainly relied on epidemiological investigations, which are often limited by insufficient sample size, incomplete information, imperfect follow-up, and biases in investigation. The new data processing and analysis technologies can integrate not only clinical data but also data from multiple sources including genomic data, environmental exposure, daily life habits, geographic location, social media, and a variety of other information. This can compensate the traditional clinical research and provide a more precise and thorough understanding of cancer.

With the rapid development of the second- and third-generation sequencing technologies, our understanding of the molecular biology and pathogenesis of tumors has been gradually revolutionized. Multileveled omics data, such as whole genome, whole exome, transcriptome, proteome, DNA Methylome, and microbiome, will soon become an important basis for uncovering the mystery of tumor etiology and pathogenesis. These omics data are large in volume, complex in structure, and difficult to integrate with clinical data and to analyze. Various omics research focuses on mining these big data sets to generate new knowledge. The vast amount of medical big data available on public platforms is an excellent resource for innovative scientific research. For example, ONCOMINE (http://www.oncomine. org) is a cancer microarray database and web-based data mining platform that contains more than 4700 gene expression data sets and information of more than 80,000 cancer and normal tissue samples, with complete cancer mutation profiles, gene expression data, and related clinical information. Using the ONCOMINE, we can compare and analyze the genes differentially expressed between the most important cancer types and corresponding normal tissues, in order to discover the potential pathogenic genes and to reveal the pathogenesis of tumors. Other important omics information sharing platforms include the TCGA (The Cancer Genome Atlas, https://cancergenome.nih. gov/), GEO (NCBI Gene Expression Omnibus, https://www.ncbi.nlm.nih.gov/geo/), ICGC (International Cancer Genome Consortium, http://www.icgc.org), COSMIC (Catalogue Of Somatic Mutations In Cancer), 1000 Genomes (http://www.international genome.org/), ENCODE(Encyclopedia of DNA Elements, http://genome.ucsc.edu/ ENCODE/).

At present, there is an increasing disease burden of malignant tumors in China and this has received widespread attention from the public. Tumor screening is an important measure for cancer prevention. Through medical checkups, each person can have a record on the big data platform, and the data can be integrated and analyzed to update the tumor risk status of an individual in real time and perform timely screening for the susceptible population. For instance, big data analysis can screen and identify individuals or families at high risk for Lynch syndrome (an inherited disorder that increases the risk of many types of cancer, particularly colorectal cancer), facilitating early detection and diagnosis before progression of the disease. Therefore, with the advancement of such technologies, these individuals can receive preventive examination, treatment, and close follow-up.

2.2 The Significance of Big Data in Improving Diagnosis of Cancer

The clinical diagnosis of cancers is a prerequisite for implementation of proper treatment and to improve disease prognosis. The process includes query of medical history, physical, laboratory, imaging, and pathological examinations. The early detection and treatment of tumors are important. It can significantly alleviate the cost-related burden on the health care system, as early treatment may cost only half of that of late treatment. However, early diagnosis of tumors is difficult due to the insidious clinical signs and symptoms and lack of sensitivity to test indicators. Big data can integrate and analyze disease-related information from a large number of healthy people, which plays an important role in early diagnosis of tumors. The research team of the Sun Yat-sen University Cancer Center compared the methylation levels of nearly 500,000 CpG sites in liver cancer tissues and normal blood DNA at the genome-wide level and screened the sites with the most significant differences using big data mining and analysis techniques. These sites were further analyzed in 2000 cases of liver cancer patients and normal controls by methylation PCR amplification and sequencing, and a diagnostic model based on ten to eight of these sites was established. This model showed higher sensitivity and specificity than the current standard serum marker for liver cancer, AFP. This could be of great significance for the early diagnosis and treatment of liver cancer.

Big data platform has an important role in promoting cancer diagnosis accuracy and treatment efficacy in China. At present, the medical resources in China are isolated, and the ability of cancer diagnosis in primary hospital needs to be improved. Through an integrated data warehouse that collects a large number of medical records, imaging data, pathological results, and other types of medical data, medical institutions of multiple levels would be able to share information to improve the diagnosis and treatment of cancers in primary institutions by remote consultation. In 2016, the *Notice on Promoting the Pilot Work of Graded Diagnosis and Treatment* announced by the National Health and Family Planning Commission also emphasized that it is necessary to accelerate the informatization of medical institutions and sharing of medical resources, to form a regional medical and health information platform, continuously improving electronic health and medical records adoption and sharing information across medical institutions of different levels and categories. Meanwhile, it's also necessary to use telemedicine to improve the accessibility of high-quality medical resources and the efficiency of medical services, and to develop internet-based medical and health services with new technologies such as the internet and big data. For example, the Pan-Central South Region Oncology Specialist Alliance and Cancer Clinical Research Collaboration Network is an alliance of 50 hospitals from 12 provinces (or autonomous regions) including Guangdong, Guangxi, Fujian, Yunnan, Hubei, Hunan, Anhui, and Shaanxi. The alliance focuses on the construction and application of a big data platform in primary medical institutions to establish telemedicine platforms between affiliated hospitals. The aim is to conduct remote consultations, provide skilled training, and academic exchange, in order to enhance the services provided in medical institutions and to improve the overall quality of cancer diagnosis and treatment in China.

2.3 The Significance of Big Data in Promoting Treatment of Cancer

The progress of cancer treatment relies heavily on large multicenter RCTs, but even in the USA, the forefront country of cancer research, less than 5% of cancer patients participate in RCTs. Moreover, these studies only include patients who meet specific physiological indicators, which often do not reflect the specific conditions encountered by oncologists in their daily work. Real-World Study (RWS) has been proposed as a solution to overcome this problem. Medical workers can analyze large tumor database to improve their daily clinical practice, for example, to identify patients' vital signs, to estimate treatment cost and efficacy, to clarify the indications and verify the efficacy of antitumor drugs in real-world population, to collect adverse reaction reports in real time, to investigate rare adverse effects, to perform secondary evaluation of drugs, and to find the most clinically effective treatment for each cancer patients to provide more personalized medical care.

In December 2013, the American Society of Clinical Oncology (ASCO) launched a project to use big data to help cancer treatment, CancerLinQ, which was designed to collect data on the diagnosis and treatment of cancer patients at a large-scale to facilitate clinical decision-making and for quality assessment of clinical practice. The system collects data that doctors usually keep on their medical records, such as age, gender, medications, concomitant diseases, diagnosis, and treatment procedures until death. The project has currently recruited 87 cancer centers, more than 2500 oncologists, 15 electronic health record providers, and has collected over 1.8 million records of patient data. The program allows doctors and researchers to access and analyze the medical records of anonymous cancer patients, and almost all patients will become subjects of oncological studies. After pooling enough patient records, doctors can study the disease comprehensively at a broadscale, helping to develop treatment plans for other patients.

Cancer treatment has gradually entered the "era of precision medicine." Through integrative analysis of omic-level big data, e.g., genomics, transcriptomics, proteomics, and DNA methylation profile, researchers would be able to identify, validate, and apply biomarkers in populations with specific tumor types, thereby defining disease status and finding therapeutic targets with greater accuracy, and optimistically driving toward the goal of individualized and precise treatment of cancer.

2.4 The Significance of Big Data in the Development of New Anti-tumor Drugs

Research and development of novel anticancer drugs constitute a great part of the progress in cancer treatment. At present, the cost of drug research and development continues to soar. Relevant data shows that for the development and testing of a new drug till its approval by regulatory agencies, pharmaceutical companies spend more than $2.5 billion. Big data techniques may greatly facilitate drug research and development. Subsequently, they can help pharmaceutical companies to be informed of current drug(s) approvals and their use in the market to better investigate clinical requirements. Researchers can mine real-world clinical data to identify new molecules which have the potential to become safe and effective drugs, to validate effective targets and biomarkers, and can also to screen the follow-up on preclinical and clinical trial data. In all, the time for drug development can be shortened and cost can be lowered.

In addition, big data can also play a role in the use of new drugs in clinical trials. Medical databases can accelerate patient recruitment by identifying suitable patients in accordance with their characteristics and lifestyle, evaluating their compatibility for clinical trials of new drugs. Big data analysis can also greatly amplify the value of data obtained from clinical trials by providing clinicians with the best aid toward treatment decision-making. In general, the key value of big data for the development of new drugs is to make the trial design more reasonable, reduce the rate of trial failure, and improve the efficiency and cost-effectiveness of the clinical trials.

3 Key Technologies for the Development of Oncology Big Data

3.1 Standardized Acquisition and Standardization Techniques in Oncology Big Data

Standardized acquisition and standardization techniques in oncology big data aim to establish a series of standards and norms which integrate tumor-related clinical information dispersed on different platforms into a unified large database. The process includes data collection, cleaning, transmission, storage, integration, and strict quality control. It lays the foundation for mining and transformation of the big data.

For the construction of the standard specification system for medical big data applications, current international researches mainly focus on the ontological technology for constructing medical terminology standards and the interoperability technology for constructing information exchange standards. By constructing a controlled medical vocabulary, medical ontological technology clarifies the standard usage of clinical concepts and accumulates synonyms that express the same concept basing on their standard usage, and further clarifies the hierarchical and semantic relationships between the different clinical concepts. From the perspective of synonymous descriptions, medical ontology can be regarded as the semantic standard of medical big data integration. Developed countries in Europe and the USA attach great importance to the development of medical ontology systems. For example, the Systematized Nomenclature of Medicine-Clinical Terms (SNOMED-CT) developed by the International Health Terminology Standards Development Organization provides a comprehensive and unified medical ontology system covering 18 types of clinical information including diseases, findings, operations, microorganisms, drugs, etc. Unlike ordinary dictionaries, SNOMED's terms are not independent of each other, but are strictly organized according to the logic principle of ontological description, and it has become a standard of clinical medical terminology widely used around the world.

Interoperability describes the ability of different information systems to exchange, share, and interpret data and work together. Data exchange schema should allow data to be shared between clinicians, laboratories, hospitals, pharmacies, and patients, regardless of the application used. In order to achieve interoperability between different information systems, it is necessary to define a common data structure, document specification, transmission standard, and interface protocols. From these perspectives, interoperability technology is essentially the terminology and information sharing standards of communication between medical information systems. At present, the most widely used health information exchange standard in the world is the HL7 (Health Level 7). HL7 regulates the specifications of clinical medicine, manages information formats, and develops the standard format of a hospital's data transfer protocol and integration interfaces used by manufacturers to design application software. These technologies enable information sharing and exchange between various health information systems by translating locally formatted data into a standard XML format.

Based on the standard specifications, the collection and integration of medical big data rely on a set of information technology (IT) architectures that include the data Extraction, Transform, Load (ETL), that is to extract out the required data from the data source, transform it into a standard format for data cleaning, and load it into the large database according to the predefined data model, so that the original heterogeneous data can be unified. For the collection and integration system of the medical big data apart from the ETL process, other aspects must also be considered, including the convenience to maintain the system, the quality control of the data, and the extensibility of customized application of the system. In the field of big data collection and integration, the current internationally mainstream products include the data stage by Ascential plc and power center by Informatica Company. Meanwhile, in the field of medical big data, the most representative system is the Informatics for Integrating Biology and the Bedside (I2B2) project from the Harvard Medical School, which has developed a complete technical framework for medical big data collection, and integration. It has standardized a large number of clinical electronic medical records based on international medicine terminology standards, such as ICD-10, SNOMED-CT, LOINC, RxNorm, etc., and has established a patient-centered unified data model for storage, integration, and management of clinical information. This system has a number of advantages, including that it is an open-source license system and has a user-friendly database designed for efficient data retrieval so that doctors and researchers can quickly query the collection of patient queues with different combinations of characteristics, thus providing solid data support for subsequent translational medical research.

China has developed a large number of data set standards for medical information exchange, comparable to that of international information exchange standards. The Statistical Information Center of National Health and Family Planning Commission has launched a series of studies on the national health information standards and its technical application. Up to now, it has established the basic framework of China's health information system, which includes 277 health information standards for electronic health records, electronic medical records, regional health information platforms, and telemedicine, and 231 of the standards have been officially released. In the field of medical ontology, domestic researchers have built an ontology library with linguistics characteristics of Chinese and compatible with international medical ontology systems. The ontology library includes the standard usage of medical terminology, synonyms, hierarchical structure, semantic tags, etc. At present, the Chinese version of Logical Observation Identifiers Names and Codes (LOINC) and Human Phenomenological Terminology (HPO) has been developed and provides public inquiry services through their respective websites. The National 863 Program named as the "Big Data Analysis and Application Research for Common Malignant Tumors" has established six data standards, including "Basic Architecture and Data Standards for Electronic Medical records of Tumor Diseases," "The Basic Data Set of Tumor Diseases," "CDA Document Template for Follow-up of Tumor Disease," "CDA Document Template for Chemotherapy Medical Record," "CDA Document Template for Radiotherapy Medical Record," "Controlled Vocabulary CMV for Tumor Diseases"; based on which a number of data standards for nasopharyngeal carcinoma, colorectal cancer, etc. have been established for speeding the integration of heterogeneous data sources.

3.2 Search Engine and Cross-Database Retrieval Technology for Big Data in Oncology

After the collection and integration of medical big data, how to quickly and accurately obtain the desired information from massive data is a challenging. In addition to the challenges of query efficiency, the challenges of big data search also include the following aspects. First, due to the multimodal of big data, often including characters, numbers, text, sequences, network (pathways), and images, it is necessary to construct different index for multimodal data to support comprehensive search request. Secondly, in the case of multimodal search, it is necessary to integrate the correlation of different modal data to improve the rationality of search ranking. To improve the easy-to-use interface of search results, besides the form-based retrieval, interactive and visual retrieval are also needed. The interactive and transparent data browse and exploration based on the data directory should support queries-based patient, disease, pathway, genes, and drugs. In particular, with the establishment of regional multicenter medical big data platform, the development of a cross-center big data search engine has become a major problem.

Universal big data query and analysis is one of the core issues in cloud computing. To create a medical big data search engine with good user experience and cross-database search capacity, one would need not only to support on general-purpose big data architecture and technology but also to be familiar with the characteristics of medical terminologies and be able to integrate existing technologist into medical big data and practical requirements. Specifically, the medical big data search engine and cross-database search involve the following key technologies: big data and cloud computing technologies represented by (Hadoop Distributed File System (HDFS), MapReduce, HBase, Hive, Spark, etc., multimodal medical data indexing and joint query technology, distributed big data retrieval technology, interactive visual retrieval technology, multiview retrieval technology, and search ranking algorithm. At present, many general-purpose technologies in the field of big data are led by international IT giants, such as Google, Apache, etc., while other countries are mainly learning and applying these corresponding technologies in specific businesses.

In 2009, the Harvard University launched the Shared Health Research Information Network (SHRINE) project with five hospitals and research institutes, to explore an IT architecture that could facilitate the cross-database search of more than 1 billion clinical data items from 6 million patients across the five organizations. One of the SHRINE project aims is to develop an interactive, visualized search engine that could make the search process clearer and more intuitive. In addition, based on the characteristics of clinical information, SHRINE could also optimize the search for comprehensive disease-level data, such as laboratory test indicators, to provide a more accurate search.

Many Chinese institutions have done excellent works to establish medical big data platform and have created relevant technical standards. In 2017, the big data alliance for colorectal cancer (BACC) of the Chinese Society of Clinical Oncology (CSCO) was established for creating terminology standards for colorectal cancer. The BACC cooperate with Yidu Cloud (Beijing, China) Technology Co., Ltd. to generate a medical data platform, which provides a cloud-based service to collect, clean, store, and integrate isolated clinical information from different hospitals. The original clinical information was further processed by using advanced machine learning and artificial intelligence technologies, such as natural language processing, structuration, and enterprise master patient index (EMPI), which allows electronic health record (EHR) data for further use in clinical research.

3.3 Security and Privacy Protection Technology in Oncology Big Data

A patient's personal privacy information is the first priority issue to be considered in the process of using medical big data. If the information security and privacy protection issues are not properly handled, it may not only cause harm to the patients but also damage the development of the entire medical big data industry. Therefore, we must continuously strengthen the awareness of information security and privacy protection. In 2000, the USA specifically set up a Privacy Rule in the Health Insurance Portability and Accountability Act (HIPAA) to protect patients' private information. The HIPAA ensures the confidentiality, consistency, and availability of information systems from the perspectives of management processes, physical protection, technical security services and mechanisms, and data access. Requirements for de-identification of patient data are also very specific. In the HIPAA, it is clearly stated that 18 types of sensitive health and privacy information must be stripped off before clinical records can be entered in industry–university research applications.

In addition to institutional protection, it is also critical to improve information security and privacy protection from a technical perspective. Anonymity and perturbation are commonly used technologies to publish patients' data with not only patient privacy protected but also valid information preserved. In recent years, privacy protection technologies of medical big data have been preferentially investigated, including data masking-based medical data desensitization, automatic data leakage risk assessment, desensitization for unstructured medical data, and risk assessment and early warning for individuals being identified during joint application of multiple databases.

Information security technology is a very large technical system which mainly includes physical security policy, information encryption strategy, access control policy, fire wall control policy, and network security management strategy. It is designed to ensure security, confidentiality, integrity and authenticity of data during storage, transmission, exchange, and use. In the cloud computing era, the mainstream technologies of storage encryption include hybrid encryption, homomorphic encryption, BLS short signature POR model, DPDP, Knox, and so on. In the area of the data access control, current main technologies include spatiotemporal role-based access control, attribute-based encryption access control, and hierarchical attribute-based access control. The main technologies of information security application in the field of medical big data include the following, medical data classification, hierarchically management, and safety protection strategies; authentication and access control technology for data application; medical data encryption and decryption technology in distributed computing architecture; detection and blocking of illegal data access technology; and technologies that ensure authenticity and integrity of health and medical big data. The establishment of the information security and privacy protection system for medical big data could be achieved by systemic integration of regulations, technologies, and management in all processes (data storage, access, application, etc.).

Compared with developed countries, China's information security industry is still in its infancy. The existing privacy protection system is still problematic, but with the awakening of information security and privacy awareness in the era of big data, China's information security industry will inevitably grow stronger. At present, many institutions in China are trying to build a solid foundation in the fields of medical big data security and privacy protection. Many universities in China have a solid background in the fields of network attack and defense, evolutionary cryptography, trusted computing, and information security. They have made important achievements with the support of some important scientific research projects such as the national defense research projects, national security major basic research projects, and provincial research projects. They have also participated in the formulation of relevant technical standards for trusted computing for Chinese government and military, published a series of high-quality papers in authoritative journals and important academic conferences, and developed a series of high-tech security products which were successfully applied to the actual information warfare. These institutions play an important role in the scientific research, national economic construction, and national defense in the field of information security construction, and can protect information security of the medical big data platform at technical levels. If we can take full advantage of the existing superiority of each research institution in China and integrate their efforts in medical big data security and privacy protection, then industry will be able to reach international advanced level in the foreseeable future.

3.4 Integrated Analysis of Oncological Clinical Data and Omics Data

With the continuous development of high-throughput sequencing technology and the increasing accumulation of omics data, the fusion of oncological clinical data and omics data has become an unstoppable trend. Based on the clinical big data and omics big data of large-scale population cohort, the omics characteristics of a disease can be extracted by integrating multiple levels of omics data, including genomic, immunomic, transcriptomic, proteomic, metabolomic, and phenomic data. Then the correlations of the omics characteristics with early screening, classification, efficacy evaluation, prognosis assessment, and other clinical applications can be explored, providing the basis for accurate diagnosis and precise treatment of cancer patients. This area has now become a hot topic in the field of oncological big data and translational medicine. MD Anderson Cancer Center launched a project called Adaptive Patient-Oriented Longitudinal Learning and Optimization (APOLLO) in 2012 to continuously track cancer patients and dynamically collect omics data, clinical diagnosis, and treatment data before, during, and after treatment, in order to reveal the inherent law in the process of disease occurrence and development. The MD Anderson Cancer Center further collaborated with IBM to develop and train the medical artificial intelligence named Watson, which learns from the MD Anderson Cancer Center more than 600,000 pieces of medical evidence about cancer treatment, more than 300 medical journals in the field of oncology, more than 200 textbooks, which altogether consists of nearly 15 million pages of text. The IBM Watson for Oncology can help doctors provide patients with better-personalized treatment options, advice on drug selection and medication regimens based on the patient omics data and clinical data. From the perspective of international progress, integrated analysis of clinical data and omics data is a key technology to deeply analyze the molecular mechanism of disease phenotype and the clinical application of precision medicine. Therefore, it is an indispensable component of medical big data application.

Benefiting from the rapid development of China's natural sciences in integrated analysis of clinical data and omics data, Chinese scientists have been able to achieve significant progress in the field of medical big data analysis and mining. In the study "Identification of Novel Markers for Prognosis and Treatment of Human Liver Cancer," which was selected as one of China's top ten scientific advances in 2011, small RNA profiles of normal human liver, liver with cirrhosis, liver with viral hepatitis, and liver cancer were determined by deep sequencing technology. It was found that the expression level of miR-199 was closely related to prognosis of liver cancer patients, and it was proved that miR-199 could suppress the tumor-promoting kinase PAK4 and significantly inhibit growth of liver cancer, thus providing a new prognostic marker and a potential target for targeted therapy of liver cancer. This work aimed for the prevention and treatment of major diseases in China and translational medical research. It is the result of cross-collaboration between many institutions and disciplines including basic research, biotechnology, clinical specimen management, and patient data analysis, and has laid the foundation for studies on the role of RNA in the liver physiology and liver diseases. This study is a typical case of integrated analysis of clinical data and omics data.

3.5 Big Data Phenotyping Technology for Clinical Data

Extracting structured phenomic information such as signs, symptoms, clinical tests, diagnosis, surgery, drugs, etc. from unstructured texts such as clinical records, can facilitate complex disease classification, search of similar patients, and treatment plan recommendation. Therefore, phenotyping technology is critical for the analysis and mining of medical big data and transformative applications. According to used the methodologies, commonly used phenotyping technology can be divided into three types, namely ontology-based, rule-based and machine learning-based. Among them, ontology-based phenotyping technology mainly compares the medical terms and semantic type of the clinical text with the structured Classes of the ontology language and finds interrelationship of the concept through the Properties of the ontology language. The ontology-based phenotyping technology depends heavily on the descriptive power of the ontology, and the comprehensive ontology construction is a complicated project that needs to be iterated continuously, so this type of phenotyping technology is rarely used alone. Rule-based phenotyping technology uses natural language processing technology to extract text elements such as nouns, adjectives, quantifiers, etc. from training text samples, and then extracted phenotypic information and derived rules in description of associated attributes, wherever possible. The extracted phenotypic information and descriptive rules were further applied to a larger extent of text samples for information extraction, which is highly accurate. However, due to the complexity of clinical texts, the formulation and maintenance of extraction rules require a lot of manpower and time, so it is not suitable for large-scale phenome. Machine learning-based phenotyping technology uses expert annotation as the training set to construct a machine model that learns medical Classes (i.e., concepts), Individuals within Classes (i.e., concrete entities of the concepts), and Properties (definitions of the interrelationship among Classes and the interrelationship of their Individuals). This technology has the characteristics of relatively low labor cost and high efficiency. Moreover, compared with the above two phenotyping technologies, the whole machine learning process can be easily migrated from one disease to another, so it is becoming a research focus in this field.

Although machine learning-based phenotyping technology has gradually become the mainstream, traditional machine learning technology still has many limitations for its application in medical big data. First, the construction of such machine learning model relies heavily on high-quality corpus set labeled by experts at very high cost. Therefore, how to automatically construct such models with less expert input is a very important issue. Second, the selection of features has a great impact on the machine learning model's efficiency and generalization capacity. Traditional feature selection depends on expert experience and knowledge, which is time-consuming and laborious. Finally, the training and text processing speed of machine learning are relatively slow, which greatly limits its ability to process massive medical big data. These three major problems directly point to the "throughput" problem of phenotyping technology in the medical big data era, drawing desired features of an efficient, fast, and accurate machine learning model. This has also led to a new direction of machine learning-based phenotyping technology, namely the high-throughput phenotyping algorithm based on active learning or self-taught learning. Development of deep learning technology and its continuous integration with big data and cloud computing technology in recent years has further strengthened this trend. The phenotype extraction model can use a small amount of labeled terms as seeds to automatically learn effective features from massive data by the usage of deep learning technology. After that, the phenotype extraction model will conduct rapid machine learning on text feature description with cloud computing or distributed computing architecture and would be continuously optimized by constructing an interaction system in response to outer feedback.

Machine learning and natural language processing are the two key supporting technologies for the development of medical big data phenotyping technology. In the field of medical natural language processing, countries like Europe and the USA are very active in academic research and have formed many mature and practical medical natural language processing frameworks or tools. For example, there are the Named Entity Recognizer developed by Stanford University, the cTakes system developed by Apache for clinical text analysis and knowledge extraction, and the MedEX system developed by the University of Texas for drug information extraction, and more.

Meanwhile, due to the particularity of clinical text terms, grammar, and the complexity of Chinese language, the general natural language processing tools do not often perform well in medical text processing; which greatly restricts the development of phenomic construction technologies. Despite such, many research institutions in China have long been committed to the use of natural language processing and knowledge expression technologies. They have used these technologies for medical data analysis, modeling, and mining practices with unstructured texts such as electronic medical records. For instance, the National Cancer Big Data Center conducted a preliminary exploration of clinical terminology application and high-throughput phenotyping technology for a large number of medical record texts, developing extraction algorithms for the phenotype (i.e., symptoms, signs, drug administration, and clinical prognosis) of lung cancer, gastric cancer, liver cancer, esophageal cancer, colorectal cancer, prostate cancer, breast cancer, etc. The National Cancer Big Data Center also further explores molecular typing and new molecular markers of common tumors and performs preliminary phenome-wide association study (PheWAS) of some selective tumors.

4 Development of Big Data Platforms for Oncology

4.1 Framework of Big Data Platforms for Oncology

Health information is stored in different systems from hospital, including health information system (HIS), laboratory information system (LIS), and picture archiving and communication system (PACS). These systems should be integrated into a unified and structured format so they could be secondarily used for upper-layer application. The raw data from these systems were firstly extracted, transformed, and loaded into a data warehouse. With natural language processing and machine learning technique, the information was normalized and structured for the applications in management, clinical research, clinical decision support system, and risk assessment system. Data standards and keyword dictionaries can be used as query specification, so that the users can quickly retrieve, review, and understand the location of the data for efficient use. The comprehensive query function introduces large, full-text indexing technology on the basis of traditional structured data query, capable of joint query of traditional structured fields and large text unstructured fields. This function uses online big data analysis technology to enable researchers to quickly understand and explore the data. The framework of the big data analysis and application system for oncology is shown in Fig. 1.

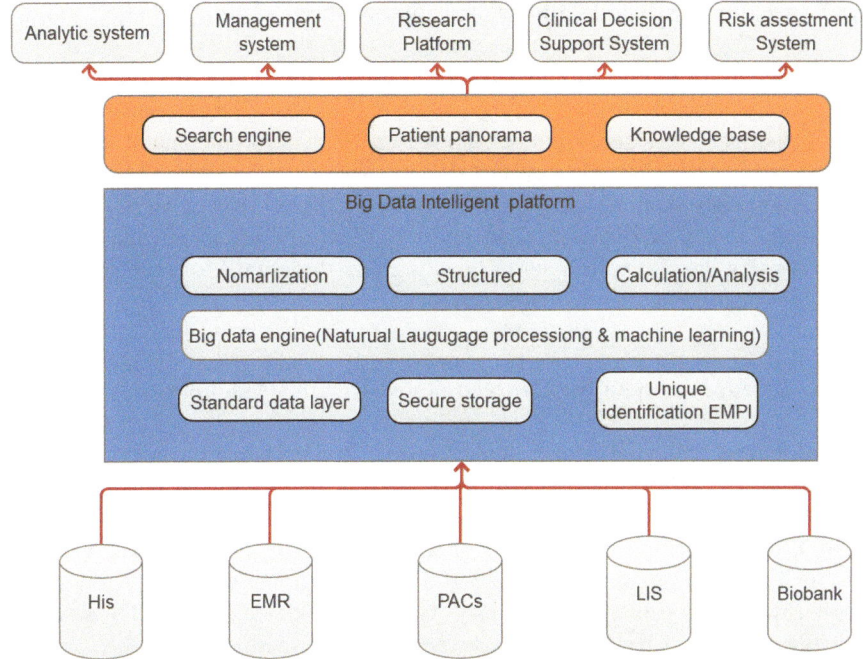

Fig. 1 Framework of the big data analysis and application system for oncology

4.1.1 Big Data Analytic System for Oncology

The big data analytic system for oncology incorporates data exploration, statistical analysis, and data mining. It integrates the popular methods of medical statistics to provide an all-round support for scientific research. By using the big data analysis system in the oncological field, researchers can get rid of the data preparation tasks which are used to be time-consuming and do not need to export the data before using professional applications such as SAS/SPSS since all of the work can be efficiently performed online.

4.1.2 Clinical Research System

The single-disease scientific research management system provides a powerful support for single-disease researchers throughout the workflow, from collecting scientific research data, filling out CRF forms, to subsequenting 360-degree views, statistical analysis, and multidimensional analysis of patients. The system establishes a plurality of single-disease databases of tumor or specific hospital department. Each database accumulates cases according to the inclusion criteria designated by the researcher and provides the doctors with access for scientific research purposes.

4.1.3   Health and Disease Risk Assessment System

The Health and Disease Risk Assessment System studies the health and disease risks from a perspective of cancer prevention and control. The correlation between historical cancer screening test data and the cancer-related outcome of the patient can be analyzed to find the key factors in cancer screening test and to provide decision-making support for early evaluation of disease risks. Meanwhile, risk assessment model can be integrated into the system based on previous research results so that the system can automatically assess disease risk of each patient and provide cancer prevention recommendations.

4.1.4   Research System for Cancer-Related Health Economics

Cancer-related health economics studies factors that influence medical cost of the patients to provide references for medical insurance payment and efficiency evaluation. At present, the general version of the China Diagnosis-Related Groups (CN-DRGs), a classification criterion used for grouping patients according to their principal diagnoses, has shown poor performance for tumor diseases. The financial burden on a tumor patient is more closely correlated with the patient's TNM staging and other variables that indicate the severity of the disease. While these variables are not included in the general version of CN-DRGs grouping model, the model should be optimized.

4.1.5   Clinical Decision Support System

Clinical decision support system (CDSS) for oncology is designed to assist physicians and other health professionals for clinical decision-making, by linking health observations with health knowledge. The clinical decision support system can be categorized into knowledge-based and nonknowledge-based. Knowledge-based contains rules and associations most often in the form of "IF ..., THEN ...." Most of these types of CDSS are guideline-based systems which integrate guidelines into EHR for clinical reminder, prevention, etc. The nonknowledge-based employs machine learning to review past experiences and/or find patterns in clinical data. They raise checkpoints and suggest clinicians to investigate in more depth.

4.2   Construction of a Cross-Regional Big Data Platform for Oncology

4.2.1   Establishing a Data Standard System

After the clinical data of different hospital systems are integrated, and after model reconstruction, data cleaning, data standardization, data structuring, and data normalization have been performed, the analysis of a patient diagnosis and treatment model based on a complete data source can be established. The patient diagnosis and treatment model cover all the serving systems of the hospital and are flexible to add new module. For example, the model can be designed to accompany the new top-grade radiotherapy system for patients requiring radiotherapy. Based on the patient's diagnosis and treatment model, the data standard system can be formulated jointly with the oncologist according to the International Classification of Diseases Tenth Revision (ICD-10) standard, the Health Level 7 (HL7) Clinical Document Architecture (CDA) standard, the Ministry of Health electronic medical record standards, and other clinical special standards, such as the National Comprehensive Cancer Network (NCCN) guidelines, American Joint Committee on Cancer (AJCC) staging manual, etc.

Then, according to the characteristics of the clinical diagnosis and treatment of different tumor types, the data set of each type of tumor can be formulated (Fig. 2). Clinical oncology research can use tumor type-specific data sets to perform big data analysis and to answer related questions. The established data standard system can

Demographic information	Chief complaint	History of present illness	Past History	Family History
• Name • Gender • Age • Nationality · Race • Occupation • ...	• Symptom • Course • Diarrhea • Vomitting · Abdominal pain • Weight change • ...	• Symptom • Course • Obstruction • Abdominal Pain • Weight change • ...	• Previous health status • Immunization • Allergen • Trauma • Surgery • ...	• Family history of tumor • Family history of mental illness • Family history of other disease • ...

Physical examination	Diagnosis	Laboratory examination	Imaging examination	Endoscopy
• Height • Weight • Blood Pressure • ECOG score • Rectal digital examination • ...	• Clinical diagnosis • Imaging diagnosis • Pathological diagnosis • ...	• Liver function • Kidney function • Complete Blood count • Urine teat • ...	• Abdominal CT scan • Abdominal MRI • PET-CT • ...	• Colonoscopy • Gastroscopy • Endoscopic ultrasonography • ...

Pathological examination	Genetic profile	Operation procedure	Chemotherapy	Targeted therapy
• Pathologic type • Differentiation • IHC • Frozen section examination • ...	• BRAF mutation • NRAS mutation • KRAS mutation • Microsatellite Instability • ...	• Date • Surgery • Surgeon • Findings • Process • ...	• Date • Regimen • Dosage • Administration • Treatment cycle • ...	• Date • Regimen • Dosage • Administration • Treatment cycle • ...

Immunotherapy	Interventional therapy	Radiation therapy	Efficacy evaluation	Follow-up
• Date • Regimen • Dosage • Administration • Treatment cycle • ...	• Date • Name of procedure • Process • Agents • ...	• Gross Target Volume • Clinical Target Volume • Planning Target Volume • ...	• Complete response • Partial response • Stable disease • Progression disease • ...	• Date • Status • Averse event • Recurrence • Metastasis • ...

Fig. 2 Colorectal cancer vocabularies

be used as a standard for data conversion between hospitals in different regions to achieve the integration of heterogeneous data sources.

4.2.2 Data Security and Privacy Protection of the Tumor Big Data Platform

Data leakage is always a threat during data collection, processing, and application of medical data. On the one hand, medical data can be accessed by multiple nodes in the hospital's internal procedure, leading to privacy leakage within the hospital. On the other hand, data leakage can also occur during transmission among different institutions of the platform, which may occur during scientific research and regional data exchange within the platform. Data leakage can jeopardize the privacy of patients and may lead to a series of issues such as fraud. The tumor big data platform must place the highest priority on data safety and privacy protection, and perform multilevel data security management through definition and desensitization of sensitive information, and authorization management. The big data platform should establish a strict and flexible authorization management system in line with national norms for clinical data management, which allows access according to the different identity of the users, different scenes, etc., so that cooperative hospitals can better mobilize clinicians and departments to participate in related clinical research, management, and teaching, as shown in Fig. 3.

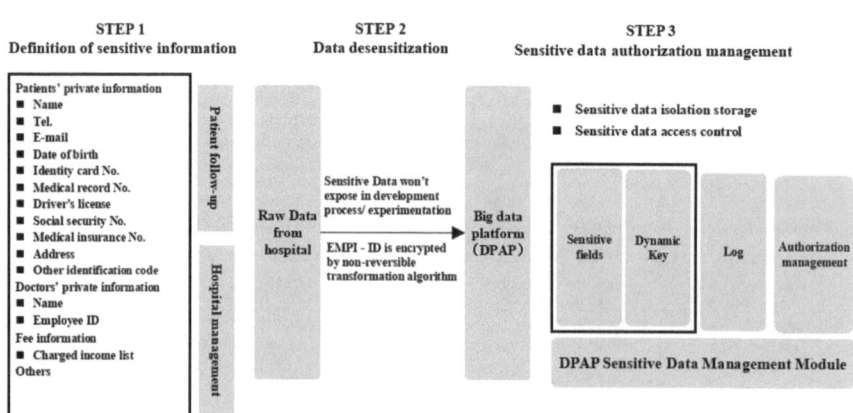

Fig. 3 Multilevel data desensitization in the cancer big data platform

4.3 Significance of Cancer Big Data Platform for Oncology Research

4.3.1 Cancer Big Data Platform Supports Efficient and Flexible Medical Record Retrieval

The cross-regional tumor big data platform can integrate all the data of multiple hospitals and provide a complete medical record search engine. The search engine of the tumor big data platform is built on the entire distributed computing framework, based on efficient memory and storage media. The search engine achieves fast retrieval of global multifield or full-text data by data fragmentation technology. To ensure the validity and accuracy of the search results, the big data platform applies post-structuration and normalization process of raw data and applies correlation algorithm and a variety of sorting filtering strategies. The establishment of search engines and indexes can achieve queries in millions of medical records with seconds, perform intelligent retrieval of keywords within full-texts, and can support accurate search of complex logical relationships with multiple conditional combinations to meet the requirements of different scenarios. Based on scientific research data marts, tumor big data platform can integrate filters that apply inclusion and exclusion criteria, and flexibly define the indicators and screening logic, so that the investigating subjects and corresponding parameters that meet the inclusion criteria can be included and be represented. In addition, the tumor big data platform also supports visual search of relative-time events, with flexible definition of relative-time events. The limitation of conventional data platform is being time-consuming, labor-intensive, and not being able to flexibly covering various relative-time events. While the big data platform can support a search based on the unstructured clinical basic fields covering laboratory examination, imaging examinations, symptoms, signs, family history, marriage and childbirth, surgery procedures, drug administrations, etc.

4.3.2 Cancer Big Data Platform Provides a Patient-Centered Global View

As a chronic disease, tumor patients often have repeated long-term visits which pose a major challenge to data analysis. For patient data that correlated with complex conditions and long treatment term, the timeline model would be a good choice to demonstrate the overall diagnosis and treatment procedure, and the changes in the patient's condition. The key characteristics of the patient, diagnosis, and treatment of the disease can be presented graphically in a manner that allows comparison and association, so that doctors can quickly obtain significant information which would be of great value for both case analysis and scientific research. The timeline model demonstrates the patient's full treatment cycle and records the patient's diagnosis, medication, physical data, examinations, treatment, surgery, and other information at each time point. In the past, doctors usually needed to memorize the abnormal values and time of the examinations. Despite abnormal values shown in the reports, they may not reflect the fluctuation trend, nor will they reflect their indications to the treatment- or end point-related events. However, using the database, multiple test results can be shown graphically on the timeline, and the abnormal situation can be identified at a glance. The timeline model reveals the hospital's routine diagnosis and treatment path as well as the specific patient's personalized program. This information together with evaluation of the patients' outcome can be used as a knowledge base to provide patients with a more evidence-based treatment plan.

4.3.3   Cancer Big Data Platform Allows Data Analysis and Mining

Knowledge base content can be mined from the tumor big data platform through prefabricated analysis and mining models. The analysis and mining models analyze each piece of data in the system and establish a relationship of it with a standard vocabulary, forming a knowledge base that contains all the hospital's actual data. The platform also maintains the mapping relationship between the hospital's actual data and the standard big data vocabulary and records the data time, version number, the original value, word frequency, corresponding standard value, and the time of iterative update of the algorithm. The model is continuously updated through background machine learning plus manual annotation. Using the data mining model, the baseline data and main observation indicators of the research objects in the research project can be statistically described and presented in the form of data charts. The result clinical data can be exported to medical statistical software for statistical analysis or to be used for construction of big data analysis and mining model to deal with massive and high-dimensional data based on the treating doctors' needs. Through data mining, the correlation between various data can be revealed, new scientific findings can be proposed, and better treatment options can be provided to the patients.

4.3.4   Cancer Big Data Platform Supports Real-World Research

Real-World Study (RWS) is based on the full-scale patient data collected in the real world without intervention in the collection process and uses statistical methods to extract evidence of relevance and validity from the real world. Such study involves no blinding, randomization, control, or placebo, and the results can be directly applied to clinical practice (patient-centered outcome study, in which the study site and intervention conditions follow real clinical settings and patients were included without restrictions). Single/multiple factor correlations found in the RWS can then be tested by randomized control trials to find the causative relations. Through the combination of these two research methods, the deviation in statistics can be effectively eliminated, and the efficiency of researches can be greatly improved.

Based on the tumor big data platform, the pooled data can be fully mined and analyzed and used to support RCTs and RWSs that comply with Good Clinical Practice (GCP) specifications.

5 Opportunities and Challenges Facing the Development of Cancer Big Data

5.1 Current Status and Challenges Facing Cancer Big Data in China

The development of China's medical and health business has entered the era of big data, and medical big data has become an important strategic resource for the country. Deep integration of the internet and big data technologies with the medical industry is needed, to deepen medical and health system reform, to realize the "Healthy China" dream, to optimize medical and health resources, to improve accuracy of medical technology services and public health policies, and to improve the fairness and accessibility of medical services. In recent years, the government has developed a series of policies to promote the research and development of medical big data application. The National Medical and Health Service System Planning Outline pointed out several problems existing in medical system, including the insufficiency of the medical resources and medical service quality in China, the irrationality and fragmentation of the structure of medical system, and the excessively expanding of some public hospitals. The rapid development of information technology, such as cloud computing, internet of things (IoT), mobile internet, big data, etc., provides opportunities for optimizing the efficiency of medical service, and will certainly promote a profound transformation of the management models for medical and health services. The continuous deepening of medical reform has also put forward new requirements for the number of public hospitals and the optimal allocation of medical resources. Other policies include the "Outline for the Promotion of Big Data Development," "Notice on Organizing and Implementing Major Projects for Promoting Big Data Development," and "Guiding Opinions on Promoting and Regulating the Development of Big Data Applications in Health and Medicine." It was proposed to build an electronic health and medical record database include all data related to public health, medical services, medical security, drug supply, family planning, and integrated management services. Medical and health big data applications were considered as the key direction of development in the future. These policies also stated that the government shall promote construction of medical big data platform for application innovation, and support technology research for medical big data integration management, big data sharing, big data analysis and retrieval, standard specifications, privacy protection. The policies also required that relevant organizations shall conform to the development trend of emerging information technologies to standardize and promote integration and sharing of big data application in health and medicine. With the approval of the State Council, it has been clear that the guiding ideology for the development of medical big data is to improve the medical policy system, to innovate the working mechanism, to share the health medical information and public medical data, eliminating information islands and actively creating an environment that promotes development of innovative applications of health and medical big data. Through the "internet plus healthcare," relevant organizations shall explore new modes of medical service, provide satisfactory medical and health services for the people, and provide strong support for a "Healthy China".

The emergence of the Internet and big data technology has greatly promoted the development of medical science, making "internet plus big data plus medicine" a new trend in the medical field. Through comprehensive and intelligent data collection, standardized data management, deep data analysis, and patient-centric transformation applications, big data-based medical revolutions are gradually overturning traditional medical models.

5.2   The Future of Cancer Big Data

Medical big data has wide application prospects in oncology. Medical quality supervision, clinical auxiliary diagnosis, health economic analysis, public health decision-making, precision medical research, and other tumor big data-based applications will significantly promote the integration of social medical resources and improvement of the standard of the medical system. Such techniques would ultimately lead to the improvement of the overall health of the people and generate enormous potential social and economic value.

However, sharing of tumor big data has become a global problem, due to the following reasons: complexity of oncological medical services, diversity of interest among different participants in medical industry, heterogeneity of informatization of different medical institutions, and the unstructured nature of the medical data. With the rapid development of information technology, the massive data generated by medical informatization has become the basis for promoting medical innovation, but the effective aggregation and sharing of medical big data are the prerequisite for its usage. It is necessary to establish an integrated system for standardized collection, storage, management, regulation, exchange, and retrieval of tumor big data. And the integrated system described above in combination with standard medical terminology system, resource catalog system, data quality control system, and safety system could provide the medical institutions with a big data interconnection and sharing platform. In this way, medical data from different sources can be pooled and integrated, and mutual security certification between different medical institutions can be achieved. A series of medical database can be established, and interconnection and sharing of tumor big data can be realized on such a platform.

Meanwhile, concerted efforts of hospitals, research institutions, data users, and parties from the information technology industry would be required to promote transformation of medical research progresses into application. The seamless cooperation of all relevant parties will effectively promote the transformation and industrialization of research and development (R&D) results. The technology platform will become the foundation of R&D, and the accumulated experience of numerous medical IT companies will also provide strong support for industrialization of R&D results. In addition, the technological breakthroughs, standard outputs, and application demonstrations based on the medical big data application platform will also directly affect the diagnosis, treatment plan, and prognosis evaluation in oncology.

The rapid development of oncology big data depends on breakthroughs in standardized big data collection technology, big data security technology, oncological ontology terminology services, medical big data search engine, oncology phenome construction, integrated analysis of clinical data and omics data, and other bottlenecks in medical big data technology. Breakthrough in these technologies will greatly promote the general proficiency of medical big data in integration management, standard specification, privacy protection, search analysis, and mining application. R&D-generated standards and models would be verified and form a series of standard specifications in the tumor big data industry, which would promote improvement of the overall proficiency of the tumor big data field in China. Application of tumor big data depends on the platform for innovation of big data application technology. A series of data models and analysis subsystems shall be built for medical quality supervision, clinical auxiliary treatment, health economic analysis, public health policy evaluation, and precision medical support, which will greatly enhance medical big data-based decision-making and R&D, and help to solve the current problems that have plagued China's medical system. Therefore, it is necessary to strengthen the construction of medical big data innovation platform to support relevant applications and improve medical informatization and the application of medical big data in China.

The continued development of cancer big data requires the construction of a talented team in this field. China's medical institutions and scientific research institutions need to firstly carry out more close academic cooperation, secondly continuously cultivate high-level professional talents and modern laboratory management talents with a prospective point of view, and thirdly construct high-level national engineering realization teams and research teams for the development and application of China's health industry. In order to actively promote the ability of the clinicians to the tumor big data platform to solve clinical problems, it is also necessary to establish a data mining and analysis training program for the clinicians, and train them to be better oriented for medical big data applications.

Acknowledgements We gratefully thank Sharvesh Raj Seeruttun from *Cancer communication* and staffs in the Department of Medical Oncology and GI Surgery Oncology at Sun Yat-sen University Cancer Center for their suggestion and assistance.

References

1. Toga AW, Foster I, Kesselman C, et al. Big biomedical data as the key resource for discovery science. Journal of the American Medical Informatics Association: JAMIA, 2015, 22:1126–31.
2. Beyer MA DL. The Importance of 'Big Data': A Definition. http://www.gartnercom/it-glossary/big-data/2012.
3. Dinov ID. Methodological challenges and analytic opportunities for modeling and interpreting Big Healthcare Data, GigaScience, 2016, 5:12.
4. Baro E, Degoul S, Beuscart R, Chazard E. Toward a Literature-Driven Definition of Big Data in Healthcare. BioMed research international, 2015, 2015:639021.
5. Vicini P, Fields O, Lai E, et al. Precision medicine in the age of big data: The present and future role of large-scale unbiased sequencing in drug discovery and development. Clinical pharmacology and therapeutics 2016;99:198–207.
6. Collins FS, Varmus H. A new initiative on precision medicine. The New England journal of medicine 2015;372:793–5.
7. Hochster HS, Niedzwiecki D. Big Data, Small Effects. Journal of clinical oncology: official journal of the American Society of Clinical Oncology 2016; pii: JCO658161. [Epub ahead of print].
8. Papolos A, Narula J, Bavishi C, Chaudhry FA, Sengupta PP. U.S. Hospital Use of Echocardiography: Insights From the Nationwide Inpatient Sample. Journal of the American College of Cardiology 2016;67:502–11.
9. Minikel EV, Vallabh SM, Lek M, et al. Quantifying prion disease penetrance using large population control cohorts. Science translational medicine 2016;8:322ra9.
10. Yan X, Chu JH, Gomez J, et al. Noninvasive analysis of the sputum transcriptome discriminates clinical phenotypes of asthma. American journal of respiratory and critical care medicine 2015;191:1116–25.
11. Xu RH, Wei W, Krawczyk M, et al. Circulating tumour DNA methylation markers for diagnosis and prognosis of hepatocellular carcinoma. Nat Mater. 2017 Nov;16(11): 1155–1161.
12. Bourne PE, Bonazzi V, Dunn M, et al. The NIH Big Data to Knowledge (BD2K) initiative. Journal of the American Medical Informatics Association: JAMIA 2015;22:1114.
13. Nikpay M, Goel A, Won HH, et al. A comprehensive 1,000 Genomes-based genome-wide association meta-analysis of coronary artery disease. Nature genetics, 2015, 47:1121–30.
14. Consortium EP. The ENCODE (ENCyclopedia Of DNA Elements) Project. Science 2004;306:636–40.
15. Cancer Genome Atlas Research N, Weinstein JN, Collisson EA, et al. The Cancer Genome Atlas Pan-Cancer analysis project. Nature genetics 2013;45:1113–20.
16. Mateo J, Carreira S, Sandhu S, et al. DNA-Repair Defects and Olaparib in Metastatic Prostate Cancer. The New England journal of medicine 2015;373:1697–708.
17. Jiang P, Liu XS. Big data mining yields novel insights on cancer. Nature genetics 2015;47:103–4.
18. Fehrmann RS, Karjalainen JM, Krajewska M, et al. Gene expression analysis identifies global gene dosage sensitivity in cancer. Nature genetics 2015;47:115–25.
19. Cyranoski D. China embraces precision medicine on a massive scale. Nature 2016;529:9–10.
20. Edgar R, Domrachev M, Lash AE. Gene Expression Omnibus: NCBI gene expression and hybridization array data repository. Nucleic Acids Res. 2002 Jan 1;30(1):207–10.
21. Barrett T, Wilhite SE, Ledoux P, et al. NCBI GEO: archive for functional genomics data sets–update. Nucleic Acids Res. 2013 Jan;41(Database issue):D991–5.

22. Abecasis GR, Altshuler D, Auton A, et al. A map of human genome variation from population-scale sequencing. Nature 467, 1061–1073.
23. Chin L, Andersen JN, Futreal PA. Cancer genomics: from discovery science to personalized medicine. Nat Med. 2011;17:297–303.

Ruihua Xu, M.D., Ph.D. Director of Cancer Center of Sun Yat-sen University, Director of State Key Laboratory of Oncology in South China, Director of National New Drug (Anti-Tumor Drugs) Clinical Trial Center. Professor, doctoral tutor, State Council special allowance expert, 100 talents of Guangdong. Member of the Department of Biology and Medicine of the Ministry of Education, Vice Chairman of the China AntiCancer Association, Vice Chairman of the China Society of Clinical Oncology (CSCO), Vice Chairman of the China Association of Pharmaceutical Biotechnology, Chairman of the Guangdong Anticancer Association, Chairman of Targeted Therapy Professional Committee of China Anticancer Association. Editor-in-Chief of Chinese Journal of Cancer. Consultant expert of China Food and Drug Administration (CFDA) for drug review. Editor-in-Chief of *Cancer Communication.*

Dr. Ruihua Xu has published more than 250 peer-reviewed papers on SCI collected journals, among which he as the first author or corresponding author of about 120 papers, including some in renowned journals such as *Nature Materials, Lancet Oncology, Journal of Clinical Oncology, Hepatology, Cancer Research, Leukemia, Clin Caner Research and Cancer.*

Application of Information Technology in Medical Ultrasound Engineering

Siping Chen, Xin Chen, Yuanyuan Shen, Yanrong Guo, Xiaonian He and Huiying Wen

Abstract Based on the engineering techniques such as electronic technology, computer technology and signal processing method, medical ultrasound engineering uses ultrasound as a carrier or source of energy for information detection. It utilizes acoustic characteristics and biological effects, which are induced by the interaction between ultrasound and human tissues, to diagnose or treat diseases. For example, the development of medical ultrasound imaging extends from the original one-dimensional A-mode information to the two-dimensional B-mode image, and then to the color Doppler imaging associated with blood flow. This development history consists of continuous discovery of the implicit information from the ultrasound echo. In recent years, researchers have used a variety of interdisciplinary and comprehensive information to promote the sustainable development of medical ultrasound engineering. On the one hand, more information is incorporated into transmitted ultrasound. Coded ultrasound imaging is a good example. On the other hand, ultrasound is combined with other physical technology to induce more information in ultrasound echo. Ultrasound elastography, magneto-acoustic tomography, photoacoustic imaging, and microbubble contrast imaging are such examples. These new technologies are promoting medical ultrasound engineering in the directions of functional diagnosis and precision medicine.

Keywords Medical ultrasound engineering · Information technology · Interdiscipline

1 Background

Medical ultrasonic engineering is a combination of biology, medicine, acoustics and engineering technology. The theoretical basis for medical ultrasonic engineering is acoustics, particularly the principle of ultrasound, while its realization

S. Chen (✉) · X. Chen · Y. Shen · Y. Guo · X. He · H. Wen
National-Regional Key Technology Engineering Laboratory for Medical Ultrasound,
School of Biomedical Engineering, Shenzhen University, Shenzhen, China

© Publishing House of Electronics Industry 2020
China's e-Science Blue Book 2018,
https://doi.org/10.1007/978-981-13-9390-7_20

means are built upon modern engineering technology, especially electronic, computer and information technology. Medical ultrasound usually refers to the acoustic wave in the frequency range of 0.1–50 MHz. As a mechanical wave, it has the general wave properties such as reflection, scattering, and diffraction when it propagates in the biological tissue. Acoustic information including sound velocity, sound attenuation, backscattering coefficient, Doppler shift, and nonlinear parameters of the tissue can be obtained from the echo to assess and identify the physiological and pathological states of the human tissue for disease diagnosis. Each discovery of new type of implicit information from ultrasonic echo has generated the breakthrough of the medical ultrasound imaging technology. The development of medical ultrasound is a history of continuous discoveries of the implicit information [1]. Medical ultrasound has been expanded from the first A-mode one-dimensional information to the B-mode two-dimensional information, and the color Doppler for blood flow.

In recent years, with rapid progresses of various disciplines in electronics, computer science, material science, and biology, more and more implicit information has been developed and there are much more room for advancements of medical ultrasound engineering. Particularly, the main developments are concentrated on the following three aspects. (1) Combined with other physical methods, more quantitative functional information can be found to provide a comprehensive picture of the target lesion. Traditional ultrasound imaging extracts the anatomical morphology of human tissues as well as the velocity and direction of blood flow from ultrasonic echo. These qualitative or semi quantitative information can hardly meet the requirements of clinical practices. A large number of researches have made great efforts for the search of new imaging parameters or tissue characterization methods, promoting ultrasonic diagnosis from anatomical imaging, qualitative analysis to functional, quantitative and molecular imaging. The latest technologies, such as ultrasound elastography, magneto-acoustic tomography, magneto-thermo-acoustic imaging, magneto-acoustic-electrical tomography, photo-acoustic imaging, have been commercially implemented after more than 20 years of laboratory research. (2) The imaging resolution and imaging frame rate are improved to find lesions on smaller scales, or to observe tissue movement more clearly. In recent years, Ultrafast Ultrasound Imaging (UUI) technology has been developed based on an innovative excitation-acquisition mode, resulting a frame rate of 1000 frames per second. At present, UUI has been implemented on the ultrasound system of the SuperSonic Imagine Company in France. Based on UUI, super-resolution ultrasound imaging is developed with the injection of a small amount of ultrasound contrast microbubble into the blood to image the rat brain at 500 frames per second. High-resolution images of rat blood vessels can be synthesized by locating the positions of the ultrasound microbubbles and superimposing thousands of images taken over a period of time. The results have been published in Nature in 2015 and the relevant system is currently in the laboratory stage. (3) The mechanical and thermal effects of ultrasound are used to obtain more implicit information. Ultrasound plays an important role not only in the field of diagnosis but also in the field of treatment. Ultrasound therapy covers many aspects, such as physical therapy, drug penetration, ultrasonic atomization, ultrasonic lithotripsy, pha-coemulsification, ultrasound thrombolysis, high intensity focused ultrasound. With the development of ultrasound microbubble, low-frequency low-intensity focused ultrasound combined with microbubble-mediated drug delivery are receiving more and more attentions from researchers, particularly in the drug delivery for brain diseases. In addition, the research on ultrasound neuromodulation has been grad-ually developed in recent years, enhancing the role of medical ultrasound in the frontier fields of brain central nervous system diseases and neuromodulation.

2 Application of Information Technology in Medical Ultrasound Engineering

For each of the above-described aspects, medical ultrasound engineering has made new progresses in relevant principles, technologies and applications. In the fol-lowing, we will introduce the developments achieved by domestic and oversea researchers together with the particular work carried out by our team.

2.1 Extraction of Viscoelastic Information Using Ultrasonic Mechanical Effects

It has been found in clinical practice that the information of hardness or elasticity of biological tissues is often related closely to the degree of tissue damage. For instance, the degree of stiffness for tissues with breast and prostate cancer or other common malignant diseases is significantly greater than that for the normal tissues [2]. Doctors mainly use palpation methods to assess the hardness of lesion tissues for the differentiating of benign and malignant tumors. However, palpation relies heavily on the subjective experience of the doctor and it is not suitable for the detection in deep tissues.

Ultrasound elastography is a new type of ultrasound imaging technology emerging in recent years. It uses ultrasound to detect the mechanical properties of soft tissue, which has the advantages for non-invasive, rapid, and repeatable detection. The research team in the University of Texas in the United States led by Prof. Ophir firstly reported the internal strain distribution map obtained by quasi-static pressure in 1991, and proposed the concept of ultrasound elastography [3]. This groundbreaking study has a profound impact, breaking through the lim-itations of ultrasound for anatomical imaging and blood flow imaging, and demonstrating that ultrasound can also image the mechanical properties of tissue. The Japanese Hitachi company released an ultrasound scanner in 2005, which firstly provided the function of ultrasound elastography called real-time tissue elastography. Based on the low-frequency vibration excitation method, the French

Echosens company developed the FibroScan product in 2001, which was special-ized for liver hardness measurement and had been used in a wide range ranges of clinical applications. More elastography methods utilize acoustic radiation to locally excite the internal tissues of the human body, and to detect elasticity information by the ultrasonic mechanical effects. The resulting products include Siemens ACUSON S2000 system and SuperSonic Imagine Aixplorer system [4].

Ultrasound elastography has been widely used in the detection of diseases in liver, breast and prostate [5]. However, this technology is still evolving, and various new methods and new applications are emerging. Our team, with the support of the National Natural Science Foundation of China (61031003), focuses on in-depth research of key technologies in elastography and tries to extract more information. Human tissue has both elasticity and viscosity, which are closely related to human physiological and pathological features; meanwhile viscosity depends on the excitation frequency. Usually the excitation frequency of the existing ultrasound elastography technology is restrained to a single frequency (about 50 Hz), and only the elastic information is detected. For the first time, our team applied orthogonal coding technology to ultrasound elastography and extended the excitation fre-quency range to 1000 Hz so that more information for elastography can be revealed [6]. Our team also applied new techniques to animal models of liver disease, such as liver fibrosis and steatosis, in order to study changes in liver viscoelasticity during disease progression [7]. Figure 1 shows the changes of viscoelasticity in rat liver during fibrosis development. The results were published in the peer-reviewed journals of medical ultrasound, including Ultrasound in Medicine and Biology, Ultrasonics, IEEE Trans UFFC, etc.

Our team also carried out researches on the test method of ultrasound elastog-raphy equipment, and formulated the pharmaceutical industry standard of China: Test method for performance of ultrasonic elastography equipment based on acoustic radiation force (YY/T 1480-2016). This standard fills the gap in the test method for the elastography equipment. Our team also tried to promote the application of the standard in the industry. Many companies have used the standard

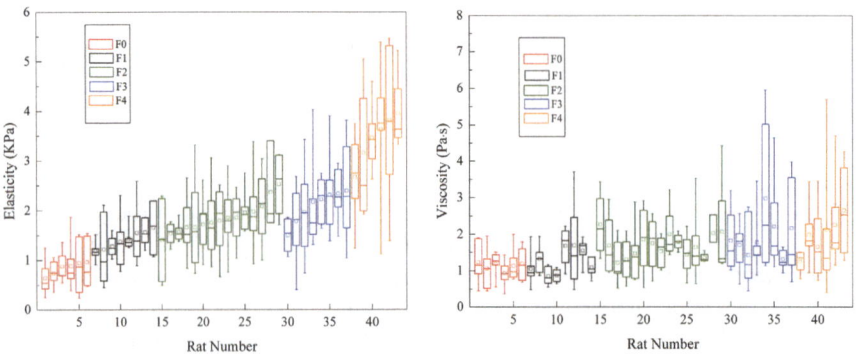

Fig. 1 Changes in liver viscoelasticity during liver fibrosis development

for the development of elastic imaging functions and performance evaluation of related products. This is of great significance for the development of China's medical device industry as well as its international influence.

2.2 Extraction of Biomechanical Information Using Rheological Techniques

Biorheology is the study of the response of biological tissues to mechanical loads. This response is usually characterized by a constitutive equation describing the mechanical properties of the tissue [8]. The parameters in the constitutive equation are often analyzed based on a large amount of experimental data, which reflects the physiological state of the biological tissue to some extent. For example, when arthritis occurs, the permeability coefficient and the modulus of aggregation of the articular cartilage will change significantly. Therefore, the occurrence of certain diseases can be predicted in some way by measuring the mechanical parameters of biological tissues.

Fung et al. [9] have proposed the representation models, namely viscoelastic models, through extensive experiments. They consider that biological soft tissue is essentially a viscoelastic material that combines elastic solid properties with viscous fluid properties. For the mechanical behavior of materials under steady-state resonance conditions, vibration test methods are usually used to study the corresponding response of viscoelastic bodies under alternating stress. The physical quantities related to dynamic viscoelastic properties, such as complex modulus, storage modulus, loss modulus and loss factor, can be obtained. Some researchers have applied rheological mechanics techniques for the evaluation of human tissues including brain tissue, liver, bones, etc. [10–12].

Our team utilized the dynamic mechanical analysis (DMA) technique to measure the rheological properties of tissues using the commercial system (DMA, Boss, ElectroForce 3200, USA). An overview of the experimental system is shown in Fig. 2.

An alternating stress $\sigma(t) = \sigma_0 e^{j(\omega t + \delta)}$ is applied to a biological tissue sample by the shear wave oscillation mode, and produces a corresponding alternating strain $\varepsilon(t) = \varepsilon_0 e^{j\omega t}$. The complex modulus G^* of the tissue is defined as the ratio of stress to strain

$$G^*(\omega) = \frac{\sigma(t)}{\varepsilon(t)} = \frac{\sigma_0 e^{j(\omega t + \delta)}}{\varepsilon_0 e^{j\omega t}} = \frac{\sigma_0}{\varepsilon_0}(\cos \delta + j \sin \delta) = G'(\omega) + jG''(\omega), \quad (1)$$

where G' is the storage modulus and G'' is the loss modulus. By assuming a mechanical model of the biological structure, the storage modulus and the loss modulus can be linked to the viscoelasticity of the tissue. For example, if the tissue conforms to the Voigt model, then

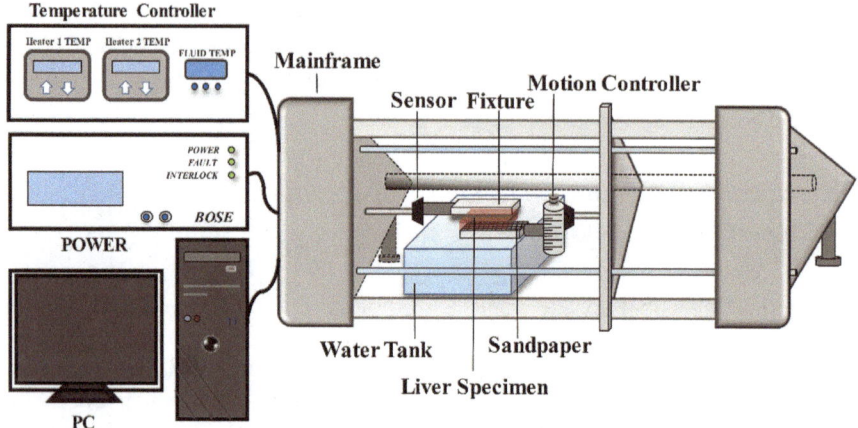

Fig. 2 Dynamic mechanical analysis test experiment diagram

$$G' = \mu, \quad G'' = \omega\eta, \tag{2}$$

where μ and η are the elastic coefficient and the viscosity coefficient, respectively.

Our team applied DMA techniques to animal models of liver disease, such as liver fibrosis and steatosis, and compared the results of DMA with the results of ultrasound elastography measurements to evaluate the measurement accuracy of viscoelasticity by ultrasound elastography. Figure 3 shows the experimental results of shear wave elastography (SWE) and DMA in an animal model of steatosis. Similar elasticity results were obtained between SWE and DMA for different steatosis severity (S0–S4). However there was a deviation between the viscosity coefficients of SWE and DMA. The main reason is that SWE and DMA use different frequencies of vibration, and the viscosity coefficient is more dependent on frequency.

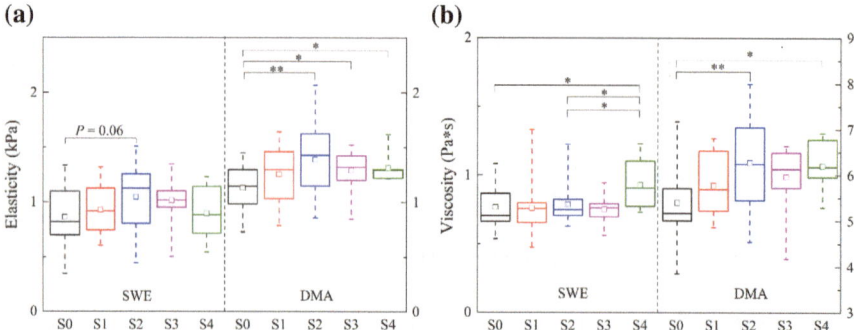

Fig. 3 Comparison results between shear wave elastography and dynamic mechanical analysis

2.3 Extraction of Conductivity Information Using MAET Technology

State changes within a biological tissue, such as tissue type, water content, temperature, may cause significant changes in electrical conductivity. Meanwhile changes of pathological and physiological within the target tissue will change the permeability of the cell membrane and the concentration of the cell fluid, thereby affecting the electrical impedance characteristics. Consequently, it is possible to detect pathological and physiological abnormalities by detecting the variations of the electrical properties of biological tissues which provide useful and valuable information for diagnosis of early-stage cancer [13]. There are various methods to detect electrical characteristics, such as Electrical Impedance Tomography (EIT), Magneto-Acoustic Tomography (MAT) and Magneto-Acoustic-Electrical Tomography (MAET). In the three methods, the EIT method applies a safe driving current/voltage into a target biological tissue, and the internal conductivity distribution image of the target sample is reconstructed by detecting the corresponding response information of the driving current or voltage. In addition, both the MAT and MAET methods are imaging methods based on the coupling of sound field, electromagnetic field. The two methods use two different implementation modalities [14]. One implementation is to apply an alternating magnetic field to the biological tissue placed in a static magnetic field, thereby causing biological tissue vibration. The conductivity reconstruction of the tissue is achieved by receiving the acoustic signal. The other implementation is to apply an acoustic radiation force to the biological tissue placed in static magnetic field, causing local vibration of the tissue to generate an alternating current. An image of the electrical conductivity is reconstructed by detecting the electrical signal. Recently, experimental platforms and methods of MAT and MAET are still in the laboratory stage, and there is still a long way to go before the two imaging modalities are applied in clinics.

The team of Professor Liu Guoqiang in the Institute of Electrical Engineering of the Chinese Academy of Sciences proposed a MAET imaging method using the Linearly Frequency-Modulated (LFM) theory, which employed the relationship between distance and frequency offset and converted the resolution of conductivity spatial detection into the resolution of the spectrum. It is also a potential medical imaging modality with high resolution. This team simulated the imaging procedure by COMSOL software and designed a MAET apparatus for detecting the interface of conductivity variations [15]. Furthermore, experiments were performed on pork tissue [15], which demonstrated that the conductivity change interface with the axial resolution of 1 mm can be clearly detected. The experimental results were shown in Fig. 4. Arrow ① ② ③ represented the interfaces between the fat and lean meat of pork, and the conductivity interfaces of the reconstructed conductivity image were highly consistent with the interfaces of pork.

With the support of the National Natural Science Foundation of China (61427806), our team cooperated with the team of Professor Liu and proposed a new multimodal imaging method: acousto-electromagnetic dual-mode imaging

Fig. 4 **a** Cross section of the
pork sample. **b** Image of the
conductivity parameter of the
pork

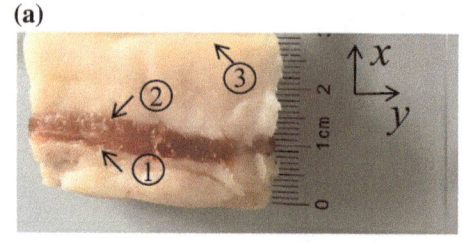

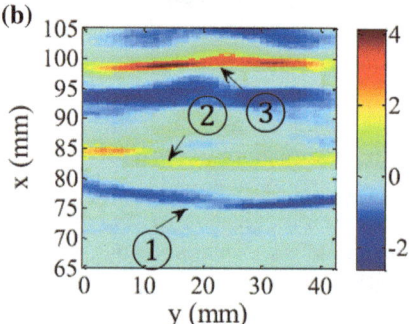

method based on acoustic radiation force and Lorentz force. The core of the method
is to combine the ultrasound and electromagnetic imaging modalities through
acoustic radiation force to obtain the morphological, mechanical and electrical
properties of tissues.

The proposed MAET system consisted of three parts, (1) ultrasonic excitation
source; (2) motion control and detection platform; and (3) Verasonics acquisition
and control platform. As the major part of the system, Verasonics acquisition and
control platform was used for a 14-bit analog-to-digital conversion (ADC),
amplifying, band-pass filtering, demodulating (mixing, low-pass filtering), and
system timing control. The ultrasonic excitation source is composed of a signal
generator, a power splitter, and a power amplifier for amplifying the excitation
signal. The motion control and detection platform contains a high-power excitation
probe, a C-shaped permanent magnet with two cube magnets and its support
structure, a pair of silver-plated copper electrodes and a detection tank inserted into
the C-shaped static magnet. In addition, the static magnet space size is 10×10
4 cm and the MAET system diagram is shown in Fig. 5.

2.4 Improving Imaging Resolution with Coded Excitation Technology

The developments of interdisciplinary researches have continuously promoted the
development of medical ultrasound engineering. Coded excitation has been used

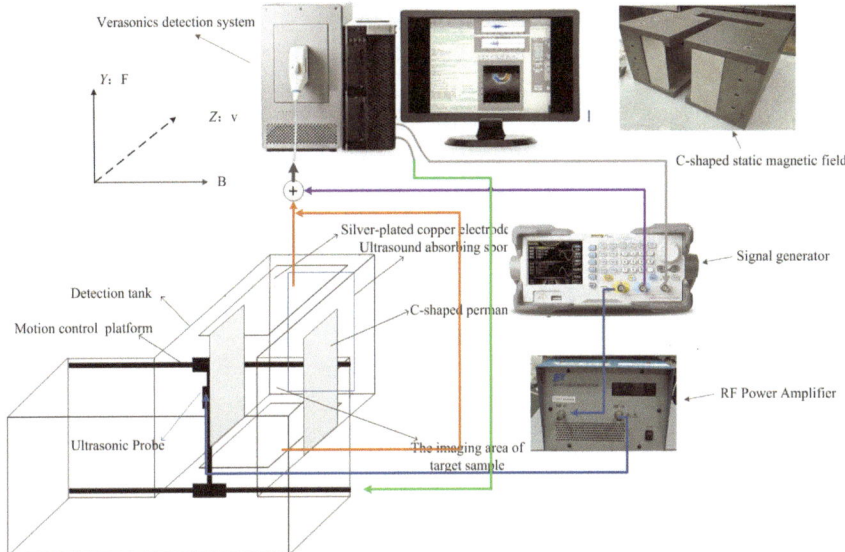

Fig. 5 Connection diagram of conductivity imaging system

successfully in many fields such as radar, communication, and medical ultrasound imaging. O'Donnell found that coded excitation can help to improve the system signal-to-noise ratio (SNR) with 15–20 dB while maintaining the same resolution level [16]. The coding excitation technology has been used in some commercial ultrasound equipment. For example, the LFM-encoded signal was selected for the Siemens system and the Golay codes was used as the transmit signal for the GE system. However, the effectiveness of this technique in ultrasound elastography has not been fully studied.

One of the main difficulties in ultrasound elastography is that the amplitude of the shear wave induced in the tissue is weak. The propagation distance of shear wave in human tissue is limited to millimeter magnitude because of the high attenuation, especially for the high frequency components. Therefore, ultrasound elastography suffers from poor SNR and shallow detection depth. The results of Walker and Trahey show that the SNR of the shear wave vibration signal is determined by the SNR of the ultrasonic echo signal [17]. Some researchers have applied coded excitation in ultrasound elastography to improve the SNR of shear wave [18].

Our team applied coded excitation in the transient elastography (TE) technique to improve the imaging performance [19]. A commercial elasticity phantom (Model 049, CIRS Inc., Norfolk, VA, USA) was used to evaluate the performance of the coded excitation for shear wave detection. In this experiment, a fresh piece of excised pork belly with a thickness of about 2.0 cm was placed between the transducer and the phantom surface to simulate the attenuation. The center

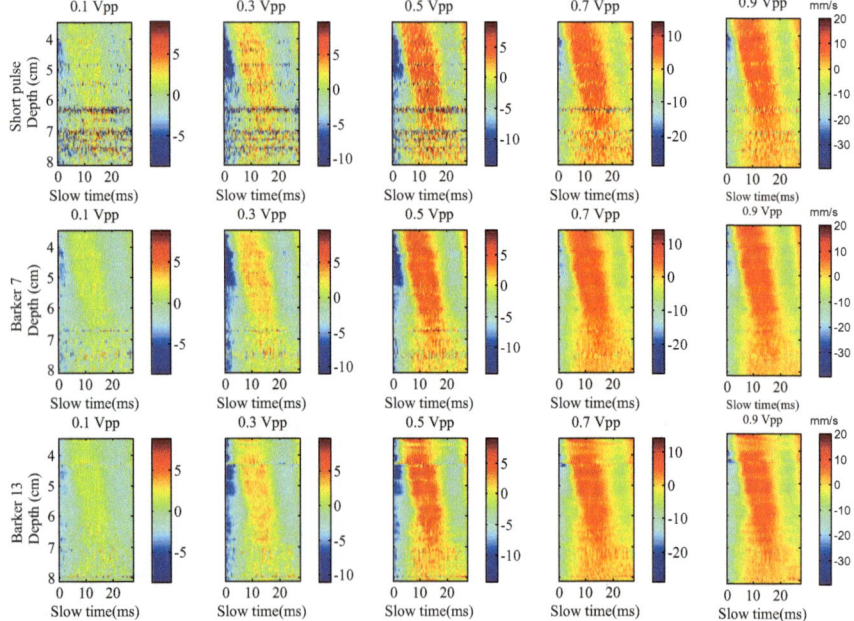

Fig. 6 The shear wave particle velocity images after adding a piece of pork belly, which are obtained from the 3 detected pulses by gradually increasing vibration input voltage from 0.1 to 0.9 Vpp with a interval of 0.2 Vpp. Note different color scales with units of mm/s are used for different input voltages

frequency of the low frequency (LF) excitation is 50 Hz and the voltage of the LF signal was gradually turned up from 0.1 to 0.9 Vpp with a 0.2 Vpp interval. At each voltage, the shear wave motion was detected by short pulse, Barker 7, and Barker 13 respectively (3 different detection pulses).

Figure 6 shows the time-distance images of the shear wave. At 0.1 Vpp input voltage, the image obtained by the short pulse was polluted by severe noise and the shear wave motion trajectory can be hardly recognized. For the coded excitation, the trajectory was better delineated with significantly less background noise than the short pulse.

2.5 Study on Blood–Brain Barrier Opening by Using Nonlinear Acoustic Information of Ultrasonic Cavitation Effect

Ultrasound contrast agent has been used as an ultrasonic molecular probe in ultrasound molecular imaging, which can display inflammation, thrombosis, tumor, etc. at the molecular level, and can carry drugs/genes for targeted therapy. Since the early 1990s, great progresses have been seen in the preparation, theory and application of ultrasound contrast agents. Contrast microbubbles have complex physical and chemical properties in the ultrasonic field. The contrast-enhanced echo signal contains not only the resonant frequency (i.e. the fundamental frequency), but also the energy components of higher-order harmonics and subharmonics and superharmonics. When the sound pressure rises to a certain critical point, the microbubbles will rupture, and a short high-intensity broadband signal will be released instantaneously [20].

In addition to contrast enhancement, ultrasound microbubbles can also be used in targeted therapies, among which the targeted, highly efficient delivery of therapeutic drug molecules into the brain is a huge challenge. The blood–brain barrier (BBB) can protect the stability of internal environment of brain. However, it also hinders the delivery of therapeutic agents into the central nervous system, affecting the treatment efficiency for many intracranial diseases. In recent years, domestic and foreign animal experiments have found that low-frequency focused ultrasound combined with microbubbles can temporarily promote Evans blue and helium ions across the BBB through various forms, and pathological sections show no damage to neuronal cells, avoiding the limitations and side effects of the traditional BBB opening methods. The results suggest that the method of low-frequency focused ultrasound combined with microbubbles can reversibly and partially open the targeted BBB, opening up a new way to improve the targeted delivery and treatment of central nervous system diseases [21–23].

Low-frequency low-energy ultrasound combined with microbubbles can mediate a variety of macromolecular drugs into brain tissue, such as anti-tumor drugs, antibody drugs, etc. The effectiveness of the method has been confirmed in many animal models (glioma model, Alzheimer) [24–26]. In November 2015, scientists and doctors at the Sunnybrook Health Science Center in Canada conducted the first clinical trial of glioma patients for the first time by delivering the anticancer drug doxorubicin into brain tissue.

In our preliminary experiment, the BBB could be opened successfully and non-invasively without any hemorrhage and neuron damaging effects at a peak rare-factional pressure of 0.53 MPa and with a microbubble dosage of 0.1 mL/g [27]. In each group, one mouse was injected intravenously with Evans blue, which could bind to albumin (MW: 67 kDa) once entering circulation. Evans blue-tagged albumin extravasated from vasculature into the brain parenchyma only when the BBB was opened. Thus, blue coloration of the brain indicated successful BBB opening by FUS with microbubbles. In this study, the FUS was applied to the left hemisphere. As shown in Fig. 7, Evans blue extravasation was observed on the left hemisphere, while it was invisible on the unsonicated right hemisphere. Moreover, the degree and area of Evans blue dispersion increased when the acoustic pressure rises from 0.53 to 0.64 MPa, as shown in Fig. 7a, c. When the microbubble dose was raised from 0.1 to 0.5 mL/g, the Evans blue extravasation was enhanced gradually, as shown in Fig. 7.

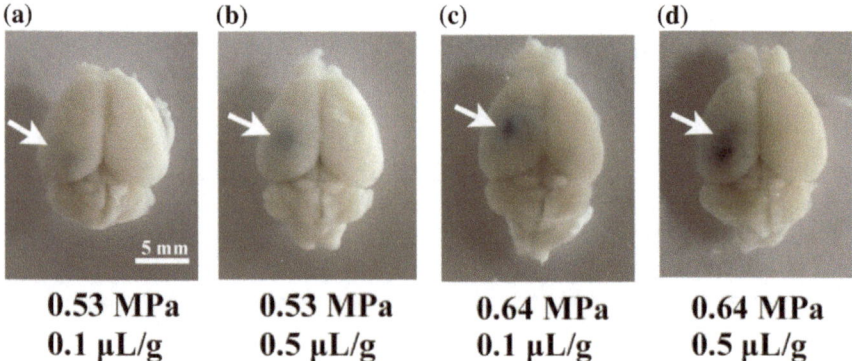

Fig. 7 Images of mice brains injected with Evans blue dye after sonication with different peak pressures and microbubble dosages

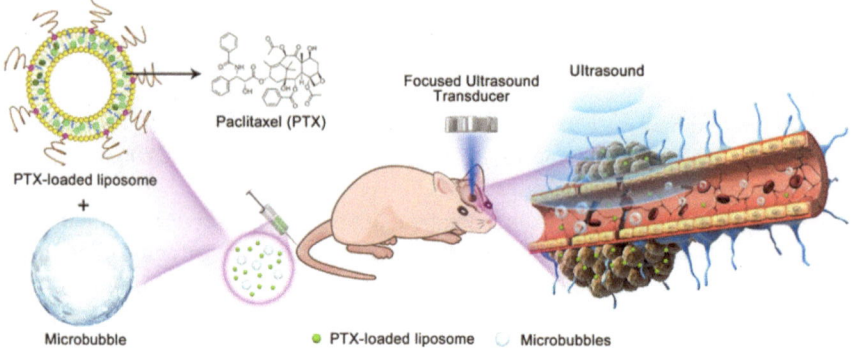

Fig. 8 Schematic illustration outlining the delivery of PTX-LIPO using FUS exposure in the presence of circulating MBs

We further investigated the feasibility of enhancing the delivery of PTX-LIPO using pulsed low-intensity FUS exposure in the presence of circulating microbubbles as well as improving therapeutic efficacy in an intracranial glioblastoma nude mice model [28], as shown in Fig. 8.

The experimental flowchart is presented in Fig. 9a. Tumor-bearing mice were randomly divided into four groups: Untreated Control group (Control); Only for ultrasound microbubbles (FUS + MB); Only paclitaxel liposome injection (PTX-LIPO); FUS sonication following injection of microbubbles and paclitaxel liposome (FUS + MB + PTX-LIPO). Treatment starting from the 7th day after implantation, once treatment every two days, three times in total. The concentration of PTX was analyzed using high-performance liquid chromatography. Longitudinal MRI analysis illustrated that the intracranial glioblastoma in nude mice treated with PTX-LIPO delivered via FUS with MBs was

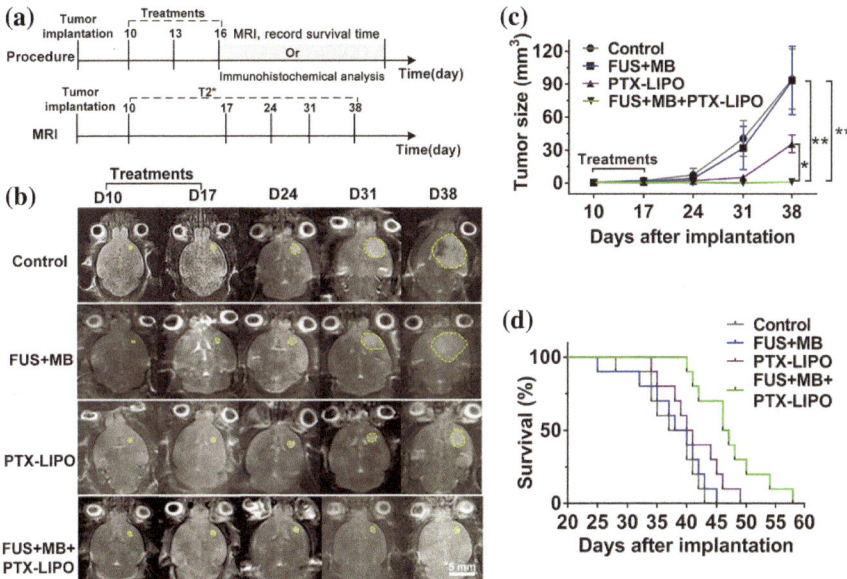

Fig. 9 In vivo anti-glioma efficacy in nude mice bearing intracranial glioblastoma in different treatment groups: no treatment as the CONTROL, FUS exposure with MBs but without PTX-LIPO as FUS + MB, PTX-LIPO injection only and PTX-LIPO delivery using FUS with MBs as FUS + MB + PTX-LIPO

suppressed consistently for four weeks compared to the untreated group. Immunohistochemical analysis further confirmed the anti-proliferation effect and cell apoptosis induction. Our study demonstrated that noninvasive low-intensity FUS with MBs can be used as an effective approach to deliver PTX-LIPO in order to improve their chemotherapy efficacy toward glioblastoma.

2.6 Extraction and Application of Other Information on Medical Ultrasound Features

Ultrasound has wave effect, mechanical effect and thermal effect, and these effects have yielded different applications in biomedicine field. Ultrasound imaging for diagnosis such as B-ultrasound and color ultrasound can be realized by the wave effect of ultrasound. In addition to elastic imaging, the mechanical effects of ultrasound can be used for acoustic manipulation and targeted drug delivery. The ultrasonic field is used to drive the intravascular drug particles to accumulate at targeted position. Then the ultrasonic radiation force and the cavitation effect of the microbubbles can be used to crush the drug-loaded ultrasonic microbubbles to release the drug. This technology is expected to increase the drug concentration in the target site, reduce the toxic side effects of the drug on the normal tissues of the body, and improve the therapeutic effect. In recent years, researchers have also achieved neuromodulation using acoustic radiation force. The central nervous system of the target site is stimulated or inhibited by ultrasound with different intensities, frequencies, pulse repetition frequencies, pulse widths, durations to induce reversible changes in the bidirectional regulation of neural function. The thermal effect of ultrasound is caused by abortion of the scattered ultrasound energy by the surrounding tissue when ultrasound propagates through the tissue and it may result in an increase in tissue temperature. Thermal ablation of the tumor can be achieved using the thermal effects of ultrasound. High-intensity focused ultrasound (HIFU) concentrates energy in the target area and forms a transient high-temperature up to 65–100 °C in the target area. HIFU can kill the tumor cells and cause coagulative necrosis of tumor tissue. Studies have shown that HIFU treatment can significantly enhance the host cell immune function and promote the body to produce specific anti-tumor immunity. More applications of ultrasound properties in biological tissues are yet to be explored, and information technology plays an important role in this discovery process.

3 Summary and Outlook

Biological tissues contain a large amount of information associated with their physiological pathology, such as the viscoelastic and conductivity information. The propagation of ultrasound in biological tissues is complex. Through interaction with tissues, ultrasound echo signals carry a wealth of information specifying the biological characteristics of tissues. Information technology promotes the deep exploration of the physical interaction mechanism between ultrasound and human tissue. It is expected to extract more accurate and valuable medical diagnosis information to provide a scientific basis for the treatment and diagnosis of diseases.

References

1. Siping Chen, 10000 Scientific Problems, Volume of Information Science, Science Press, 2011: 256–259.
2. Garra BS, Cespedes EI, Ophir J, et al. Elastography of breast lesions: initial clinical results. Radiology, 1997, 202(1): 79–86.
3. Ophir J, Céspedes I, Ponnekanti H, et al. Elastography: A quantitative method for imaging the elasticity of biological tissues. Ultrasonic Imaging, 1991, 13(2): 111–134.
4. Shiina T, Nightingale KR, Palmeri ML, et al. WFUMB guidelines and recommendations for clinical use of ultrasound elastography: Part 1: Basic principles and terminology. Ultrasound in Medicine and Biology, 2015, 41(5): 1126–1147.
5. Piscaglia F, Marinelli S, Bota S, et al. The role of ultrasound elastographic techniques in chronic liver disease: current status and future perspectives. European Journal of Radiology, 2014, 83(3):450–415.

6. Zheng Y, Yao A, Chen S, et al. Ultrasound vibrometry using orthogonal- frequency-based vibration pulses. IEEE transactions on Ultrasonics, Ferroelectrics, and Frequency Control, 2013, 60(11): 2359–2370.

7. Chen X, Shen Y, Zheng Y, et al. Quantification of liver viscoelasticity with acoustic radiation force: a study of hepatic fibrosis in a rat model. Ultrasound in Medicine and Biology, 2013, 39 (11): 2091–2102.

8. Tingqing Yang, Theory and Application of Viscoelasticity, Science Press, 2004.

9. Fung Y. Biomechanics: mechanical properties of living tissues. Springer, 1993.

10. Atay SM, Kroenke CD, Sabet A, et al. Measurement of the Dynamic Shear Modulus of Mouse Brain Tissue In Vivo By Magnetic Resonance Elastography. Journal of Biomechanical Engineering, 2008, 130(2):021013.

11. Ocal S, Ozcan MU, Basdogan I, et al. Effect of preservation period on the viscoelastic material properties of soft tissues with implications for liver transplantation. Journal of Biomechanical Engineering, 2010, 132(10):101007.

12. Freed AD, Diethelm K. Fractional calculus in biomechanics: a 3D viscoelastic model using regularized fractional derivative kernels with application to the human calcaneal fat pad. Biomechanics and Modeling in Mechanobiology, 2006, 5(4):203–215.

13. Joines WT, Zhang Y, Li C, et al. The measured electrical properties of normal and malignant human tissues from 50 to 900 MHz. Medical Physics, 1994, 21(4):547.

14. Guoqiang Liu, Magneto-Acoustic Tomography Technology, Science Press, 2016.

15. Li Y, Liu G, Xia H, et al. Numerical Simulations and Experimental Study of Magneto-Acousto-Electrical Tomography With Plane Transducer. IEEE Transactions on Magnetics, PP(99):1–4.

16. O'Donnell M. Coded excitation system for improving the penetration of real-time phased-array imaging systems. IEEE Transactions on Ultrasonics, Ferroelectrics, and Frequency Control, 1992, 39(3):341–351.

17. Walker WF, Trahey GE. A fundamental limit on delay estimation using partially correlated speckle signals. IEEE Transactions on Ultrasonics Ferroelectrics and Frequency Control, 2002, 42(2):301–308.

18. Song P, Urban MW, Manduca A, et al. Coded excitation plane wave imaging for shear wave motion detection. IEEE Transactions on Ultrasonics Ferroelectrics and Frequency Control, 2015, 62(7):1356–1372.

19. He XN, Diao XF, Lin HM, et al. Improved shear wave motion detection using coded excitation for transient elastography. Scientific Reports, 2017, 7:44483.

20. Zhigang Wang, Ultrasonic molecular imaging, Science Press, 2016.

21. Beccaria K, Canney M, Goldwirt L, et al. Opening of the blood-brain barrier with an unfocused ultrasound device in rabbits. Journal of Neurosurgery. 2013, 119(4): 887–98.

22. Choi JJ, Pernot M, Small SA, et al. Noninvasive, transcranial and localized opening of the blood-brain barrier using focused ultrasound in mice. Ultrasound in Medicine and Biology. 2007, 33(1): 95–104.

23. Hynynen K, Mcdannold N, Vykhodtseva N, et al. Noninvasive MR imaging-guided focal opening of the blood-brain barrier in rabbits. Radiology. 2001, 220(3): 640–646.

24. Aryal M, Arvanitis CD, Alexander PM, et al. Ultrasound-mediated blood-brain barrier disruption for targeted drug delivery in the central nervous system. Advanced Drug Delivery Reviews. 2014, 72: 94–109.

25. Treat LH, Mcdannold N, Zhang Y, et al. Improved Anti-Tumor Effect of Liposomal Doxorubicin after Targeted Blood-Brain Barrier Disruption by Mri-Guided Focused Ultrasound in Rat Glioma. Ultrasound in Medicine and Biology. 2012, 38(10): 1716–1725.

26. Yang FY, Wong TT, Teng MC, et al. Focused ultrasound and interleukin-4 receptor-targeted liposomal doxorubicin for enhanced targeted drug delivery and antitumor effect in glioblastoma multiforme. Journal of Controlled Release. 2012, 160(3): 652–658.

27. Shen Y, Guo J, Chen G, et al. Delivery of Liposomes with Different Sizes to Mice Brain after Sonication by Focused Ultrasound in the Presence of Microbubbles. Ultrasound in Medicine and Biology, 2016, 42(7):1499–1511.
28. Shen Y, Pi Z, Yan F, et al. Enhanced delivery of paclitaxel liposomes using focused ultrasound with microbubbles for treating nude mice bearing intracranial glioblastoma xenografts. International Journal of Nanomedicine, 2017, 12:5613–5629.

Siping Chen, Professor and Ph.D. supervisor, School of Biomedical Engineering, Shenzhen University. He once served as the chief engineer of Anke Company and vice president of Shenzhen University. He has been engaged in biomedical engineering research and medical ultrasound engineering and devoted to industrialization of scientific and technological achievements. He and his team developed first color Doppler scanner and promoted the industrialization of it in China in 1989. He was the principle investigator of many national key projects such as the key project of the National Natural Science Foundation of China. He published more than 100 peer-reviewed journal papers and three books. He established the Biomedical Engineering program, the National Engineering laboratory and Provincial Key Laboratory in Shenzhen University. He was awarded many prizes, including National Science and Technology Progress Award, the Special Allowance Expert of the State Council, the National Young Experts with Outstanding Contributions. Currently, he is the deputy director of National Technical Committee on Medical Electrical Equipment of Standardization Administration of China, chairman of the Instrument Engineering Committee of the Chinese Academy of Ultrasound Medical Engineering, expert reviewer of the National Basic Research Program of China.

The Applications of the e-Science in National R&D Program for Major Research Instruments

Yalin Lu, Mengmeng Yang, Fang Jiang, Zhengping Fu, Zhigang He
and Gongfa Liu

Abstract The development of major scientific instruments has indispensable impacts on the modern science and technology, however, it is a complex and difficult mission. Comparing to common scientific projects, the development of major scientific instruments not only requires progresses in relevant theories and technologies, but also requires efficient projects management. In this chapter, quoting the Special Fund for Research on National Major Research Instruments as an example, we will discuss the applications of the information for the program of developing major scientific instruments, and comment on their importance and necessity. Our discussion and comments are made on the basis of following five parts, including sources of project ideas, the process of developing instruments, the integration of instrument systems, the implementation of the projects management, and sharing information after the completion of instruments.

Keywords Information · Major scientific instruments · Management

1 Introduction

"e-Science" was firstly proposed in UK at the end of the twentieth century, and is about "global collaboration in key areas of science and the next generation of infrastructure that will enable it" [1, 2]. An official concept of "e-Science" by the UK e-Science programme is as follows: "e-Science is the increasing distributed global collaboration and the typical characteristic of "e-Science" is that scientists can enter immense database, digital network and high-quality visualization system" [3].

The essence of "e-Science" is to promote scientific research through information technology, including the adoption of new information technologies, the con-

Y. Lu (✉) · M. Yang · F. Jiang · Z. He · G. Liu
National Synchrotron Radiation Laboratory, Hefei, China

Y. Lu · Z. Fu
Hefei National Laboratory for Physical Sciences at the Microscale, Hefei, China

© Publishing House of Electronics Industry 2020
China's e-Science Blue Book 2018,
https://doi.org/10.1007/978-981-13-9390-7_21

struction of new information facilities, scientific research based on such infrastructure and supporting technologies, and activities of scientists in such environment [2]. Open study, resource sharing and collaborative research are the characteristics of "e-Science". The Infrastructure includes a variety of computing resources, data resources, network communication resources, and scientific research equipment. Internet, digital library, supercomputing, database, collaborative software, video conference, and sensor technology are all belonging to the category of "e-Science".

"e-Science" enables global, interdisciplinary, mass collaboration. It also enables resource sharing and collaborative work across time, space and physical barriers, which will change the methods and modes of scientific research. "e-Science" can greatly promote communication and cooperation, promote the development of scientific research, significantly change the organization of scientific research and activity model, and promote the science and technology transition. "e-Science" is now triggering the transition of science and engineering in the twenty-first century.

National R&D Program for Major Research Instruments focus on the frontier of science and national demand. Towards the scientific goals, these programs pay attention to the top-level design, clarify the development emphasis, encourage and support the development of exploratory scientific research equipment with creative ideas, thus innovative tools can be provided for scientific research to comprehensively enhance China's initial innovation capability.

National R&D Program for Major Research Instruments has the basic characteristics of emphasizing national goals, highlighting integrated innovation, balancing software and hardware development, emphasizing user participation, the complexity of organization and management and high risk [4]. These characteristics determine the importance and necessity of "e-Science" for National R&D Program for Major Research Instruments, and the application of "e-Science" will also run through all stages of the Program from establishment to acceptance.

2 "e-Science" is the Important Ideological Origin of National R&D Program for Major Research Instruments

The idea of National R&D Program for Major Research Instruments origins from the existing research practice of the applicant on the one hand and on the other hand from the acquisition of massive scientific research information. Taking one of the National R&D Programs for Major Research Instruments—"Near-field terahertz scanning tip system for high-throughput material characterization" as an example, Prof. Lu Yalin, the project leader, proposed a microwave near-field microscope firstly in 1997 [5]. The imaging system can increase the spatial resolution to one millionth ($10^{-6} \lambda$) of the wavelength of microwave used in the system, which becomes the main thought for the "Near-field terahertz scanning tip system for high-throughput material characterization" project. Based on this, with the addition of analysis of a large amount of scientific research information, Prof. Lu keenly captured the unsubstitutability of THz on the detection of the collective behavior of carriers in functional materials. Meanwhile, during the implementation of Materials Genome Initiative, High-flux characterization techniques are still in urgent need. In addition, multi-physical field (e.g. temperature field, magnetic field, electric field, light field) are necessary means to regulate and reveal the physical properties and mechanisms of functional materials. Based on the above information, Prof. Lu and his team propose the specific goals of the project and the technical framework required to achieve these goals. Similarly, based on the existing research practice of the team, a number of innovative technologies have been developed through broadly absorption of scientific and technological information in related fields, including compact free-electron laser terahertz sources, dual-mode (high-precision and wide-range) scanning probes, large-diameter vector magnets, high-quality light paths, etc. Based on these innovative technologies, a significantly original idea of the Major Research Instruments has been ultimately formed.

Given the above, it is shown that the idea of the R&D of the Major Research Instruments is inseparable from the data resources included in "e-Science". In addition, the extraction, analysis, integration and reprocessing of information will greatly help enhance the significance of R&D of major research instruments.

3 The Developing Process of Major Research Instrument Calls for "e-Science"

3.1 "e-Science" is the Basic Condition for Integrated Innovation

Compared to other scientific research programs, R&D Program for Major Research Instruments is often more complicated, and highly relying on the integration of cross-disciplines. These programs need overall integration of multi-discipline and comprehensive application of a variety of new technologies. Therefore, strengthening multi-department cooperation and integrated innovation is a key factor to the success of these programs.

Taking the "Near-field terahertz scanning tip system for high-throughput material characterization" project as an example (the following discussion in this chapter is all based on this project), the system is oriented to the forefront of functional materials research, so it is necessary to keep up closely with the development of this field. In addition, the system requires integration of technologies such as optics, mechanics, electricity, cryogenics, magnetics, automatic control, precision machining, etc. Some of these fields are developing so rapidly that new technologies are emerging. Therefore, it is critical to strengthen the information sharing, promote the communication and cooperation of scientist in various research fields and integrate kinds of new technologies. As was mentioned above, this kind of wide-ranging and high-level interacted information sharing and cooperation extremely relies on the Internet technology.

The project team built a variety of discussion groups based on internet, including groups based on e-mail and social media. The main organization form of groups based on e-mail is mail group, and groups based-on social media including QQ group, QQ discussion group and WeChat group.

Mail groups are mostly used for formal discussions, in order to retain the data for later review. Meanwhile, webmail-based conference initiation is gradually put into operation. Figure 1 shows a webmail-based conference initiation. It is shown that in addition to regular parameters such as time, location, and agenda, notifications and reminders are also set to memorize team members.

Social media-based discussion groups are often used for real-time discussions, including article sharing and discussion about requirements. Discussion groups based on social media has the advantages in terms of group discussion and real-time feedback, which greatly improve the information interaction and group cohesiveness between team members. Figure 2 shows a discussion group based on the famous social software QQ.

Team members can promptly share the results of their research via the discussion groups based on Internet, which greatly improve the efficiency of information transfer of the project team.

3.2 "e-Science" Helps to Improve the Practicality of the Instrument Under Research

The R&D Program for Major Research Instruments is oriented not only to the R&D, but also the application of the instruments [6]. R&D is focused on the construction of the instrument according to the program idea and R&D plan.

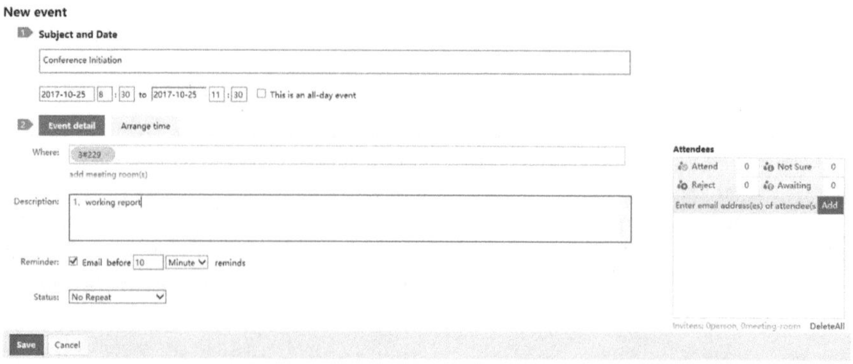

Fig. 1 Webmail-based conference initiation

Fig. 2 Discussion group based on QQ

What's more, the application is emphasized on how the instrument works after the program completes. At the kick-off meeting of the project "Near-field terahertz scanning tip system for high-throughput material characterization", the expert group mentioned that the project team should pay more attention to the practicality of the instrument. They also pointed that the best way is to let future users of the instrument participate in the whole process of the project, in order that their demand can be considered during the construction of the instrument.

The project team did exactly what the experts recommend. As future users may exist in various fields and different regions, it is not realistic to concentrate future users at the same time or in the same location. In fact, feedback and discussion are basically completed through information-based means, such as Internet.

For this purpose, the project team specially set up a "scientific research and technology development" sub-team, which is responsible for docking with potential users of the instrument to be researched. The members of this sub-team have sought the demand of potential users many times via e-mail or video discussion. The project team also organize times of discussion between R&D members and researchers, in order to integrate the demands of users and on this basis guide the project R&D process and appropriately adjust the project R&D plan. Thus, the practicality of the instrument could be maximized after the successful R&D of the instrument in the future.

3.3 "e-Science" Can Effectively Reduce the Risks of the Project

The R&D Program for Major Research Instruments involve the knowledge and skills from the frontier of various research fields, and the risks caused by the uncertainty of technology are more prominent. During the execution of the R&D Program for Major Research Instruments, it is especially important to do excellent risk management. Through fully utilization of kinds of means such as pre-assemble and simulation, it is beneficial to reduce risks sufficiently, and the data acquisition, data processing, large-scale calculations involved in the simulation process are also attached to "e-Science".

During the execution of the "Near-field terahertz scanning tip system for high-throughput material characterization", the primary goal is to verify the rationality of the physical design. Due to the lack of a complete technical solution or a prototype, the simulation is the only reasonable way for verification. The project carried out a large number of simulations, by which the technical design was guided. Figures 3 and 4 are schematic diagrams of the simulation of one key technology for the project.

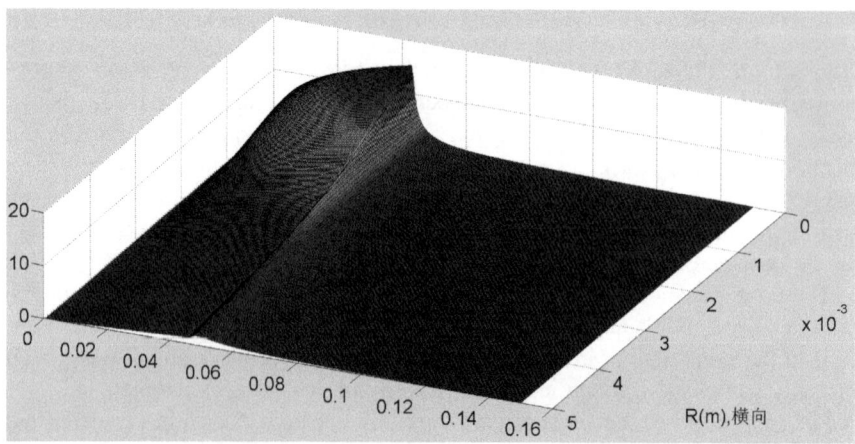

Fig. 3 Schematic diagram of the simulation for the project (1)

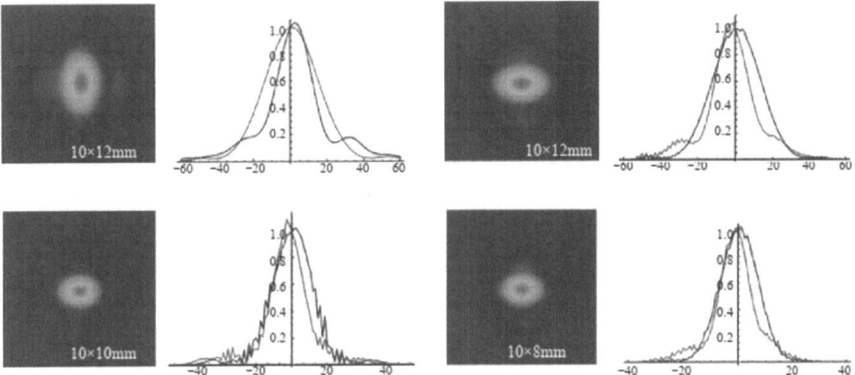

Fig. 4 Schematic diagram of the simulation for the project (2)

4 Informatization of the Instrument Under Research

From the perspective of the innovation chain, the general chain of technological innovation is: basic research—application research—product development—market, while the scientific equipment innovation chain is: basic research—application research—product development—application development—market. Compared with general technological innovation, scientific equipment innovation has an additional link of application development. The R&D Program for Major Research Instruments include not only the R&D of instruments, but also the development of related databases, software and the basis of control systems. This is a unique requirement for scientific equipment innovation [4]. Therefore, for the R&D of major research instruments, not only the execution is inseparable from informationization, but also the instrument itself requires a very high informationization level.

Take the "Near-field terahertz scanning tip system for high-throughput material characterization" project as an example, the system includes not only the construction of test platform and test environment, but also the integration of various devices and various types of interfaces. In addition, the collection and analysis of a large amount of complex data is also an essential function of the system. Therefore, the development of related software and control systems is indispensable for the completion of the project, and this is also inseparable from "e-Science" including technologies such as the database, communication technology etc.

Figure 5 is a schematic diagram of the central control hardware structure of the above-mentioned instrument, including the management layer, data communication layer and front controller layer. The management layer includes a supervisory control computer, a VM server platform, etc., in order to implement distribute control. The data communication layer is applied for connection between the management layer and the front controller layer, and also data acquisition. The front controller layer is used for the overall control of many sub-sections.

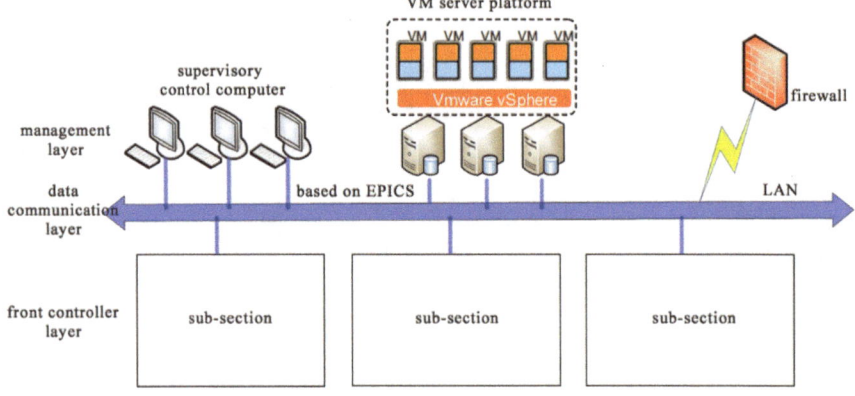

Fig. 5 A schematic diagram of the central control hardware structure of the above-mentioned instrument

The central control software system of the above-mentioned instrument is based on EPICS, a large-scale control software system with the full name of "Experimental Physics and Industrial Control System", which was jointly developed by the Los Alamos National Laboratory (LANL) and Argonne National Laboratory (ANL) in the early 1990s. The EPICS based on distributed standards consists of two parts: the IOC layer and the OPI layer. The IOC layer can be subdivided into 6 layers: Channel Access (CA) Server Interface, Database Access Interface, Dynamic Database, Record Support Module, Device Support Module, Device Driver. The OPI layer can be subdivided into two layers: Channel Access (CA) client interface and application software. The two basic mechanisms in the EPICS software system are channel access and distributed dynamic databases.

Obviously, in order to improve the availability and maintainability of the system, the integration process of the R&D Program for Major Research Instruments requires applications of information technology in an all-round way.

5 Informatization is an Effective Way to Improve Project Organization and Management

The R&D Program for Major Research Instruments needs repeated attempts. It is necessary to sustainably learn from the experience and lessons of the previous stage, and promote R&D activities through continuous feedback and interaction [6]. The organization and management of these programs involve a series of rules, norms and procedures, as are directly related to each participant in the program. It is necessary to organize and coordinate efficiently to achieve effective resource allocation. Through information-based collaborative management, it can effectively help to improve the organization and coordination of the program.

The organization and management of scientific research programs mainly include personnel management, organization management, process management, fund management, achievement management, information release, etc. [7]. The following is detailed explanation for the application of informatization in the above six aspects in our project.

5.1 Talent Introduction Based on Internet

Human resources are the core of social activities. The introduction of high-level scientific research talents plays an extremely important role in effectively promoting the progress of the program and filling the talent gap in several areas of the program. The "Near-field terahertz scanning tip system for high-throughput material characterization" project places great emphasis on talent introduction, making full use of various social media and portal websites, such as Acabridge, "www.gaox-iaojob.com". Clear recruitment information is shown in order to introduce personnel professional and urgently-needed. Figure 6 shows the recruitment information of the project published on "www.gaoxiaojob.com".

Meanwhile, in order to greatly accelerate the process of talent introduction, in the interview section the work team fully utilizes the Internet and flexibly adopts the form of online interviews to provide convenience for those who are not able to participate in traditional on-site interviews.

5.2 Using Information-Based Means to Improve the Coordination Efficiency

In the execution of the project, the application of information-based means has greatly improved the efficiency of system coordination. For example, when doing task assignment, the task assigner and the task bearer use email, SMS, telephone or other means based on social-media to carry out an adequate discussion on the task content, and a formal task assignment notice was formed after an agreement is reached, which efficiently avoid the possibility of subsequent changes.

5.3 Process Management Based on Internet and Database

The document management during the execution is of great significance for putting forward the program, especially for R&D program of Major Research Instrument. As the number of documents involved is often huge in those programs, if the document management is chaotic, it will bring great inconvenience to project check

Fig. 6 The recruitment information of the project published on "www.gaoxiaojob.com"

and subsequent application of the instrument. A complete and normal technical record covering all the steps as design drawings, material selection, component processing and integration helps ensure that the technology is solidified and easy to inherit and develop. In addition, during the program execution, the establishment of rules, regulations, and document standards related to personnel, procurement, schedule control, inter-departmental communication is also convenient for summaries and reference [6].

During the execution of the "Near-field terahertz scanning tip system for high-throughput material characterization" project, a complete document specification was established. These document specifications were sent to the related persons' mailbox in the form of electronic documents, avoiding the accumulation of a large number of blank paper documents and ensuring that the relevant documents are easy to find when needed. The project also established an archive database, which is continuously updated and maintained. This not only facilitates the management of the process documents, but also facilitates the smooth execution of the file check work in the future and as a basis for traceability. Figure 7 is a schematic diagram of the project's file directory.

The Schematic Diagram for "Near-field terahertz scanning tip system for high-throughput material characterization"

No.	Document number	Title for Archive			No. of cases	No. of pages	beginning and ending dates	retention period	security classification	comment
		File types	Sub types	Title						
1	NFTHZ-ZG01-01	integrated management	proposal		1	59				
2	NFTHZ-ZG01-02	integrated management	proposal		?	89				
3	NFTHZ-ZG01-03	integrated management	proposal		2	/				
4	NFTHZ-ZG01-04	integrated management	proposal		4	44				
5	NFTHZ-ZG02-01	integrated management	management		9	43				
6	NFTHZ-ZG02-02	integrated management	management		2	10				
7	NFTHZ-ZG02-03	integrated management	management		1	63				

Fig. 7 A schematic diagram of the project's file directory

In addition, for the safe storage of the files and avoidance of accidental data loss, the project also adopted the cloud storage and back up important project documents in the cloud, bringing "double insurance" to the project files.

5.4   Using Information-Based Means to Supervise the Use of Funds

The use of funds is always the focus of research projects, and self-monitoring of the funds is particularly important in the current environment in which the country continues to relax the use of research funding. In the implementation process of the "Near-field terahertz scanning tip system for high-throughput material characterization" project, in addition to standardizing the use of funds, the project team also established a self-inspection table for revenue and expenditure. As shown in Fig. 8, as contrasted with the budget declaration form, detailed records including the purchase date, the amount of the expenditure, the responsible person etc. are set up.

In addition, the semi-annual comparison with the data in the financial system of the project supporting institution is continuously carried out for self-examination and self-correction of the difference to ensure the legal use of the funds.

	Sub Section	Sub System	Name of Equipment	Type and Specification	Unit	No.	Unit cost (RMB)	Cost (RMB)	Recommended Vendor	Altered or not	standard or not	Proposer	Proposed cost	Currency	In terms of RMB	Name of application for purchase	Date of Application	Choice of purchasing	Final supplier
1																			
2																			
3																			
4																			
6																			
7																			
8																			
9																			
10																			
11																			

Fig. 8 The self-inspection table for revenue and expenditure of the project

5.5 Establish a Database to Fully Manage Scientific Research Achievements

The management of scientific research achievements is also very important for the implementation of the project, and supporting data can be provided very conveniently for the preparation of various summaries and annual reports. For this project, we have established a database to comprehensively record the publication of articles and patent applications (shown in Fig. 9). Meanwhile, we use the patent management functions provided by the patent search website such as "www.baiten.cn" to maintain intellectual property rights.

In addition, the project team also used document management software such as endnote, combined with a variety of literature search engines, to establish a literature tracking database, and closely follow with the literature related to the project. According to this, a continuously optimizing of the instrument design is carried out to ensure that the instrument design is oriented to the frontier of science.

5.6 Establishing the Project Portal Website and Posting Information Timely

The project portal website is an external window of the project and is the platform of information disclosure and communication. The portal website of the "Near-field terahertz scanning tip system for high-throughput material characterization" project (in preparation) consists of brief introduction, research trends, research achievements, recruitment information, etc. The contents of the website can be dynamically updated.

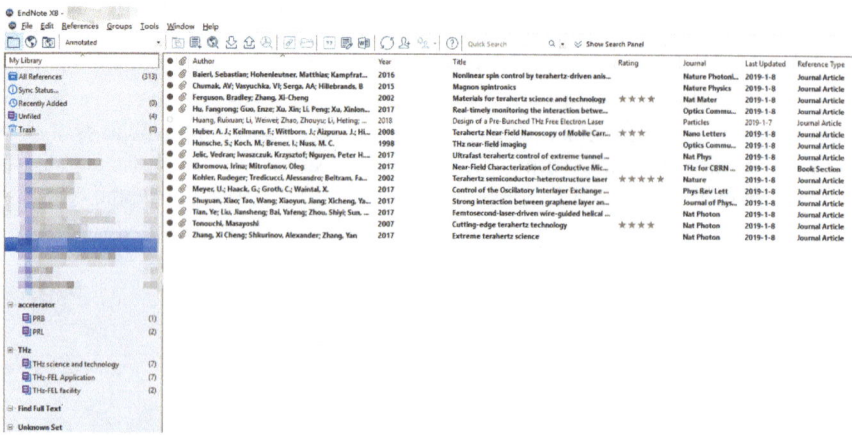

Fig. 9 A database to comprehensively record the publication of articles and patent applications

Fig. 10 News of the project

In addition, the "Near-field terahertz scanning tip system for high-throughput material characterization" project focuses on news communication. Once there are important developments, such as kick-off meetings, seminars or major achievements, etc., related reports will be pressed on relevant websites (such as the portal website of the project supporting institution). The release of relevant information continues to attract attention and talents, and the talents can promote the smooth progress of the project, thus forming a good cycle (shown in Fig. 10).

6 Management, Opening and Sharing of the Instrument After R&D Requires Informationization

The practicality of the instrument has been mentioned above. In addition, the practicality of the instrument also includes the utilization rate of the instrument after successful R&D. A high utilization rate depends on fully opening and sharing. An information management platform helps the instrument information to be fully disclosed and can effectively increase the opening and sharing rate.

After the "Near-field terahertz scanning tip system for high-throughput material characterization" project check, opening and sharing will be considered in the following ways:

(1) Incorporating into SAMP

Figure 11 shows the homepage of SAMP. Through selecting the regional center, research institute, instrument category, instrument class and instrument category in the quick appointment function area, users can quickly query the intended instrument. Users can make a reservation in the result interface according to the reservation status and reservation type of the relevant instrument, as shown in Fig. 12.

After the construction of the instrument, if it is incorporated into SAMP, it will belong to the University of Science and Technology of Hefei, a large-scale regional center for strategic energy and material science, and the instrument category

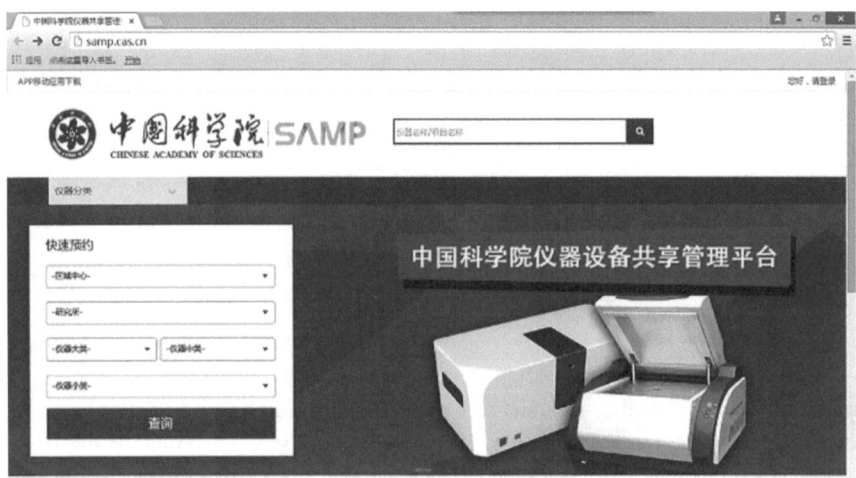

Fig. 11 Homepage of SAMP

Fig. 12 The reservation interface in SAMP

Fig. 13 A general view of the category of the instrument after construction if incorporated into SAMP

Fig. 14 The file card of the instrument

belongs to the indoor analysis and test equipment, as shown in Fig. 13. Meanwhile, by establishing the file card as shown in Fig. 14, the application scope and main functions of the instrument are fully clarified, and highly specialized and domain-oriented opening and sharing is realized.

(2) Establish an opening and sharing platform like SAMP and manage it autonomously and provide management platform portal links on the project portal website, websites of supporting institutions and other related websites.

Through information-based means, the instrument's opening and sharing management, open data resource management, and results achievements management are included. Combined with popular science propaganda, the public influence of the instrument is enhanced. Meanwhile, high-quality resources are mined, potential users are cultivated and instrument applications are expanded. Thus, the instrument can much better serve China's scientific research.

7 Conclusion

For National R&D Program for Major Research Instruments, informatization is necessary in all stages of the program. We should fully understand the needs of informatization for those programs and improve the application of informatization in the implementation of those programs. A continuous innovation should be carried out in the program management. Meanwhile, we should strengthen coordination and communication, and improve risk prevention capabilities, in order to promote the smooth execution of those programs, in addition to truly contribute to improve the independent innovation capability of China's scientific instruments and equipment.

References

1. Taylor J. e-Science definition. Retrieved March, 2002, 28: 2006.
2. Mianheng Jiang. E-Science. e-Science Technology & Application, 2008(1): 4–11.
3. Changjiang Qin. The Influence of E-Science on Modern Science. Science & Technology Progress and Policy, 2008, 8: 41.
4. Jiaxi Wu, Zhongqing Yu. Discussion on Management Mode of Major Scientific Instruments and Equipment R&D Projects. Project Management Technology, 2011, 9(12): 56–60.
5. Lu Y, Wei T, Duewer F, et al. Nondestructive imaging of dielectric-constant profiles and ferroelectric domains with a scanning-tip microwave near-field microscope. Science, 1997, 276(5321): 2004–2006.
6. Kunchao Bai, Peiwen Ji, Shouzhu Zhang. Thoughts on the management of National Major Scientific Research Equipment. Bulletin of National Natural Science Foundation of China, 2017(4): 380–383.
7. Yu Wang. Network-based Information Management of Science Research at College & University. Technology And Innovation Management, 2007, 28(2): 34–36.

Yalin Lu, Professor in the University of Science and Technology of China, Director of the National Synchrotron Radiation Laboratory. He was selected into the first batch of "Thousand Talents Program" in 2008 and was the recipient of China National Award for Natural Science (first class) in 2006. His research focuses on quantum functional materials, photonic science and advanced spectroscopy. He has published more than 260 high-quality papers in journals such as Science and Nature, and has nearly 50 Chinese or US invention patents. He has received more than 90 invitation reports and conference reports, and edited 5 album monographs. He is serving as one of the expert group for the "National Science and Technology Infrastructure Program" of the Ministry of Science and Technology, also an expert of Future Development Strategic Planning Research Committee of CAS large research infrastructures, and directors of the Chinese Physical Society and the Society of Silicate, and deputy editor and editorial board of more than ten domestic or foreign academic journals, and also third and fourth Hong Kong Qiushi Technology Foundation review expert. He is the director of the "Near-field terahertz scanning tip system for high-throughput material characterization" program and "a new generation of high-contrast low-dose X-ray phase contrast CT device", also chief scientist of one National Key Science Projects Program.

Current Situation and Prospect of CERNET

Ying Liu

Abstract China Education and Research Network (CERNET) is the first nation-wide education and research computer network in China. It has become an important public infrastructure greatly supports China's high education, science and technology development. CNGI-CERNET2/6IX is the largest and only academic one of the network backbones of CNGI. It is the largest pure IPv6 Internet backbone around the world. CNGI-CERNET2 has become an important infrastructure for promoting the Chinese next-generation Internet strategy. This chapter introduces the development progress of CERNET and CNGI-CERNET2 since 2015.

Keywords CERNET · CNGI-CERNET2/6IX · National research and education network

1 Overview

The national academic computer network-CERNET was built in 1994. It was supported and constructed by the Chinese Government; the Ministry of education is responsible for the management. The Tsinghua University and other top universities are responsible for the construction and operation of the national backbone. CERNET is one of the major Internet backbone networks with international export rights in China. After more than two decades of efforts, CERNET has connected over 3000 universities, educational institutions and scientific research units, and over 30 million users. The world's advanced 100G CERNET backbone network was built in 2013. At present, CERNET has developed into the world's largest national academic network, the important infrastructure of Chinese education informationization and the important composition part of national informationization infrastructure.

Y. Liu (✉)
CERNET Network Center, Beijing, China

© Publishing House of Electronics Industry 2020
China's e-Science Blue Book 2018,
https://doi.org/10.1007/978-981-13-9390-7_22

Since 2003, CERNET has united more than 100 universities to participate in the China next-generation Internet demonstration project-CNGI which has been approved by the State Council and organized jointly by the National Development and Reform Commission (NDRC) and other eight ministries and commissions, building the largest pure IPv6 core network of CNGI-CNGI-CERNET2/6IX, making important breakthrough in the key technologies of the next-generation Internet. CNGI-CERNET2/6IX has become an important infrastructure supporting next-generation Internet technology researches, application development and industry evolution.

In June 2016, the NDRC approved the Internet+ major project support project "IPv6 demonstration network oriented to the field of education (referred to as CERNET2 Phase II)." Based on CNGI-CERNET2/6IX, the core network of CNGI demonstration network, which has been built and put into operation for over 10 years, a large-scale, high-performance IPv6 next-generation Internet demonstration network oriented to the field of education has been built, with 41 core nodes of the backbone network, 100G bandwidth and over 10 million IPv6 users. Moreover, CERNET carried out "Internet +" technology testing and application demonstration, IPv6 transition technology, real source address validation technology and other technical tests, provided test verification platform for the national "Internet +" action plan, to promote China's accelerated development IPv6 next-generation Internet, enhance the national network space security capacity, play a demonstration role in supporting "Internet +" action plan and making the layout of next-generation Internet ahead of time.

The following introduces the recent developments of CERNET since 2015 from three aspects, including the CERNET backbone network, CNGI-CERNET2 backbone network and CERNET/CNGI interconnection center.

2   CERNET Backbone Network

2.1   Construction and Operation of CERNET

Through the "211 Project" third phase CERNET construction project which completed in 2013, the backbone bandwidth of CERNET is $10 \sim 100$ Gbps with 38 10 Gbps core nodes, 23 100 Gbps core nodes, the total bandwidth of the backbone network is above 3.15 Tbps, and each core node has an exchange capacity of more than 2×1.6 Tbps. CERNET provides a high-quality information infrastructure for higher education and scientific and technological innovation.

The topology of the CERNET backbone network is shown in Fig. 1.

As of December 2016, CERNET backbone network has about 14.59 million IPv4 addresses (equivalent to 227B+9C) and 6937 EDU.CN domain names. As the global IPv4 addresses run out, the number of IPv4 addresses has barely changed since 2011.

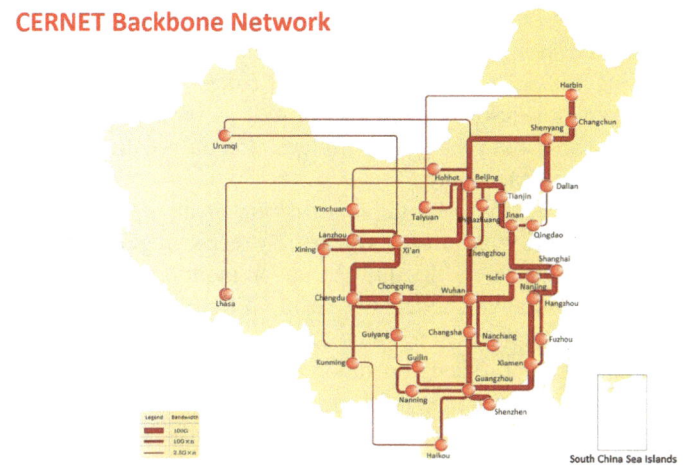

Fig. 1 CERNET backbone network topology

2.2 CERNET Backbone Network Access

Thirty-eight core nodes of the CERNET backbone network have established high-speed access systems, having devices with high-speed access capacity of 100 Mbps/1 Gbps/10 Gbps; they can provide over 100 Mbps high-speed access capabilities for more than 2000 colleges and universities and form the ability to provide 10 Gbps access to 500 of these colleges and universities. At present, a total of more than 3000 units have accessed CERNET.

2.3 CERNET Supports and Application

In the past more than 20 years, CERNET has supported and promoted a large number of network application innovation services constructed and completed the basic support systems for public network application, including network security service system and video service system. For the education system, digital certificate services were provided for over 100 applications; CERNET also set up a video service center and a high-definition video conferencing service platform and management system distributed in 38 core nodes, which provide a convenient environment for domestic and international academic exchanges between colleges and universities, and they are an important supporting platform for universities to carry out international cooperation and exchanges. Moreover, CERNET has completed the construction and promotion of key subject information service system, established distributed information service nodes in the CERNET network center and Beijing, Shanghai and Guangzhou and completed the construction of 54 key subject information resource systems, as well as formed a distributed information service system for key subjects of colleges and universities, covering 11 key subjects and with large capacity.

CERNET has further improved the construction of high-performance network management and security guarantee system. By deploying distributed network management system, backbone harmful behavior monitoring system and emergency response collaborative service system in CERNET network center and 38 core nodes, the network operation status can be "informed and controllable" to ensure the safe, stable and reliable operation of CERNET backbone network.

CERNET actively supports national basic and frontier scientific research. In May 2017, CERNET provided 100G bandwidth connectivity to Wuxi National Supercomputing Center (Sunway TaihuLight). Relying on the network and technology advantages of CERNET and the resource advantages of Wuxi national supercomputing center, it extensively serves major education and scientific research projects and promotes research and innovation in the field of education. Thus, visit users of the National Supercomputing Center in Wuxi via CERNET will get the best online experience.

In November 2017, CERNET and Pilot National Laboratory for Marine Science and Technology (Qingdao) (PNLMST) signed the "Supercomputing Internet Strategic Cooperation Agreement." CERNET and PNLMST will jointly construct a high-speed dedicated communication network for marine scientific research and are committed to building an international supercomputing ocean big data center and constructing an ocean big data monitoring network with the widest range of global data perception. CERNET will provide a 100G dedicated network for the Intelligent Supercomputing and Big Data Lab, connecting three supercomputing scientific devices put in PNLMST, Jinan Center and National Laboratory Supercomputing Center, respectively, to form a set of supercomputing large scientific device groups, which constitute a highly interactive, wide coverage, rapid feedback and informative Internet system, and ensure the interaction and analysis of big data in laboratory marine research.

3   CNGI-CERNET2 Backbone Network

3.1   Status of CNGI-CERNET2 Basic Network

As of December 2016, the CNGI-CERNET2 backbone network was distributed in 25 core nodes in 20 cities, and the bandwidth between the core nodes is 2.5 G/

10 Gbps; the connection bandwidth between the core device and the access device is 2.5–10 Gbps. The total bandwidth of the backbone network has reached 127.5Gbps.

By the end of December 2016, there were 19/32 IPv6 addresses, 1100 IPv6 address allocation records (including 11*/32, 1089*/48). The number of IPv6 BGP prefixes of CNGI-CERNET2 backbone network reached 37,000. The peak flow flowing in the CNGI-CERNET2 backbone network reached 59.82 Gbps, and the peak flow flowing out of the CNGI-CERNET2 backbone network reached 61.28 Gbps. As of 2016, there were over 700 units accessed to the CNGI-CERNET2 backbone network.

3.2 Construction of CERNET2 Phase II Projects

In June 2016, NDRC approved the second batch of "Internet +" major project support project "IPv6 demonstration network for the field of education (CERNET2 Phase II project)." These projects are undertaken by 41 universities, including CERNET network center and Tsinghua University.

The construction goals of these projects include: to build a large-scale IPv6 next-generation Internet demonstration network oriented to the field of education, based on CNGI-CERNET2/6IX, which has been built and put into operation for more than 10 years, and the core nodes of the backbone network are 41 and the bandwidth reaches 100 G, with over 10 million IPv6 users; launch an "Internet +" technology experiment and application demonstration, to provide a pilot verification platform for the implementation of the "Internet +" action plan in the country. To promote China's accelerated development of the IPv6 next-generation Internet, enhance the national network space security capabilities, as well as play a demonstration role in supporting "Internet +" action plan, making the layout of the next-generation Internet ahead of time (Table 1).

After the completion of project construction, the number of core nodes of the backbone network increased from 26 to 41, enlarging to 31 province/autonomous

Table 1 Comparison of CERNET2 backbone networks before and after construction

Content	Before project construction	After project construction
Network protocol	IPv6	IPv6
Number of core nodes	25	41
Core node coverage	20 provinces and cities	31 provinces and cities
Backbone line bandwidth	2.5 G/10 G	10 G/100 G
Total bandwidth of the backbone network	127.5 G	2950 G
Number of IPv6 users	5,000,000	Over 10,000,000

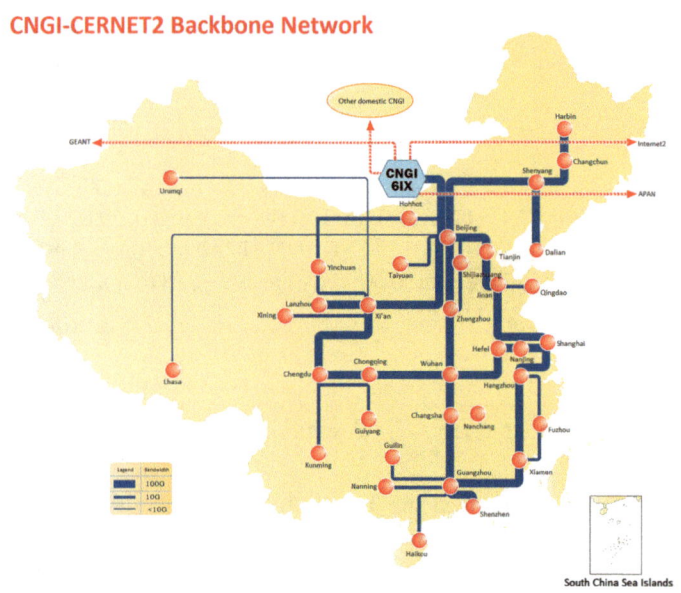

CNGI-CERNET2 Backbone Network

Fig. 2 CERNET2 backbone network topology (CERNET2 Phase II project)

regions nationwide from the original 20 ones; the backbone bandwidth was upgraded from the 2.5 G/10 G to 10 G/100 G. The total bandwidth of the backbone network was upgraded from 127.5 to 2950 G, and the number of IPv6 users has increased from 3 million to over 10 million (Fig. 2).

4 Status of CERNET Connectivity

4.1 Basic Situation of CERNET/CNGI Interconnection Center

CERNET is responsible for the construction and operation of CERNET Beijing Interconnection Center (CERNET-IX), the CNGI Beijing Interconnection Center (CNGI-6IX) and CERNET Hong Kong Interconnection Center (CERNET-HKIX) located in Beijing and Hong Kong, respectively. They connect to China's Science and Technology Network (CSTNET), China Telecom, China Unicom, China Mobile and other domestic Internet and next-generation Internet test networks in high speed and have realized high-speed interconnection with the international next-generation Internet academic networks, including the US Internet2, Europe GEANT2 and Asia-Pacific APAN, etc.

4.2 Domestic Interconnection Situation

In 2013, the Ministry of MIIT agreed to add seven regional central cities as the direct link points for the Internet backbone, and in 2016, with the help and cooperation of various interconnection units, CERNET achieved the growth of connectivity bandwidth and the increase of backbone network traffic. By the end of 2016, CERNET's domestic connectivity bandwidth had grown to 287 G, an increase of nearly three times compared to that of 2014. Among them, the bandwidth of CERNET to China Telecom increased by 38 G, to China Unicom increased by 20 G, to China Mobile increased by 50 G and to CSTNET increased by 50 G.

4.3 International Interconnection Situation

Since 1995, CERNET has set up its only international export in Beijing, connecting directly with academic networks, such as the US Internet2, the European GEANT, the Asia-Pacific APAN, and providing dedicated international access to education and scientific research. By the end of 2016, the total bandwidth of CERNET international exports exceeded 65 G.

Since 2004, Tsinghua University has obtained the operational management rights of the cross-Eurasian Information Network TEIN through international bidding, providing cross-intercontinental network connection operation services for the interconnection of academic networks of different countries and regions (such as Southeast Asia and South Asia) with the European Academic Network and supporting international cooperation in education and scientific researches among Eurasian countries. In 2016, Tsinghua University continued to be designated as a direct participant in the TEIN project and its successor Asi@connect project, and would remain responsible for the operation of the TEIN NOC; Asi@connect NOC service upgrades were under planning and preparation.

In 2015, CERNET and the European Pan-European Academic Network renewed a 10-year-long-term cooperation agreement.

5 Summary and Prospect

5.1 Summary

In general, CERNET and CNGI-CERNET2 got stable and booming development in 2015–2016. The contrast changes in CERNET interconnection bandwidth, IPv4 address, EDU.CN domain name and the number of access units can be seen from Table 2 below. The data show that CERNET's domestic interconnection bandwidth, number of access units and others are in an increasing trend.

Table 2 CERNET infrastructure construction and application status

Time	Backbone bandwidth (Gbps)	Domestic interconnection bandwidth (Gbps)	International interconnection bandwidth (Gbps)	IPv4 address (million)	EDU. CN domain name	Access units
2014	10–100	90	65	14,59	4321	3000
2015	10–100	198	65	14,59	6871	3000
2016	10–100	287	65	14,59	6937	3200

Table 3 CNGI-CERNET2 infrastructure construction and application status

Time	Backbone bandwidth (Gbps)	IPv6 address (a)	IPv6 BGP prefixes (10,000)	IPv6 AS	Access unit (a)	Number of users (10,000)
2014	2.5–10	902	2.1	51	500	300
2015	2.5–10	1000	2.8	51	600	400
2016	2.5–10	1100	3.7	51	700	500

Table 3 lists the comparison changes of CNGI-CERNET2 bandwidth, number of IPv6 addresses, number of access units and number of users from 2015 to 2016. The data show that CERNET2's access units continue to grow, becoming an important demonstration network infrastructure for the country to break the deadlock in IPv6 development and to prepare for the CERNET2 Phase II project.

5.2 Thoughts on the Future Development

Since 2012, with the completion of IPv4 addresses distribution, the world has entered the rapid development period of IPv6 next-generation Internet. At present, the US, Germany and other developed countries have more than 20% IPv6 users, forming a complete next-generation Internet industry chain, Google, Facebook and other well-known Internet enterprises are promoting and leading the development of world's next-generation Internet. On the contrary, the development of China's IPv6 next-generation Internet is deadlocked, the number of IPv6 users since 2012 has remained at millions of orders, compared with developed countries, and the main reasons for this situation include: (1) Lack of business drivers, although operators have upgraded the backbone network to a dual stack; however, due to the lack of IPv6 infrastructure and IPv6 applications that extend to users, the absence of a clear profit model, the lack of business drivers, the high cost of IPv6 upgrades and the existence of NAT alternatives reduce the incentives of telecom operators, Internet enterprises to adopt IPv6; some terminal equipment, such as mobile phone, home gateway and IoT device cannot effectively support IPv6, network, terminal, applications remain circular waiting state, resulting in slow development of the industry. (2) Lack of awareness of IPv6 by local governments. The local government's understanding of IPv6 is not in place, and the IPv6 organically does not combine with the application of all walks of life in the construction of the smart city in time. (3) The operational IPv4–IPv6 interchange technology has not been effectively applied. IPv4 and IPv6 will coexist. In the 12th Five-Year period, China had been trying a variety of IPv4–IPv6 interoperability technology, but large-scale operation of IPv4–IPv6 interoperability technology had not been effectively promoted and used, resulting in the existence of a large number of IPv4 end users cannot smoothly use the growing IPv6 applications, and growing IPv6 users are unable to use the vast array of IPv4 applications that exist. (4) The synergy of industrial chain development has not yet been formed. All links of the next-generation Internet industry chain are lack of coordination and have not a formed joint force. It is necessary to give full play to the integration, promotion and coordination role of social organizations, such as industrial alliances, with the guidance and support of national authorities, to better promote the development of the whole industrial chain of IPv6.

An effective way to break the deadlock in the development of IPv6 is to form a benign ecological environment with a long complement of IPv6 users, IPv6 networks and IPv6 applications. It is necessary to build a large-scale IPv6 demonstration network, develop large-scale users and attract well-known Internet companies to provide IPv6 information services based on the core network CNGI-CERNET2/6IX and the maintenance of millions of IPv6 users in the construction and operation by university groups. Under this demonstration role, it leads China's network operators, information service providers and other enterprises to jointly promote the next-generation Internet development, so as to consolidate the "Internet +" infrastructure.

CERNET will unite colleges and universities, seize the core technologies of Internet architecture, adhere to the technical route of Internet evolution and innovation and well construct CERNET2 Phase II projects. Moreover, CERNET will play a demonstration leading role for China breaking the IPv6 development deadlock as soon as possible, making layout of the next-generation Internet ahead of time, further guide the application and content providers to adopt IPv6, drive network operators to deploy IPv6, lead network equipment manufacturers to break through the core technologies, form a benign development cycle, promote the formation of ecological environment for IPv6 benign development as soon as possible. In the future, CERNET should implement the network power strategy, strengthen independent innovation, popularize and promote the key technologies of the next-generation Internet and strive to obtain more voice in the international competition of cyberspace for China.

Liu Ying Associate Researcher, Institute of Network Science and Cyberspace, Tsinghua University. She was Secretary General of the Internet Professional Committee of the China Computer Society. Ms. Liu's main research interests are the next-generation interconnection development planning, network architecture, multicast routing algorithm research, multicast routing protocol design, high-performance router architecture. As a project and subject person in charge, Ms. Liu has undertaken and participated in a number of national, provincial and ministerial key scientific research projects.